CAMBRIDGE TRACTS IN
MATHEMATICS

General Editors
B. BOLLOBAS, P. SARNAK, C.T.C. WALL

101 Infinite electrical networks

This is the first book to present the salient features of the general theory of infinite electrical networks in a coherent exposition. Using the basic tools of functional analysis and graph theory, the author presents the fundamental developments of the past two decades and discusses applications to other areas of mathematics and engineering.

The jump in complexity from finite electrical networks to infinite ones is comparable to the jump in complexity from finite-dimensional to infinite-dimensional spaces. Many of the questions conventionally asked about finite networks are currently unanswerable for infinite networks, while questions that are meaningless for finite networks crop up for infinite ones and lead to surprising results, such as the occasional collapse of Kirchhoff's laws in infinite regimes. Some central concepts have no counterpart in the finite case, for example, the extremities of an infinite network, the perceptibility of infinity, and the connections at infinity.

The first half of the book presents existence and uniqueness theorems for both infinite-power and finite-power voltage-current regimes, and the second half discusses methods for solving problems in infinite cascades and grids. A notable feature is the recent invention of transfinite networks, roughly analogous to Cantor's extension of the natural numbers to the transfinite ordinals. The last chapter is a survey of applications to exterior problems of partial differential equations, random walks on infinite graphs, and networks of operators on Hilbert spaces.

ARMEN H. ZEMANIAN
Electrical Engineering Department
State University of New York at Stony Brook

Infinite electrical networks

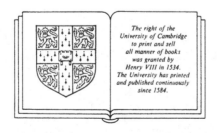

The right of the
University of Cambridge
to print and sell
all manner of books
was granted by
Henry VIII in 1534.
The University has printed
and published continuously
since 1584.

CAMBRIDGE UNIVERSITY PRESS
Cambridge
New York   Port Chester
Melbourne   Sydney

CAMBRIDGE UNIVERSITY PRESS
Cambridge, New York, Melbourne, Madrid, Cape Town, Singapore, São Paulo

Cambridge University Press
The Edinburgh Building, Cambridge CB2 8RU, UK

Published in the United States of America by Cambridge University Press, New York

www.cambridge.org
Information on this title: www.cambridge.org/9780521401531

First published 1991
This digitally printed version 2008

A catalogue record for this publication is available from the British Library

Library of Congress Cataloguing in Publication data
Zemanian, A. H. (Armen H.)
Infinite electrical networks / Armen H. Zemanian.
p. cm. – (Cambridge tracts in mathematics; 101)
Includes bibliographical references and indexes.
ISBN 0-521-40153-4
1. Electrical circuit analysis. I. Title. II. Series.
TK454.2.Z43   1991
621.319′2 – dc20   91–4491
                                              CIP

ISBN 978-0-521-40153-1 hardback
ISBN 978-0-521-06339-5 paperback

Contents

Preface		*page* **vii**
Acknowledgments		**xii**
1	**Introduction**	**1**
1.1	Notations and Terminology	1
1.2	Countable Graphs	4
1.3	0-Graphs	7
1.4	Electrical Networks	9
1.5	Kirchhoff's Laws	12
1.6	Curiouser and Curiouser	13
1.7	Electrical Analogs for Some Differential Equations	20
1.8	The Transient Behavior of Linear RLC Networks	27
2	**Infinite-power Regimes**	**30**
2.1	An Example	31
2.2	The Chainlike Structure	32
2.3	Halin's Result for Locally Finite Graphs	35
2.4	An Extension to Countably Infinite Graphs	38
2.5	Limbs	41
2.6	Current Regimes Satisfying Kirchhoff's Current Law	44
2.7	Joints and Chords	46
2.8	The Equations for a Limb Analysis	47
2.9	Chord Dominance	50
2.10	Limb Analysis, Summarized	52
2.11	Nonlinear Networks	53
2.12	A Contraction Mapping Result	55
2.13	A More General Fixed Point Theorem	58
3	**Finite-power Regimes: The Linear Case**	**61**
3.1	Flanders' Theorems	62
3.2	Connections at Infinity: 1-Graphs	66
3.3	Existence and Uniqueness	74
3.4	The Validity of Kirchhoff's Laws	80
3.5	A Dual Analysis	83
3.6	Transferring Pure Sources	87
3.7	Networks with Pure Sources	95
3.8	Thomson's Least Power Principle	99
3.9	The Concavity of Driving-point Resistances	100
3.10	Nondisconnectable 0-Tips	103
3.11	Effectively Shorted 0-Tips	105
4	**Finite-power Regimes: The Nonlinear Case**	**108**
4.1	Regular Networks	109
4.2	The Fundamental Operator	116
4.3	The Modular Sequence Space l_M	120
4.4	The Space $l_M^\#$	122
4.5	The Space c_M	125

4.6	Some Uniform-variation Conditions	127
4.7	Current Regimes in l_M	130
4.8	A Minimization Principle	133

5   Transfinite Electrical Networks   **139**
5.1	p-Graphs	140
5.2	ω-Graphs	146
5.3	Graphs of Still Higher Ranks	149
5.4	(k, q)-Paths and Terminal Behavior	151
5.5	k-Networks	153

6   Cascades   **158**
6.1	Linear Uniform Cascades	161
6.2	Nonlinear Uniform Cascades	167
6.3	Backward Mappings of the Axes	171
6.4	Trajectories near the Origin	175
6.5	Characteristic Immittances for Nonlinear Uniform Cascades	179
6.6	Nonlinear Uniform Lattice Cascades	186
6.7	Nonuniform Cascades: Infinity Imperceptible	193
6.8	Nonuniform Cascades: Infinity Perceptible	197
6.9	Input–Output Mappings	201
6.10	ω^p-Cascades	204
6.11	Loaded Cascades	209

7   Grids   **215**
7.1	Laurent Operators and the Fourier-Series Transformation	217
7.2	n-Dimensional Rectangular Grids	220
7.3	A Nodal Analysis for Uniform Grids	223
7.4	One-dimensional Nonuniformity	229
7.5	Solving Grounded Semi-infinite Grids: Infinity Imperceptible	233
7.6	Solving Ungrounded Semi-infinite Grids: Infinity Imperceptible	240
7.7	Semi-infinite Grids: Infinity Perceptible	245
7.8	Forward and Backward Mappings	250
7.9	Solving Semi-infinite Grids: Infinity Perceptible	255
7.10	Transfinite Grids	260
7.11	Grids with Two-dimensionally Transfinite Nonuniformities	266

8   Applications   **267**
8.1	Surface Operators	268
8.2	Domain Contractions	274
8.3	Random Walks	285
8.4	Operator Networks	290

Bibliography	**294**
Index of Symbols	**303**
Index	**305**

Preface

*... accumulations of isolated facts and measurements which
lie as a sort of dead weight on the scientific stomach, and which
must remain undigested until theory supplies a more powerful
solvent ... .*

Lord Rayleigh

The theory of electrical networks became fully launched, it seems
fair to say, when Gustav Kirchhoff published his voltage and current laws in 1847 [72]. Since then, a massive literature on electrical
networks has accumulated, but almost all of it is devoted to finite networks. Infinite networks received scant attention, and what
they did receive was devoted primarily to ladders, grids, and other
infinite networks having periodic graphs and uniform element values. Only during the past two decades has a general theory for
infinite electrical networks with unrestricted graphs and variable
element values been developing. The simpler case of purely resistive
networks possesses the larger body of results. Nonetheless, much
has also been achieved with regard to RLC networks. Enough now
exists in the research literature to warrant a book that gathers the
salient features of the subject into a coherent exposition.

As might well be expected, the jump in complexity from finite
electrical networks to infinite ones is comparable to the jump in
complexity from finite-dimensional spaces to infinite-dimensional
spaces. Many of the questions we conventionally ask and answer
about finite networks are unanswerable for infinite networks – at
least at the present time. On the other hand, questions, which
are meaningless for finite networks, crop up about infinite ones
and lead to novel attributes, which often jar the habits of thought
conditioned by finite networks. A case in point is the occasional
collapse of such fundamentals as Kirchhoff's laws. Indeed, Kirchhoff's current law need not hold at a node with an infinity of incident branches, and his voltage law may fail around an infinite loop.
Moreover, the use of an infinite loop is itself an issue, and different
voltage-current regimes can arise depending on whether or not infinite loops are allowed. The theory of infinite electrical networks
is perforce very different from that of finite networks. Concepts
that have no counterparts in the finite case are fundamental to the

subject, for example, the extremities of an infinite network, the perceptibility of infinity, the restraining or nonrestraining character of an infinite node, and the connections at infinity. This perhaps is one of the reasons why it took so long for the subject to become a distinctive research area.

But why bother, one may ask, to examine it at all? One answer is that it is intellectually challenging. Homo sapiens has the ability to use his/her mind in cognitive endeavors that do not directly relate to his/her physical needs – and does so. Here is one more arena in which to exercise that proficiency. To put this another way, circuit theory is intrinsically a mathematical discipline and its open problems will be attacked.

Another answer is that the subject does have practical applications. For example, a variety of partial differential equations such as Poisson's equation, the heat equation, the acoustic wave equation, and polarized forms of Maxwell's equations, have partial difference approximations that are realized by electrical networks. Moreover, the domains in which those equations are to hold are at times appropriately viewed as being infinite in extent, for instance, when a fringing problem is at hand or when a wave is propagating into an exterior region. In these cases, the discretized models are infinite electrical networks. To be sure, one may truncate the domain to obtain a finite network, but a search for a solution to the infinite network would avoid the additional error imposed by the truncation. It should be admitted, however, that much of infinite network theory does not possess engineering or physical significance – as yet. Nonetheless, though the intellectual challenge of infinite networks is an attraction, it is not only curiosity that leads one into circuit abstractions. It is conducive to the growth of that discipline to pursue, at least by some of us, circuit theory wherever it may lead.

Another explanation for the relative inattention paid to this stimulating subject is its apparent isolation from the established disciplines. It seems to be too much like electrical engineering to attract mathematicians and too much like mathematics to attract engineers. Nonetheless, it has been developing apace in recent years, and it is accessible to anyone with some knowledge of circuit theory and elements of functional analysis. Moreover, the subject is in its puberty; there are many open problems and undoubtedly much still to be discovered. Theoreticians looking for new research horizons may well consider infinite electrical networks.

We assume that the reader has a basic knowledge of graph theory and functional analysis. Nonetheless, we explicitly state the principal results from these subjects that we use and provide references to textbooks for expositions of them. A summary of most of the needed standard results from functional analysis is given in the appendices of [154]. On the other hand, since an electrical network consists of a certain analytical structure imposed upon a graph, the graphs we encounter in this book are almost entirely infinite ones. We will restrict our attention exclusively to countable graphs, that is, to those having finite or denumerable sets of branches. Thus, the graph theory we shall employ is not the standard theory encountered in the customary courses on finite circuits. Finally, it would be helpful to the reader to have some prior exposure to electrical circuit theory; such knowledge would enhance a comprehension of this book's subject matter, but is not essential because we shall define and derive all the concepts and results in circuit theory that we use. Actually, this book is addressed to two different audiences: engineers and mathematicians. Hence, background information that is well-known to one group may be belabored for the sake of the other. Moreover, we strive for a rigorous exposition and, on the other hand, present many examples to illuminate this radical extension of circuit theory.

The topics are organized as follows. The first chapter starts with some fundamental concepts and definitions. It continues with a variety of examples, which illustrate the peculiarities and paradoxes that distinguish infinite networks from finite ones. It then presents two examples to illustrate how infinite networks arise as discretizations of physical phenomena in exterior domains. It ends by pointing out how the transient behavior of a linear RLC network can be obtained from the analysis of a resistive network.

There is a dichotomy in the theory of infinite networks, which arises from the fact that Ohm's law and Kirchhoff's voltage and current laws do not by themselves determine a unique voltage-current regime except in certain trivial cases. This leads to two divergent ways of studying infinite electrical networks. One way is to impose those laws alone and to examine the whole class of different voltage-current regimes that the network can have. This is done in Chapter 2. In general, the power dissipated in the network under such a regime is infinite. Chapter 2 is not essential to the rest of the book and can be skipped if one is interested only in finite-power regimes.

The second way, which is examined in Chapter 3 for linear networks and in Chapter 4 for nonlinear ones, is motivated by the following question. What additional conditions must be added to Ohm's law and Kirchhoff's laws to ensure a unique voltage-current regime? One conspicuous requirement that suggests itself is finiteness of the total dissipated power. This suffices for some networks but not for all. In general, what is occurring at infinity has to be specified as well. For certain networks infinity is imperceptible from any particular point of the network, and so the connections at infinity can remain unspecified, but in other cases it is necessary to know what is connected to the network at infinity if a unique voltage-current regime is to be obtained.

In the latter case, what is connected out at infinity may be another infinite network. Thus, we are led naturally to the idea of a transfinite network, one wherein two nodes may be connected through a transfinite path but not through a finite one. Actually, this process of constructing a network that "extends beyond infinity" can be repeated by connecting infinite collections of infinite networks at their extremities to obtain hierarchies of transfinite networks. Roughly speaking, we might say that the process is analogous to the construction of an ordinary infinite network by connecting an infinity of branches at various nodes; in the transfinite case, branches are replaced by infinite networks and the nodes of a branch are replaced by the extremities of an infinite network. This generalization of infinite networks is explored in Chapter 5.

Chapters 2 through 5 are concerned almost entirely with existence and uniqueness theorems. To be sure, some of the proofs therein are constructive so that solutions can in principle be calculated. Nevertheless, the focus of those chapters is not on the computability of solutions. If, however, we restrict our attention to infinite networks having certain regularities, new methods become available, which in some cases lead to ways of calculating voltage-current regimes. A particularly simple regularity is the one-dimensional infinite cascade of three-terminal and two-port networks. This is the subject of Chapter 6. Chapter 7 considers various kinds of grounded and ungrounded grids. A useful application arises in this context. Since the discretizations of various partial differential equations can often be realized by electrical grids, methods for solving infinite electrical networks may be used to compute solutions for the differential equations for which the the domain of analysis is infinite in extent. In fact, such procedures may avoid the domain truncations customarily used to convert the infinite domain into an approximating finite one. Applications of this nature are surveyed in Sections 8.1 and 8.2.

It should be emphasized, however, that this book is not a comprehensive treatment of all aspects of infinite electrical networks. For instance, its thrust is toward resistive networks. To be sure, several of our analyses extend to RLC networks, and we point this out as occasions arise; but RLC networks are discussed only in this peripheral way. Another important issue is the growth of a network, something we refer to as the "perceptibility of infinity." It relates to some fundamental problems concerning random walks, Markov chains, and the classification of Riemann surfaces. A substantial body of results in this regard has been accumulating in recent years. This is very briefly covered in Section 8.3, wherein a number of references to its literature are listed. Still another context in which infinite electrical networks naturally arise is the theory of operator networks, that is, finite or infinite networks whose parameters are Hilbert-space operators. One case of this will be met in Chapter 7 wherein grids are decomposed into ladder networks of Laurent operators. Operator networks of greater generality are briefly discussed in Section 8.4, the last section.

In short, this book might be described figuratively as "road-building through the jungle of infinite electrical networks along the valley of resistive networks."

Acknowledgments

Many of the results discussed in this book were obtained during the last two decades from research sponsored by the National Science Foundation under a series of grants. That support is gratefully acknowledged. Thanks are also due to V. Belevitch, L. DeMichele, V. Dolezal, H. Flanders, I. W. Sandberg, and P. M. Soardi for their critical appraisal of various portions of the manuscript and for several suggestions that have been incorporated into the text.

Chapter 1

Introduction

The purposes of this initial chapter are to present some basic definitions about infinite electrical networks, to show by examples that their behaviors can be quite different from that of finite networks, and to indicate how they approximately represent various partial differential equations in infinite domains. Finally, we explain how the transient responses of linear RLC networks can be derived from the theory of purely resistive networks; this is of interest because most of the results of this book are established in the context of resistive networks.

1.1 Notations and Terminology

Let us start by reviewing some symbols and phraseology so as to dispel possible ambiguities in our subsequent discussions. We follow customary usage; hence, this section may be skipped and referred to only if the need arises. Also, an Index of Symbols is appended for the more commonly occurring notations in this book; it cites the pages on which they are defined.

Let X be a set. X is called denumerably infinite or just denumerable if its members can be placed in a one-to-one correspondence with all the natural numbers: $0, 1, 2, \ldots$. X is called countable if it is either finite or denumerable. In this book the set of branches of any network will always be countable.

The notation $\{x \in X : P(x)\}$, or simply $\{x : P(x)\}$ if X is understood, denotes the set of all $x \in X$ for which the proposition $P(x)$ concerning x is true. We will at times be constructing infinite sets, such as infinite paths through graphs, by selecting elements from collections of sets; when doing so, we will be tacitly invoking the axiom of choice [1]. $|X|$ denotes the cardinality of X. $\emptyset$ denotes the void set. A singleton is a set having exactly one member x and may be denoted by $\{x\}$. If Y is also a set, then $X \cup Y$ and $X \cap Y$ are the usual union and intersection sets. However, $X \subset Y$ (instead of $X \subseteq Y$) will indicate that all elements of X are members of Y, and X may equal Y; if X is smaller than Y, X is said to be a proper subset of Y. $X \backslash Y$ is the set of all members of X that are not in Y. The notation $x, y, z, \ldots \in X$ means that all the elements $x, y, z, \ldots$ are members of X; $x, y, z, \ldots \notin X$ means that none are in X. If X is a set in a metric space, X° is the interior of X, and $\bar{X}$ is the closure of X. However, if α is a complex number, $\bar{\alpha}$ is its complex conjugate. If X is a linear space, $-X = \{y \colon y = -x, x \in X\}$.

$\{x_i\}_{i \in I}$ denotes a collection of indexed elements, where the index i traverses the set I. If I is a set of integers, $\{x_i\}_{i \in I}$ is a sequence with the order induced by those integers. A sequence may be represented in several ways: If, for example, I is the set of natural numbers, we may write $\{x_i\}_{i=0}^{\infty}$ or $\{x_0, x_1, x_2, \ldots\}$ or $(x_0, x_1, x_2, \ldots)$ or just $[x_i]$. More generally, if a and b are two integers with $a < b$ and $b - a = n - 1$, $\{x_i\}_{i=a}^{b}$ is a finite sequence or synonymously an n-tuple. If $a = -\infty$ and/or $b = \infty$, then $\{x_i\}_{i=a}^{b}$ denotes an infinite sequence with I being the set of all integers from a to b; it is understood that there are no entries $x_{-\infty}$ or x_{∞}. A sequence is often called a vector. The support of a vector whose components are members of a linear space is the set of indices for its nonzero components; it may happen that those indices correspond to branches of a network, in which case we think of the support as a set of branches. Furthermore, $\cup_{i=a}^{b} X_i$ and $\cap_{i=a}^{b} X_i$ denote respectively the union and intersection of all members X_i of a sequence $\{X_i\}_{i=a}^{b}$ of sets. The Cartesian product $X_1 \times \cdots \times X_n$ of a finite number of sets is the set of all n-tuples $(x_1, \ldots, x_n)$, where $x_i \in X_i$ and $i = 1, \ldots, n$. A partition of a set X is a collection of subsets X_i of X such that every $x \in X$ is a member of one and only one X_i.

N denotes the set $\{0, 1, 2, \ldots\}$ of all natural numbers, and Z the set $\{\ldots, -1, 0, 1, \ldots\}$ of all integers. We use Greek iota for the imaginary unit: $\iota = \sqrt{-1}$. R^n and C^n denote respectively the real and complex, n-dimensional, Euclidean spaces. Thus, an arbitrary point $x \in R^n$ (or $x \in C^n$) is an n-tuple $x = (x_1, \ldots, x_n)$ of real (respectively, complex) numbers x_i. The Euclidean norm of x is $\|x\| = \left[\sum |x_i|^2\right]^{1/2}$. At times we use a different norm for x: $\|x\|_1 = |x_1| + \cdots + |x_n|$. A compact subset of R^n or C^n is a closed bounded set. A point $x \in R^n$ is called a lattice point or an n-dimensional integer if every component x_i of x is an integer. Z^n denotes the set of lattice points in R^n. Furthermore, l_{2r} and l_2 denote respectively the real and complex, infinite-dimensional, Hilbert coordinate spaces. Any member x of these spaces is an infinite sequence of real or respectively complex numbers x_i such that $\|x\| = \left[\sum |x_i|^2\right]^{1/2} < \infty$. In this case we say that the infinite sequence x is quadratically summable (instead of square summable). The inner product of $x, y \in X$ is $(x, y) = \sum x_i \bar{y}_i$. Actually, the symbol (x, y) has several meanings: an open interval in R^1, or a point in R^2, or a two-element sequence, or an inner product; we will be explicit as to which meaning we happen to be using. Also, $\text{sgn}(x)$ denotes the sign of $x \in R^1$.

A rule f that assigns one or more elements in a set Y to each element in another set X is called a relation or correspondence. Thus, f determines a subset of $X \times Y$ called the graph of f and consisting of all 2-tuples (x, y), where $y = fx$ is an element assigned to x by f. We also use the notation $f: X \rightsquigarrow Y$ as well as $f: x \mapsto y$ to denote the rule, the sets for which the rule is defined, and the typical elements related by the rule. The domain of f is the set of all elements x for which f is defined, in this case X; x is called the independent variable. The range of f is the subset of Y consisting of all elements y that are assigned by f to members of X; y is called the dependent variable.

A function f is a relation $f: X \rightsquigarrow Y$ that assigns precisely one element in Y to each member of X; in other words, f is a function if and only if the equations $y = fx$ and $w = fx$ imply that $y = w$. We may emphasize this property by referring to f as single-valued (in contradistinction to a multivalued function, which perforce is a relation). Synonymous with function are operator, mapping, and transformation. We say that $f: x \mapsto y$ maps, carries, transforms, or converts x into y and that f is a function on or from X into Y or a mapping of X into Y. We also say that f is Y-valued. When Y is R^1 or C^1, the phrase Y-valued is replaced by real-valued or complex-valued respectively, and f is sometimes called a *functional*. If W is a subset of X, the symbol $f(W)$ denotes the set $\{y \in Y : y = fx, x \in W\}$. The function g that is defined only on W and coincides with f on W is called the restriction of f to W.

A function f is said to be one-to-one or injective and is also called an injection if the equations $fx = y$ and $fu = y$ imply that $x = u$. In this case we have the inverse function $f^{-1}: y \mapsto x$, which maps the range of f into X. A function $f: X \rightsquigarrow Y$ is said to be onto or surjective and is also called a surjection if the range of f coincides with Y. If $f: X \rightsquigarrow Y$

is both injective and surjective, it is said to be bijective or a bijection.

We denote the elements in the range of f by the alternative notations $y = fx = f(x) = \langle f, x \rangle$. On occasion, it will be convenient to violate this symbolism by using $f(x)$ to denote the function f rather than its range value, as is commonly done in classical mathematics. Whenever we do so, it will be clear from the context what is meant. On still other occasions, we use the dot notation $f(\cdot) = \langle f, \cdot \rangle$ in order to indicate where the independent variable should appear.

The symbols sup and inf indicate respectively the supremum and infimum of a subset of R^1, whereas max and min indicate respectively the maximum and minimum of a finite subset of R^1. On the other hand, ess sup and ess inf indicate the essential supremum and essential infimum of the range of a measurable function. Also, lim sup and lim inf symbolize the limit superior and limit inferior.

Our notation for asymptotic behavior is standard. Let $X \subset R^n$. Let f and g be two mappings of X into R^1. Assume that $x \in X$ passes to a limit $a \in X$, and consider the ratio $f(x)/g(x)$. The notation $f(x) = o(g(x))$ means that the ratio tends to 0, $f(x) = O(g(x))$ means that the ratio remains bounded, and $f(x) \sim g(x)$ means that the ratio tends to 1.

The span of a subset X of a linear space is the set of all (finite) linear combinations of elements in X. When X and Y are Hilbert spaces, the symbol $[X; Y]$ denotes the linear space of all continuous linear mappings of X into Y; $[X; Y]$ is supplied with the uniform – synonymously, norm – topology [94, page 410].

When X and Y are Euclidean spaces, any linear mapping $f : X \rightsquigarrow Y$ has a unique representation as a matrix with respect to the natural bases in X and Y. We shall often use the same symbol to represent f and its matrix when those bases are understood. Similarly, when X and Y are the Hilbert coordinate space, a member of $[X; Y]$ and its natural matrix representation may be denoted by the same symbol.

Electrical units and their symbols are as follows: volt V, ampere A, ohm Ω, mho $\mho$, farad F, and henry H. We use some of these symbols (but in italic font) for other purposes as well. The context in which they are used should dispel any possible confusion.

♣ denotes the end of a proof or example.

1.2 Countable Graphs

In the mathematical sense an infinite electrical network is obtained by placing a certain analytical structure upon an infinite graph. With regard to graphs, we gather in this section most of the definitions that we shall employ.

Let K and J be finite or denumerably infinite index sets. Let $\mathcal{N} = \{n_k\}_{k \in K}$ be an indexed set, where n_k and n_m are distinct if $k \neq m$. The members of $\mathcal{N}$ are called *nodes*. Also, let $\mathcal{B} = \{b_j\}_{j \in J}$ be an indexed set of two-entry families $b_j = \{n_k, n_m\}$ of nodes. The members of $\mathcal{B}$ are called *branches*. That each b_j is a "family" instead of a set means that $k = m$ is allowed in $b_j = \{n_k, n_m\}$; such a branch $\{n_k, n_k\}$ will be called a *self-loop*. Also allowed is $b_i = b_j$ with $i \neq j$; in this case b_i and b_j are said to be in *parallel*. We will at times suppress the indexings of $\mathcal{N}$ and $\mathcal{B}$, but it will be understood that such indexings exist.

We say that $b_j = \{n_k, n_m\}$ is *incident to* n_k and *to* n_m and similarly n_k and n_m are *incident to* b_j. A node is *isolated* if it is not incident to any branch. A node is called *infinite* if there is an infinity of branches incident to the node; otherwise, it is called *finite*. The cardinality of the set of branches incident to a node is called the *degree* of that node.

A *countable graph* $\mathcal{G}$ is defined as the pair $\mathcal{G} = (\mathcal{B}, \mathcal{N})$. $\mathcal{G}$ is called a *finite* graph if both $\mathcal{B}$ and $\mathcal{N}$ are finite sets; otherwise, it is called *countably infinite*. $\mathcal{G}$ is also called *locally finite* if all its nodes are finite. If $n \in \mathcal{N}$ or $b \in \mathcal{B}$, we shall say that n *is in* $\mathcal{G}$ or b *is in* $\mathcal{G}$ and that $\mathcal{G}$ *has* or *contains* n or b.

This definition of $\mathcal{G}$ is quite standard. However, it is not amenable to a generalization we shall make in Chapter 5, where the idea of a transfinite graph is introduced. For that purpose we present in the next section a different definition, which will be equivalent to the present one if isolated nodes are disallowed.

Given the graph $\mathcal{G} = (\mathcal{B}, \mathcal{N})$, let $\mathcal{N}_*$ be a subset of $\mathcal{N}$ and let $\mathcal{B}_*$ be a subset of $\mathcal{B}$ such that all the nodes of all the branches of $\mathcal{B}_*$ appear in $\mathcal{N}_*$. Then, the graph $(\mathcal{N}_*, \mathcal{B}_*)$ is called a *subgraph* of $\mathcal{G}$. If $\mathcal{B}_*$ is given and $\mathcal{N}_*$ is the set of all nodes to which the branches in $\mathcal{B}_*$ are incident, then $\mathcal{G}_* = (\mathcal{N}_*, \mathcal{B}_*)$ is called the *subgraph induced by* $\mathcal{B}_*$ or a *branch-induced subgraph*. Similarly, if $\mathcal{N}_*$ is given and $\mathcal{B}_*$ is the set of all branches all of whose nodes belong to $\mathcal{N}_*$, then $\mathcal{G}_*$ is called the *subgraph induced by* $\mathcal{N}_*$ or a *node-induced subgraph*.

Let $\mathcal{H}$ and $\mathcal{K}$ be two subgraphs of a given graph $\mathcal{G}$. The *union* $\mathcal{H} \cup \mathcal{K}$ is the subgraph whose node set (branch set) is the union of the node sets (respectively, branch sets) of $\mathcal{H}$ and $\mathcal{K}$. The *intersection* $\mathcal{H} \cap \mathcal{K}$ is defined the same way except that "union" is replaced by "intersection." When $\mathcal{H} \cap \mathcal{K}$ has at least one node, we say that $\mathcal{H}$ and $\mathcal{K}$ *intersect* or *meet*. $\mathcal{H} - \mathcal{K}$ denotes the subgraph of $\mathcal{H}$ induced by all the nodes in $\mathcal{H}$ that are not in $\mathcal{K}$. At times, we replace $\mathcal{K}$ by a node set $\mathcal{S}$, in which case it is understood that $\mathcal{H} - \mathcal{S} = \mathcal{H} - \mathcal{K}$, where now $\mathcal{K}$ is the subgraph $\{\emptyset, \mathcal{S}\}$. On the other hand, if b is a branch of $\mathcal{H}$, $\mathcal{H} - b$ denotes the subgraph obtained by removing b from the branch set of $\mathcal{H}$ but leaving the node set unaltered. Finally, $\mathcal{H} \subset \mathcal{K}$ denotes that $\mathcal{H}$ is a subgraph of $\mathcal{K}$.

Let $\mathcal{M}$ be subgraph of $\mathcal{G}$. A *partition* $\{\mathcal{M}_p\}$ of $\mathcal{M}$ is a collection of subgraphs $\mathcal{M}_p$ of $\mathcal{M}$ such that every branch of $\mathcal{M}$ appears in one and only one $\mathcal{M}_p$ and $\mathcal{M} = \bigcup \mathcal{M}_p$. A branch b is said to be *adjacent* to a subgraph $\mathcal{K}$ if b is not in $\mathcal{K}$ and at least one node of b is in $\mathcal{K}$. On the other hand, a node n_0 and a subgraph $\mathcal{M}$ are called *adjacent* if n_0 is not in $\mathcal{M}$ and there exists a branch joining n_0 to a node of $\mathcal{M}$. A branch is said to *join* two subgraphs of $\mathcal{G}$ if those subgraphs do not meet and b has a node in each of the subgraphs.

A *path* is an alternating sequence of nodes n_{k_m} and branches b_{j_m}

$$\{\ldots, n_{k_m}, b_{j_m}, n_{k_{m+1}}, b_{j_{m+1}}, \ldots\} \tag{1.1}$$

wherein the indices m are restricted to the integers (as distinct from the transfinite ordinals) and number the elements of (1.1) sequentially as indicated, no node appears more than once, and each branch is incident to the two nodes immediately preceding and succeeding it in the sequence. Consequently, no branch appears more than once in the sequence. The sequence may be either finite, or one-way infinite, or two-way infinite, in which case it is called a *finite path*, or a *one-ended path*, or an *endless path* respectively. Moreover, if the sequence terminates in either direction, it is required that it terminate at a node; that node is called an *end node* of the path. A finite path with the terminal nodes n_a and n_b is called an $n_a n_b$-*path*. If the path has at least one branch, it is called *nontrivial*; a *trivial path* is a singleton $\{n_{j_m}\}$. Two paths are called *totally disjoint* if no node is contained in both paths.

A *loop* is defined exactly as is a finite nontrivial path except that its two end nodes are required to be the same node. Thus, a self-loop $\{n, n\}$ can be viewed as a special case of a loop in the following way: $\{n, \{n, n\}, n\}$.

Two nodes n_a and n_b are said to be *connected* if there is a path that contains both nodes; this is the same as saying that there is a finite path with n_a and n_b as its two end nodes. If this is true for every two nodes in $\mathcal{G}$, then $\mathcal{G}$ itself is called *connected*. Two branches are said to be *connected* if their nodes are connected. A *component* of a graph is a subgraph that is maximal with respect to connectedness. A *forest* $\mathcal{F}$ is a graph containing no loops; $\mathcal{F}$ is called a *tree* if it is also connected.

When $\mathcal{F}$ is a subgraph of $\mathcal{G}$, it is said to be *spanning in* $\mathcal{G}$ if it contains every node in $\mathcal{G}$. An *end node* of $\mathcal{F}$ is a node whose degree with respect to $\mathcal{F}$ is 1, and its incident branch is called an *end branch*.

1.3 0-Graphs

In order to facilitate our discussion of transfinite graphs in Chapter 5, we now define a special kind of (finite or infinite) graph in an unusual way. The reasons why we resort to an unconventional approach to graphs are the following. The conventional definition of a graph starts with a set of nodes and then defines the branches as unordered pairs of nodes. This is the reverse of how a physical electrical network is usually constructed. Indeed, beginning with a variety of electrical elements, each having two or more terminals, one constructs the nodes of a physical network by shorting the terminals together; in particular, starting from a set of branches one defines the nodes as shorts between various branch ends. Moreover, for our purposes there is a more forceful misgiving about the conventional definition; it arises when one tries to extend it to transfinite graphs, as was done in [177]. To accommodate connections between conventional nodes and the extremities of a transfinite graph, one is led to defining a whole hierarchy of "extended branches" involving pairs of nodes and extremities of various ranks. This results in a ponderous set of definitions.

Some simplification can be achieved if we formalize how a physical network is usually put together. We can view each terminal of an electrical element as being an "elementary tip" and define a branch as being a pair of such tips. Nodes are then sets of such tips shorted together. Next, the extremities of an ordinary infinite graph are considered to be "tips of rank zero," and a "transfinite graph of rank one" is constructed by shorting together elementary tips and tips of rank zero. This process can be continued to obtain transfinite graphs of higher ranks. In the end, branches remain as they are in an ordinary graph and nodes are merely shorts between various kinds of "tips."

Adopting this alternative approach, we now present our specialized definition of an ordinary (finite or infinite) graph. Let $\mathcal{T}$ be a finite set with an even number of elements or a denumerably infinite set. Call each element of $\mathcal{T}$ an *elementary tip* or just a *tip* for short. Partition $\mathcal{T}$ into subsets of two elementary tips each and call each subset a *branch*. Thus, a branch has the form $\{t_1, t_2\}$, where the t_1 and t_2 are its two tips. $\mathcal{B}$ denotes the set of all branches; it is countable. As before, we usually index the branches with the set J; thus, $\mathcal{B} = \{b_j : j \in J\}$.

Also, partition $\mathcal{T}$ in an arbitrary way: $\mathcal{T} = \cup n_k$. Thus, $n_k \cap n_m$ is void if $k \neq m$. The subsets n_k will be called 0-*nodes* or simply *nodes*. (Later on, we will view an n_k having two or more members as being an electrical connection that "shorts together" the tips in n_k.) As a terminology we will need in Chapters 3 and 5, we say that a 0-node *embraces* itself as well as its tips. We will often index the 0-nodes with the set K. $\mathcal{N}^0$ will denote the set of 0-nodes: $\mathcal{N}^0 = \{n_k : k \in K\}$.

A 0-*graph* is defined as the pair $\mathcal{G}^0 = (\mathcal{B}, \mathcal{N}^0)$. Note that we do not display $\mathcal{G}^0$ as the triplet $(\mathcal{T}, \mathcal{B}, \mathcal{N}^0)$ simply because $\mathcal{T}$ can be recovered as the union of all the branches. More important, observe that the present definition implies a well-defined mapping from every branch to either a single 0-node or to a pair of 0-nodes, namely, the 0-node or 0-nodes in which the two tips of the branch are found. This means that a 0-graph becomes a graph when we use that mapping to define the *incident* node or nodes for each branch. It follows that all the definitions pertaining to a graph can be transferred to 0-graphs. This we do. Later on, we shall refer to a path (1.1) in a 0-graph as a 0-*path* and will say that it *embraces* itself, all its elements, and also all the elementary tips in its branches. Also, note that there are no isolated nodes according to this unconventional definition of a graph.

However, a subgraph of a 0-graph may not be a 0-graph because, for instance, a subgraph may have isolated nodes. Nonetheless, we can modify the definition of a branch-induced subgraph to generate a 0-graph from a specified subset $\mathcal{B}_*$ of $\mathcal{B}$. Let $\mathcal{T}_*$ be the union of the branches in $\mathcal{B}_*$ (i.e., the set of all tips of the branches of $\mathcal{B}_*$). Delete from each node in $\mathcal{N}^0$ all tips that are not in $\mathcal{T}_*$. The set of all such reduced but nonvoid nodes is denoted by $\mathcal{N}_*^0$. Finally, $\mathcal{G}_*^0 = (\mathcal{B}_*, \mathcal{N}_*^0)$ is a 0-graph; we call it the *reduction of* $\mathcal{G}^0$ *induced by* $\mathcal{B}_*$ or a $\mathcal{B}_*$-*reduced* 0-graph or simply a *reduced* 0-graph.

For 0-graphs 0-*connectedness* means the same thing as connectedness. A 0-*section* of a 0-graph $\mathcal{G}^0$ is a reduction of $\mathcal{G}^0$ induced by a maximal set of branches that are pairwise 0-connected. At this stage of our definitions, a 0-section is equivalent to a component of $\mathcal{G}^0$, but in Section 3.2 we will generalize the idea of connectedness and thereby introduce a new meaning for a 0-section.

Before leaving this section, it is worth emphasizing that the introduction of 0-graphs anticipates our construction of transfinite graphs but is not essential. Another exposition of transfinite graphs is given in [182] wherein the role of the 0-graphs is borne by ordinary graphs.

1.4 Electrical Networks

An *electrical network* is defined by assigning to each branch of a given graph or 0-graph several electrical parameters connected together in a specified way. Moreover, an orientation is assigned to each branch with respect to which the direction of the branch current and branch voltage is measured. That current or voltage is positive when the orientation and direction agree and is negative when they disagree.

A *resistive branch with independent voltage and current sources* is indicated in Figure 1.1 in two entirely equivalent ways. The symbols in that diagram denote real numbers, but we will also use r, e, and h to denote the kind of parameter at hand – thereby avoiding a more cumbersome notation. The branch's orientation is the same as that indicated by the arrow for the branch current i or the plus-to-minus direction for the branch voltage v. r is the *branch resistance*, which is always a positive number (nonzero and finite); its reciprocal $g = 1/r$ is the *branch conductance*. e is the *branch voltage source* measured positively as a voltage rise (from $-$ to $+$) with respect to the branch's orientation. h is the *branch current source* measured positively in the direction opposite to the branch's orientation. These quantities are related by *Ohm's law*:

$$v + e = r(i + h) \tag{1.2}$$

or equivalently

$$i + h = g(v + e). \tag{1.3}$$

The two connections of Figure 1.1 are equivalent because e and h are both independent of any other voltage or current. The customary units of volts (V), amperes (A), ohms (Ω), and mhos ($\mho$) for voltages, currents, resistances, and conductances are adopted throughout. Actually, since r and g are real numbers which enter as multiplying factors in (1.2) and (1.3), we have in fact defined a linear element. More generally, however, r and g may be mappings of the real line R^1 into R^1; when these functions are different from multiplications by constants, we have instead a nonlinear resistance $r(\cdot)$ and a nonlinear conductance $g(\cdot)$. By rearranging (1.2) and (1.3), we can represent any branch in two more equivalent forms. *Thevenin's form* is shown in Figure 1.1(b), and *Norton's form* is shown in Figure 1.1(c). All these forms impose exactly the same relationship between v, i, e, and h.

There are other kinds of branches. A *pure voltage source* is shown in Figure 1.2(a). For this kind of branch, it is required that n_1 and n_2 be distinct nodes. Also, $v = -e$, and e is required to be nonzero; on

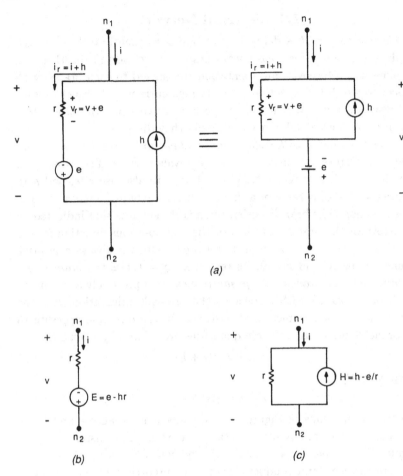

Figure 1.1. (a) Two equivalent forms of a resistive branch. The indicated quantities, other than the nodes n_1 and n_2, are real numbers and r is positive. (Also shown are two alternative symbols for the voltage source e. Both are used in this book.) (b) The Thevenin equivalent form. (c) The Norton equivalent form.

the other hand, i is determined by the network in which the branch appears. When e is set equal to zero, this branch becomes a *short circuit* or synonymously a *short*; we will not allow a short circuit to be a branch and will coalesce n_1 and n_2 into a single node by replacing n_1 and n_2 if need be.

A *pure current source* is indicated in Figure 1.2(b). Now, $i = -h$, where h is required to be nonzero; in this case, it is v that is determined by the rest of the network. When h is set equal to zero, we get an

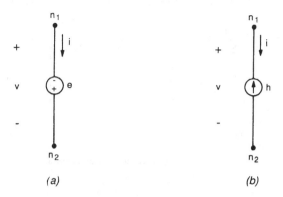

Figure 1.2. (a) A pure voltage source. n_1 and n_2 must be different nodes. Also, $v = -e \neq 0$. (b) A pure current source. Now, $i = -h \neq 0$.

open circuit or synonymously an *open*; this too is not considered to be a branch.

Reactive elements arise when currents and voltages are taken to be temporally variable and r in (1.2) and g in (1.3) are replaced by time derivatives. An *inductance l*, measured in henries (H), is indicated in Figure 1.3(a) and is defined by

$$v = l\frac{di}{dt}. \tag{1.4}$$

A *capacitance c*, measured in farads (F), is shown in Figure 1.3(b) and is defined by

$$i = c\frac{dv}{dt}. \tag{1.5}$$

These parameters too can occur in conjunction with sources e and h within a single branch; that is, r in Figure 1.1 can be replaced by l or c. When all three kinds of parameters occur among the branches of a network, that network is called an *RLC network*.

Coupling between branches occurs when the current or voltage in one branch induces a current or a voltage in another. This may be represented by a *dependent source* where e or h, instead of being a fixed real number, depends upon the voltage or current in another branch.

By the *voltage-current regime* or, for short, the *regime* of a network we shall simply mean an assignment of branch voltages and branch currents to all the branches. A typical regime is usually determined by various laws and restrictions specified in the concomitant discussion. Those laws and restrictions may change depending on the objectives of the analysis. Furthermore, the graph-theoretic definitions specified in the preceding

Figure 1.3. (a) An inductance. (b) A capacitance.

two sections are transferred to electrical networks simply by replacing "graph" by "network"; for example, we may speak of "subnetworks" instead of "subgraphs." Finally, most of this book will be focused on purely resistive networks with only independent sources. Sometimes, however, we will examine more general kinds of networks. There will be no coupling between branches unless the opposite is explicitly stated.

1.5 Kirchhoff's Laws

The fundamental relationships, which are customarily used to determine the voltage-current regime in a finite resistive network, are Ohm's law, given by (1.2), and Kirchhoff's two laws. *Kirchhoff's current law* asserts that at any given node n the algebraic sum of all branch currents flowing through n is zero, that is,

$$\sum_{(n)} \pm i_j = 0 \qquad (1.6)$$

where $\sum_{(n)}$ denotes a summation over the indices j for the branches b_j incident to n, i_j is the jth branch current, and the plus (minus) sign is used when b_j is oriented toward n (respectively, away from n). The index of an incident self-loop appears twice, and $+i_j$ and $-i_j$ both appear in the summation; this will be significant when we consider the absolute convergence of Kirchhoff's current law. As for *Kirchhoff's voltage law*, we first choose a loop L and assign to it an orientation. That law states that the algebraic sum of the branch voltages around L is zero, that is,

$$\sum_{(L)} \pm v_j = 0 \qquad (1.7)$$

where now $\sum_{(L)}$ is a summation over the indices j of the branches in

L, v_j is the jth branch voltage, and the plus (minus) sign is used if the orientations of the loop and branch agree (respectively, disagree).

These three laws truly determine a unique voltage-current regime when the resistive network has only a finite number of branches. In general, however, they fail to do so when the network is infinite. Worse yet, Kirchhoff's current law need not hold at an infinite node. These and other peculiarities of infinite networks are explored through a series of examples in the next section.

1.6 Curiouser and Curiouser

Example 1.6-1. Let us try to extend Kirchhoff's current law to an infinite node. Figure 1.4(a) indicates an infinite network wherein a 1-V voltage source in series with a 1-Ω resistor is connected to an infinite parallel circuit of 1-Ω resistors. The infinite parallel connection should be equivalent to a short circuit, according to the rule for combining parallel resistances, and so the voltage v between nodes n_1 and n_2 should be zero. Hence, the current i through the source ought to be 1 A, whereas the currents flowing through each of the purely resistive branches ought to be zero. However, calculus is unambiguous about the fact that an infinite series of zeros sums to zero. Therefore, we are led to conclude that 1 A flows toward node n_1 while 0 A flows away from it.

Perhaps our supposition that $v = 0$ is wrong; perhaps $v \neq 0$. If so, then $i = 1 - v$ A and the current flowing downward through every purely resistive branch is v A. In this case, calculus dictates that an infinite series of nonzero constants v, all identical, is infinite. So, now we have a finite current flowing toward node n_1, and an infinite current flowing away from it.

We have to conclude that Kirchhoff's current law fails at node n_1 – and at node n_2 too.

Another way of approaching this paradox is to approximate the infinite parallel circuit by a large but finite number m of parallel 1-Ω resistors, as indicated in Fig. 1.4(b). Let us now denote the currents flowing in this finite network by $i_{m,0}, i_{m,1}, \ldots$ as shown. The first subscript m is simply a parameter indicating the number of purely resistive branches. By Ohm's law and Kirchhoff's laws, $v = 1/(m+1)$, $i_{m,0} = m/(m+1)$, and $i_{m,j} = 1/(m+1)$ for all $j = 1, \ldots, m$. Moreover, Kirchhoff's current law at node n_1 is satisfied because

$$i_{m,0} = \sum_{j=1}^{m} i_{m,j}.$$

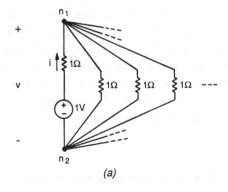

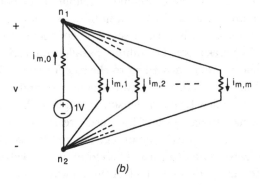

Figure 1.4. (a) An infinity of 1-Ω resistors all connected in parallel with a branch having a 1-Ω resistor and a 1-V voltage source in series. (b) A finite network that approximates the infinite network. All resistors are again 1 Ω.

There is no problem in passing to the limit as $m \to \infty$ on both sides of this equation; we obtain $1 = 1$ so long as we sum first and then take the limit on the right-hand side. However, in passing to the limit for the current $1/(m + 1)$ in each purely resistive branch, we have in effect interchanged these two processes: We have taken the limit and then summed. This is invalid, for we get $1 = 0$. It would have been valid had $\sum_{j=1}^{m} i_{m,j}$ converged uniformly with respect to all m, but this unfortunately is not the case.

(One is tempted to resolve the paradox by invoking the infinitesimals and saying that the currents in the branches to the right of the source branch in Figure 1.4(a) are infinitesimals which sum to 1 A. To be sure, nonstandard analysis [114] has rigorously resurrected the infinitesimals, but a nonstandard theory for infinite electrical networks has yet to be established.)

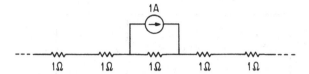

Figure 1.5. An infinite loop consisting of an infinity of 1-Ω resistors all connected in series with a branch having a 1-Ω resistor and a 1-A current source in parallel. The network extends infinitely to the left and to the right, and its "extremities at infinity" are shorted together.

In Chapter 3 we will show that for a finite-power regime Kirchhoff's current law always holds at an infinite node if the sum of the conductances for the incident branches is finite. When that sum is infinite, Kirchhoff's current law may or may not hold depending on the constraints imposed by the rest of the network. ♣

Example 1.6-2. If we try to extend Kirchhoff's voltage law to an infinite loop, that is, to an endless path whose two "extremities at infinity" are assumed to be shorted together, we run into a similar paradox. Consider the network of Figure 1.5. By interchanging voltages and currents in the arguments given in Example 1.6-1, one can show that Kirchhoff's voltage law cannot be satisfied around that infinite loop.

However, if the sum of the resistances around any infinite loop is finite, then Kirchhoff's voltage law will again hold under a finite-power regime. It may or may not hold when the sum of the resistances is infinite. This too will be shown in Chapter 3. ♣

Example 1.6-3. To see that Ohm's law and Kirchhoff's laws need not by themselves determine a unique voltage-current regime even when all nodes are finite, consider the infinite network of Figure 1.6(a). With the voltage source e and all the resistances r_k given, we might ask, "What is the value of the current i in the voltage source?" Actually, i can be any value at all – in the sense that, upon choosing i arbitrarily, we can find a voltage-current regime supporting that i and satisfying Ohm's law and Kirchhoff's laws everywhere, Kirchhoff's voltage law being applied around finite loops only. Indeed, a knowledge of i coupled with Ohm's law and Kirchhoff's voltage law determine the voltage and thereby the current for r_3. Then, Kirchhoff's current law determines the currents in r_4 and r_5. Alternately applying these laws, we can continue these manipulations to determine voltages and currents further and further toward the right. Truly the asserted voltage-current regime exists.

Moreover, the standard rule that the power $\sum e_j i_j$ delivered by the

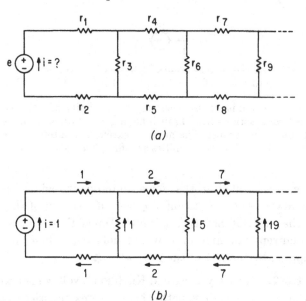

(a)

(b)

Figure 1.6. (a) An infinite network in which e and all the r_j are given. The question is, "What is i when Kirchhoff's laws and Ohm's law are satisfied?" It turns out that i can be any number at all. (b) The branch currents when $e = r_1 = r_2 = \cdots = 1$ and i is chosen to be 1 as well.

sources is equal to the power $\sum r_j i_j^2$ dissipated in the resistors, which holds for finite resistive networks, may well be violated. For instance, if $e = 1$ V and $r_j = 1$ Ω for all j and if i is chosen to be 1 A, then the branch currents take on the values shown in Figure 1.6(b). In this case, the source power is 1 watt, whereas the dissipated power is infinitely large. To restore the standard rule, we would have to say that there is a power source out at infinity pumping energy into the network. By specifying i, we indirectly specify that (possibly infinite) power source. Moreover, there is no way of directly specifying an infinite power source out at infinity; standard calculus does not allow us to distinguish one infinite source from another (but nonstandard calculus does). If on the other hand we had chosen $i = (-1 + \sqrt{3})/2$, it can be shown that the current values would tend to zero and that the power, $(-1 + \sqrt{3})/2$ watts, delivered by the source would equal the power dissipated in all the resistors so that no power source at infinity would be needed. ♣

A general procedure for determining all possible voltage-current regimes from only Ohm's and Kirchhoff's laws for arbitrary infinite networks will be established in the next chapter.

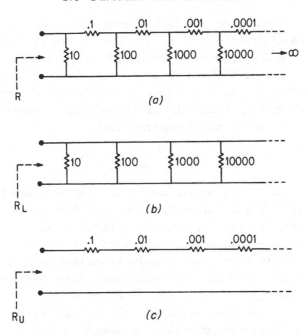

Figure 1.7. (a) An infinite ladder. All resistance values are given in ohms. R is the driving-point resistance as observed from the input terminals. (b) The infinite parallel circuit obtained by letting every horizontal resistor tend to a short circuit. An open circuit at infinity is assumed. (c) The infinite series circuit obtained by letting every vertical resistor tend to an open circuit. A short circuit at infinity is assumed.

Example 1.6-4. One might surmise from the preceding example that, if in addition to Ohm's law and Kirchhoff's laws we require that the total power dissipated in the network be finite, then a unique voltage-current regime might be achieved. This is true for some networks, such as that of Figure 1.6(b), but not for all. In particular, it is not true for the infinite ladder network of Figure 1.7(a). R of that figure denotes the input resistance, which is defined as the voltage appearing across the input terminals when a 1-A current source is impressed on those terminals. But what is R – and does it even exist as a unique quantity?

We might try to compute R as the limit of the unique input resistances of the finite networks obtained by replacing all the resistors after some point by open circuits and then moving that point out toward infinity. It can be shown that the powers dissipated in the finite networks approach a finite limit. In this case we might say that R is the input resistance when there is an open circuit at infinity. Instead of computing this R, let us establish a lower bound on it by borrowing a fundamental principle, which will be established in Chapter 3 for infinite networks. That principle states that R is not less than the value R_L it assumes when all the horizontal resistors are replaced by short circuits. This yields the circuit of Figure 1.7(b). Because of the open circuit at infinity, the impressed 1 A current must flow through the vertical resistors. Consequently, R_L ought to be the parallel sum of those resistances:

$$R_L = \frac{1}{.1 + .01 + .001 + \cdots} = 9.$$

Thus, it appears that $R \geq 9$ when there is an open circuit at infinity.

On the other hand, we might try to determine R as the limit of the input resistances for the finite networks obtained by replacing all the resistors after some point by short circuits and then moving that point out toward infinity. Once again, it can be shown that the powers dissipated in the finite networks tend toward a finite limit. Now, we might say that in the limit there is a short circuit at infinity and that R is the corresponding input resistance. The aforementioned principle also dictates that R cannot be larger than the value R_U it assumes when all the vertical resistors are replaced by open circuits as shown in Figure 1.7(c). Because of the short circuit at infinity, R_U is the series sum of the horizontal resistances, that is,

$$R_U = .1 + .01 + .001 + \cdots = 1/9.$$

Thus, $R \leq 1/9$ when there is a short circuit at infinity.

Combining the last two results, we can conclude that R depends on whether there is an open circuit or a short circuit at infinity. (In fact, a more complicated argument [179] shows that R can take on any one of the values between R_L and R_U when an appropriate load resistance is connected at infinity and a finite power regime exists.) To be sure, we have arrived at this conclusion through heuristic arguments, but they do indicate that a network's voltage-current regime, finite-powered though it may be, may well depend on what is connected out at infinity. ♣

Example 1.6-5. This example consists more of suggestions than affirmations. The affirmations will be forthcoming in subsequent chapters. As was noted parenthetically in the last paragraph, it can make sense to talk about resistive loads connected to the infinite extremities of certain ladder networks, and those loads will affect the voltage-current regimes of the networks. However, if a resistive load can be connected at infinity, so too can the input to another infinite ladder, which in turn might be

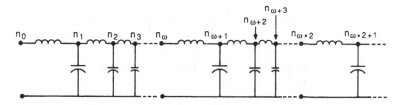

Figure 1.8. A transfinite network consisting of infinitely tapered LC ladders. Here, ω denotes the first transfinite ordinal. Node n_0 is connected to node n_ω through a transfinite path but not through a finite path.

loaded at its infinite extremity by the input to still another infinite ladder, and so forth. This leads us to the idea of a transfinite ladder, which is distinguished from the usual infinite ladder by the property that there are pairs of nodes connected by infinite paths but not by finite paths.

Another way of arriving at the concept of a transfinite ladder is to modify the customary way a distributed LC transmission line is obtained from a lumped LC ladder. Traditionally, that line is viewed as the continuous limit of a series-L-shunt-C ladder network, where the L and C lumped parameters uniformly approach infinitesimal quantities. However, we obtain a different kind of transmission line by keeping the L and C parameters as real positive numbers (not infinitesimals) but allowing them to vary with distance in such a fashion that the sequence of different L and C values approaches zero; this is illustrated in Figure 1.8. In other words, we can have a tapered ladder where the tapering progresses infinitely to a limit, at which point output terminals are taken to exist. Another such ladder may then be connected to those output terminals to obtain a transfinite ladder.

A question that arises naturally for transfinite ladders is how far can one extend a transfinite ladder; that is, if one numbers its nodes with the transfinite ordinals [1], [116] as indicated in Figure 1.8, how large can those ordinals become? It appears that they can reach any countable ordinal. We discuss this matter in Chapter 6 for nonlinear resistive ladders.

Furthermore, if ladder networks can be extended transfinitely, can grids be similarly extended? One possibility is suggested by Figure 1.9. But how about arbitrary networks? To extend these transfinitely, a proper definition of its extremities is needed – and has been formulated, albeit in a rather complicated way. Upon connecting infinite graphs at

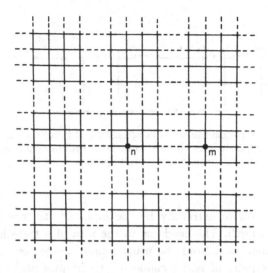

Figure 1.9. A heuristic suggestion of how an ordinary two-dimensional infinite grid might be connected at its extremities to other such grids to obtain a transfinite grid. It is implied here that there is a transfinite path connecting node n to node m, but not a finite one.

their extremities, one obtains a new construct in graph theory, namely, the transfinite graph. Moreover, once electrical parameters are assigned to the branches, a theory for transfinite electrical networks can be established in much the same way as it is for ordinary infinite networks. This is investigated in Chapter 5. ♣

1.7 Electrical Analogs for Some Differential Equations

The finite-difference approximations of various equations of mathematical physics yield partial difference equations, which can often be realized as electrical networks, in particular, grids. Kron and others have explored this analogy in some detail ([2], [28], [76]–[81], [92], [124], [135, pages 237–257], [136, pages 365–372], [138]. In certain situations it is reasonable to view the domain, in which such an equation is to hold, as being infinite in extent. This is the so-called exterior problem. In this circumstance the analogous electrical grid becomes an infinite one. The transition between a partial differential equation and its approximating electrical grid will now be explicated for two specific cases, Yukawa's potential equation and a certain polarization of Maxwell's equations.

Yukawa's Potential

The equation that governs Yukawa's potential v is

$$\nabla \cdot (\alpha \nabla v) - \beta v = f \tag{1.8}$$

where ∇ is the gradient operator, $\nabla\cdot$ is the divergence operator, α and β are given constituent parameters of the medium, and f is a given inhomogeneous term ([27], [46]). In general, α, β, and f vary spatially. Moreover, we shall require that α and β be nonnegative everywhere; this will insure that all the resistances in the analogous grid will be positive. As special cases we have Poisson's equation, which occurs when β is zero everywhere, and Laplace's equation when in addition f is zero everywhere. (Helmholtz's equation arises when β is negative everywhere, but this leads to some negative resistances in the grid.)

Let us work in a two-dimensional rectangular coordinate system (x, y). We will choose sample points of the form (x_j, y_k) where $x_j = j\Delta x$, $y_k = k\Delta y$, Δx and Δy are positive increments, and j and k are integers. The corresponding sampled values of, say, $v(x, y)$ are denoted by $v_{j,k} = v(j\Delta x, k\Delta y)$. On the other hand, we will sample $\alpha(x, y)$ at the "in between" points $(x_{j+1/2}, y_k)$ and $(x_j, y_{k+1/2})$, where $x_{j+1/2} = (j+1/2)\Delta x$ and $y_{k+1/2} = (k + 1/2)\Delta y$, and will write $\alpha_{j+1/2,k} = \alpha(x_{j+1/2}, y_k)$ and $\alpha_{j,k+1/2} = \alpha(x_j, y_{k+1/2})$. A symmetric finite-difference expression for $\alpha \partial v / \partial x$ at $(x_{j+1/2}, y_k)$ is

$$\alpha \frac{\partial v}{\partial x} \approx \frac{\alpha_{j+1/2,k}}{\Delta x} (v_{j+1,k} - v_{j,k}). \tag{1.9}$$

The next step is to subtract from (1.9) the expression obtained from it by replacing j by $j-1$ and then to divide the resulting difference by Δx. This yields a second-order, symmetric, finite-difference approximation for the x derivatives

$$\frac{\partial}{\partial x} \alpha \frac{\partial v}{\partial x}$$

in $\nabla \cdot (\alpha \nabla v)$. In addition, let us multiply that second-order approximation by $\Delta x \Delta y$. The result is

$$\frac{v_{j+1,k} - v_{j,k}}{r_{j+1/2,k}} + \frac{v_{j-1,k} - v_{j,k}}{r_{j-1/2,k}} \tag{1.10}$$

where

$$r_{j+1/2,k} = \frac{\Delta x}{\alpha_{j+1/2,k}\Delta y} \tag{1.11}$$

and

$$r_{j-1/2,k} = \frac{\Delta x}{\alpha_{j-1/2,k}\Delta y}. \tag{1.12}$$

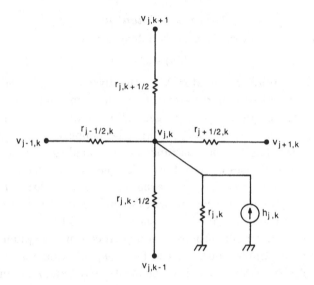

Figure 1.10. The resistors and current source incident to node (j, k) in the rectangular grid obtained from a finite-difference approximation of Yukawa's potential equation (1.8). The ground is at zero potential.

The expression (1.10) can be interpreted as the sum of two currents flowing toward a node at (x_j, y_k) through two resistors (1.11) and (1.12) connected between that node and two other nodes at (x_{j+1}, y_k) and (x_{j-1}, y_k); those nodes possess the indicated sampled potentials (i.e., node voltages), as shown in Figure 1.10.

A similar manipulation for

$$\frac{\partial}{\partial y} \alpha \frac{\partial v}{\partial y}$$

yields the sum of two currents flowing toward the node at (x_j, y_k) from the nodes at (x_j, y_{k+1}) and (x_j, y_{k-1}) through the resistors

$$r_{j,k+1/2} = \frac{\Delta y}{\alpha_{j,k+1/2} \Delta x}$$

and

$$r_{j,k-1/2} = \frac{\Delta y}{\alpha_{j,k-1/2} \Delta x}.$$

Furthermore, we may sample βv at (x_j, y_k) and then multiply by $\Delta x \Delta y$ to obtain $v_{j,k}/r_{j,k}$ where

$$r_{j,k} = \frac{1}{\Delta x \Delta y \beta_{j,k}}.$$

We can interpret $v_{j,k}/r_{j,k}$ as the current flowing from the node at (x_j, y_k) through the resistor $r_{j,k}$ to a ground node of zero potential. A similar sampling and multiplication of f yields an expression which can be viewed as a current source $h_{j,k} = \Delta x \Delta y f_{j,k}$ sending current from the ground node into the node at (x_j, y_k). These resistors and current source may be connected together as indicated in Figure 1.10, which is just that portion of a rectangular resistive grid incident to the node at (x_j, y_k).

On combining all the aforementioned currents in accordance with (1.8), we impose in effect Kirchhoff's current law upon the node at (x_j, y_k). Ohm's law is satisfied simply by the definition of our resistors. Finally, Kirchhoff's voltage law is automatically satisfied around any finite loop. Indeed, we have taken each branch voltage to be the difference between the potentials at the branch's incident nodes. Thus, on summing the branch voltages around any loop, each node potential enters the sum exactly twice – once with a plus sign and once with a minus sign. We can conclude that this finite-difference approximation of (1.8) can be realized by a resistive-grid extension of Figure 1.10. All this can be done in higher dimensions too, for each dimension simply contributes a term like (1.10) to the approximation of $\nabla \cdot (\alpha \nabla v)$.

Appropriate finite-difference expressions for the gradient and divergence operators in cylindrical or spherical coordinates lead to cylindrical or spherical resistive grids ([183], [188]). We now present an example of a physical situation that leads to a cylindrical resistive grid which is not only infinite in extent but also exhibits some additional peculiarities such as infinite nodes as well as extremities at infinity that are shorted together and also shorted to a node.

Example 1.7-1. One of the ways of ascertaining the composition of the earth, through which an oil well is being drilled, is to lower into the borehole a cylindrical, electrically insulated vehicle, called a "sonde," on which are mounted a current probe and a potential probe. See Figure 1.11(a). A current h is injected into the earth through the current probe; in the so-called "normal log" ([128, pages 776–777]) the return probe of the current source is often implanted at the surface of the earth. The borehole is filled with drilling mud, which has a high electrical conductivity. The voltages measured by the potential probe yield information about the conductivity of the earth adjacent to the sonde.

Laplace's equation $\nabla \cdot (\alpha \nabla v) = 0$ governs the electrical potential v within the earth, where now α denotes electrical conductivity. A cylindrical finite-difference approximation yields a cylindrical resistive grid centered around the borehole. A vertical cross section through the axis

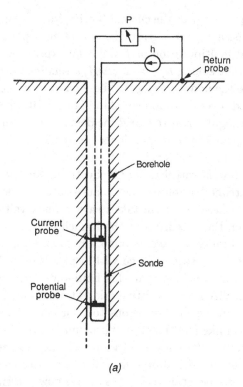

(a)

Figure 1.11(a). The normal-log configuration for electrical well logging. P denotes a voltmeter and h a current generator.

of that grid is shown in Figure 1.11(b). Since the probes are some feet apart whereas the earth's surface is perhaps a mile or so away, the earth appears to be an infinite medium. Hence, we may take it that an infinite grid is at hand and that the current source h has its grounded return probe at infinity.

Let us suppose that the earth's surface is a salt marsh whose electrical conductivity is very much larger than the conductivity deep within the earth. That salt marsh might then be viewed as an equipotential surface covering the earth. Moreover, the earth's conductivity may increase so rapidly as the earth's surface is approached that infinity is perceptible vertically upward from any point within the grid, that is, $\sum r_k < \infty$ where the r_k are the branch resistances along any one-ended vertical path extending upward indefinitely. In this case, it may well be appropriate to view the current source's return probe as being "connected to infinity in the vertically upward direction."

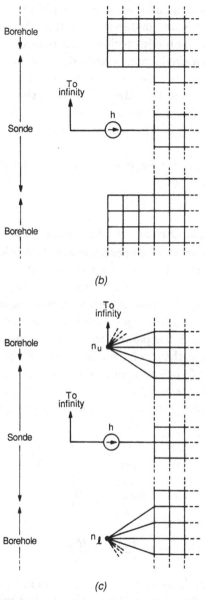

(b)

(c)

Figure 1.11(b,c). (b) A vertical cross section of the cylindrical grid for the infinite-network analog to the right of the borehole's centerline. The grid is indented in the vicinity of the insulated sonde. The return lead of the current source is at infinity in the vertically upward direction. (c) The electrical grid when the drilling mud in the borehole has infinite conductivity. The upper node n_u is shorted to infinity in the vertically upward direction.

Let us furthermore assume as an idealization that the drilling mud has effectively an infinite conductivity so that the borehole above the sonde is also an equipotential region. This reduces the borehole's grid above the sonde to a single infinite node n_u, as shown in Figure 1.11(c). Similarly, a finite or infinite node n_l represents the borehole below the sonde. If we take it that the potential of the borehole above the sonde is the same as the potential of the earth's surface, then the upper infinite node n_u is "shorted to infinity in the vertically upwards direction."

These physically motivated, heuristic ideas will be precisely defined in subsequent chapters. ♣

A Polarization of Maxwell's Equations

RLC grids arise when finite-difference approximations are taken to various polarized forms of Maxwell's equations. Let us examine one particular case. We will use an x, y, z coordinate system. E and H denote the electric and magnetic field intensity vectors respectively, which we assume are independent of the y-coordinate. Moreover, the polarization we study has it that E lies parallel to the y coordinate everywhere and that H lies parallel to the $y = 0$ plane everywhere. Thus, E has only a y-component E_y and H has only an x-component H_x and a z-component H_z. These components are assumed to vary only with respect to x, z, and the time variable t. Under these conditions Maxwell's curl equations simplify as follows:

$$\frac{\partial H_x}{\partial z} - \frac{\partial H_z}{\partial x} = \sigma E_y + \epsilon \frac{\partial E_y}{\partial t}, \tag{1.13}$$

$$\frac{\partial E_y}{\partial z} = \mu \frac{\partial H_x}{\partial t}, \tag{1.14}$$

$$\frac{\partial E_y}{\partial x} = -\frac{\partial H_z}{\partial t}. \tag{1.15}$$

The conductivity σ, dielectric permittivity ϵ, and magnetic permeability μ are assumed to vary in general with x and z but not with y and t.

We set up the same rectangular array of sample points as before except that this time y is replaced by z. Thus, $x_j = j\Delta x$ and $z_k = k\Delta z$. We could sample the spatial derivatives by using only symmetric differences. However, our notation will be simplified if we use instead a combination of forward and backward differences. Let $D_t = \partial/\partial t$,

$\sigma_{j,k} = \sigma(j\Delta x, k\Delta z)$, $\epsilon_{j,k} = \epsilon(j\Delta x, k\Delta z)$, and $\mu_{j,k} = \mu(j\Delta x, k\Delta z)$. The values of E_y, H_x, and H_z at the nodes will be denoted by

$$EY(j, k, t) = E_y(j\Delta x, k\Delta z, t),$$

$$HX(j, k, t) = H_x(j\Delta x, k\Delta z, t),$$

$$HZ(j, k, t) = H_z(j\Delta x, k\Delta z, t).$$

A replacement of the spatial derivatives in (1.13) by forward differences yields

$$\frac{HX(j, k+1, t) - HX(j, k, t)}{\Delta z} - \frac{HZ(j+1, k, t) - HZ(j, k, t)}{\Delta x}$$

$$= \sigma_{j,k} EY(j, k, t) + \epsilon_{j,k} D_t EY(j, k, t). \qquad (1.16)$$

As for the spatial derivatives in (1.14) and (1.15), we use backward differences and rearrange the incremental factors Δx and Δz to write

$$EY(j, k-1, t) - EY(j, k, t) = -(\Delta z)^2 \mu_{j,k} D_t \frac{HX(j, k, t)}{\Delta z}, \qquad (1.17)$$

$$EY(j-1, k, t) - EY(j, k, t) = (\Delta x)^2 \mu_{j,k} D_t \frac{HZ(j, k, t)}{\Delta x}. \qquad (1.18)$$

These equation can be realized by a rectangular RLC grid; those branches that are incident to node (j, k) are shown in Figure 1.12. In this representation the units of the resistances, inductances, and capacitances are in fact ohm-meters, henry-meters, and farads/meter because we have represented electric-field values by node voltages and magnetic-field values by currents times distances. The node voltages are measured with respect to a hypothetical ground node, which is at zero voltage.

If the domain in which the polarized electromagnetic field propagates is infinite in extent, we obtain once again an infinite electrical network. A transient analysis for the electromagnetic method of geophysical exploration can be based on this network [176]. Also, cylindrically and spherically polarized waves lead to cylindrical and spherical RLC grids when appropriate finite differences are taken.

1.8 The Transient Behavior of Linear RLC Networks

Most of this book is ostensibly concerned with resistive networks, that is, networks whose branches have the form shown in Figure 1.1. Existence and uniqueness theorems and computational methods are developed for them. Nonetheless, the transient behaviors of linear RLC networks are

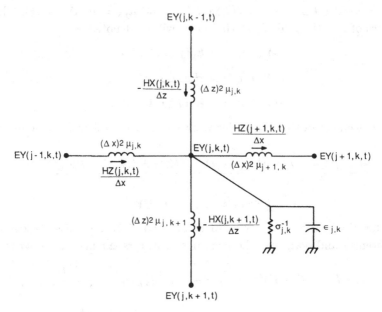

Figure 1.12. A portion of the RLC network that realizes (1.16), (1.17), and (1.18).

also implicitly encompassed within those theorems. The reason for this is that those transient responses can be analyzed by means of the Laplace transformation, which converts functions of time t into functions of the complex variable s. If those functions – or more generally Schwartz distributions – are right-sided and Laplace-transformable, their Laplace transforms may have a common right-sided half-plane of definition ([154], [156]). Furthermore, a linear inductor, which is defined by the relation $i \mapsto l\,di/dt$, is represented in the Laplace domain by the impedance ls. Similarly, a linear capacitor, defined by $v \mapsto c\,dv/dt$, is represented by the impedance $1/cs$. Thus, for each real positive value of s within the half-plane of definition, these impedances act like positive resistances. Therefore, all the existence and uniqueness theorems and computational methods can be transferred to the transformed network for each such s.

Now, by analytic continuation, a knowledge of a Laplace transform on the real axis within its half-plane of definition uniquely determines the transform throughout that half-plane, and by the uniqueness theorem of the Laplace transformation the corresponding transient function or distribution is also uniquely determined. In this way our existence and uniqueness theorems can be transferred to the transient behavior of

RLC networks so long as the functions of s can be shown to be Laplace transforms ([154, page 236]). Moreover, we need not invoke analytic continuation, for the Post–Widder formulas ([144, Chapter VII]) and their extensions to Schwartz distributions ([155]) uniquely invert Laplace transforms from their values on the real positive axis.

Actually, the Post–Widder formulas are not suitable for the computation of transient responses, but there are some more recently derived algorithms that are. They too invert Laplace transforms from values on the real axis in the s-domain. Davies and Martin [36] survey and compare a number of them and evaluate their computational efficacy.

As for the steady-state frequency-domain behavior of infinite RLC networks, some of the analyses expounded in this book can be extended for this purpose simply by replacing real parameters by complex ones. These cases will be pointed out as the occasions arise. It should be pointed out here that V. Belevitch and J. Boersma have established a variety of results concerning particular kinds of Laplace transforms for infinite RLC networks in a series of papers [16], [17], [19], [20], [21], [22]. Also, Dolezal's fundamental theories encompass infinite RLC networks [40], [41], Other works relating to infinite RLC structures are [14], [62], [112], [113], [120], [146], [157], [160], [161], [162], [164].

Chapter 2

Infinite-power Regimes

As was indicated in Example 1.6-3, the total power dissipated in the resistances by a voltage-current regime, satisfying Ohm's law, Kirchhoff's current law at finite nodes, and Kirchhoff's voltage law around finite loops, need not be finite. Moreover, these laws need not by themselves determine the regime uniquely. However, if voltage-current pairs are assigned to certain branches, the infinite-power regime may become uniquely determined. The latter result requires in addition the "nonbalancing" of various subnetworks, as is explained in the next section. In which branches the voltage-current pairs can be arbitrarily chosen and how the nonbalancing criterion can be specified are the issues resolved in this chapter. The discussion is based on a graph-theoretic decomposition of the countably infinite network into a chainlike structure, which was first discovered by Halin for locally finite graphs [63]. That result has been extended to graphs having infinite nodes [166]. The chainlike structure implies a partitioning of the network into a sequence of finite subnetworks, which can be analyzed recursively to determine the voltage-current pair for every branch. We call this a *limb analysis.*

As was mentioned before, in most of this book we restrict our attention to resistive networks. However, a limb analysis can just as readily embrace complex-valued voltages, currents, and branch parameters. In short, a limb analysis can be used for a phasor representation of an AC regime or for the complex representation of a Laplace-transformed transient regime in a linear RLC network [166]. This is why we allow complex numbers in Sections 2.8 through 2.10. Moreover, a nonlinear

RLC network is also amenable to a limb analysis when the nonlinear resistors, inductors, and capacitors are suitably restricted [167], [168].

2.1 An Example

For a simple example of a limb analysis, consider the linear, resistive, infinite network of Figure 2.1. For simplicity, we allow no voltage or current sources, even though they can be easily incorporated into the analysis. (In this section, we allude to certain subnetworks called "limbs" and "orbs" and to certain branches called "joints" and "chords." They are defined in subsequent sections, but what they are in this example should be clear.) The upper horizontal branches induce one "limb," the lower horizontal branches induce a second "limb," the single vertical branch is chosen as the one and only "joint," and the diagonal branches are the "chords." We assume that a current j flows in from infinity along the upper limb, down through the joint, and out to infinity along the lower limb. This endless path is called the "joint orb." The value of j may be chosen arbitrarily. We also assume that a current i_1 (or i_2) flows in from infinity along the upper limb, diagonally downward through the branch with resistance r_1 (respectively, r_2), and then out toward infinity along the lower limb. These paths are called "chord orbs." The values of the currents i_1 and i_2 are treated as unknowns. The branch currents resulting from the superposition of j, i_1, and i_2 satisfy Kirchhoff's current law at every node. We now write Kirchhoff's voltage law for the r_0, r_1, r_4 loop and for the r_0, r_3 r_2 loop.

$$r_1 i_1 - r_4 i_2 = (r_0 + r_4)j \tag{2.1}$$

$$-r_3 i_1 + r_2 i_2 = (r_0 + r_3)j. \tag{2.2}$$

On choosing j arbitrarily, we can solve these equations for i_1 and i_2 so long as the determinant $r_1 r_2 - r_3 r_4$ is not zero. A similar procedure can then be applied to the next four branches to the right in Figure 2.1 to solve for the currents in the next two diagonal branches. Continuing in this fashion, we can compute all the chord currents and thereby all the branch currents. The current in any limb branch is the superposition of all the joint and chord currents flowing through it.

If however, $r_1 r_2 = r_3 r_4$, then the bridge consisting of the corresponding four branches and the vertical branch is balanced, and (2.1) and (2.2) cannot be solved if $j \neq 0$. As a matter of fact, if an infinite sequence of such bridges along Figure 2.1 are balanced, we have an infinite sequence of pairs of equations like (2.1) and (2.2) that cannot be solved. Such balancing implies that the network of Figure 2.1 can carry no nonzero currents.

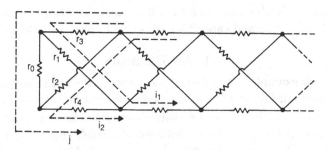

Figure 2.1. An infinite network wherein balancing may take place.

For general infinite networks, the question arises as to what conditions
can be imposed on the resistance values to ensure that nonzero currents
can flow. For Figure 2.1, the answer is that all but a finite number
of bridges are to be unbalanced. For general networks, just how the
requirement of not too much balancing is to be prescribed is problematic.
However, a sufficient condition is suggested by Figure 2.1: If the limb
branches are close enough to short circuits and the chords are close
enough to open circuits, then no balancing occurs anywhere, and limb
analysis will determine the possible current flows. This is the basic idea
we shall exploit.

First of all, however, we must examine how limbs can be chosen in an
arbitrary network. The chainlike structure of countably infinite networks
provides a means of doing so.

2.2 The Chainlike Structure

R. Halin [63] proved that every locally finite, countably infinite graph
has a certain "chainlike" structure. This result, which was extended to
all countably infinite graphs in [166], is fundamental to limb analysis,
and so we explicate it in this section and prove it in the next two. It is
illustrated in Figure 2.2.

Let $\mathcal{G}$ be a countably infinite graph and $\mathcal{M}$ a subgraph of $\mathcal{G}$. A node
of $\mathcal{M}$ is said to be $\mathcal{G}$-*infinite* (or $\mathcal{G}$-*finite*) if its degree as a node of $\mathcal{G}$
is infinite (or, respectively, finite). Thus, a node of $\mathcal{M}$ may be both
$\mathcal{M}$-finite and $\mathcal{G}$-infinite.

A countably infinite graph $\mathcal{G}$ is called *chainlike* if the following holds
true: $\mathcal{G}$ can be partitioned into a sequence of finite subgraphs $\mathcal{G}_p$:

$$\mathcal{G} = \bigcup_{p=1}^{\infty} \mathcal{G}_p \qquad (2.3)$$

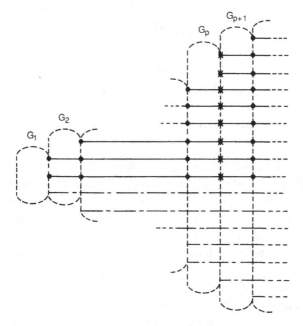

Figure 2.2. The chainlike structure of a countably infinite graph. The crossed dots represent the $\mathcal{G}$-finite nodes of $\mathcal{V}_{p+1}$.

where each branch of $\mathcal{G}$ belongs to one and only one $\mathcal{G}_p$. Moreover,

$$\mathcal{G}_p \cap \mathcal{G}_{p+1} = \mathcal{V}_{p+1} \cup \mathcal{W}_{p+1}, \quad p = 1, 2, 3, \ldots, \qquad (2.4)$$

where the following conditions are satisfied.

1. $\mathcal{V}_{p+1}$ consists of m_{p+1} $\mathcal{G}$-finite nodes (but no branches) where $m_{p+1} < \infty$, and $\mathcal{W}_{p+1}$ consists of n_{p+1} $\mathcal{G}$-infinite nodes (but no branches) where $n_{p+1} < \infty$.

2. The sequence $\{m_{p+1}\}_{p=1}^{\infty}$ is monotonic increasing but not necessarily strictly so, whereas the sequence $\{n_{p+1}\}_{p=1}^{\infty}$ need not be monotonic. Some or all of the m_{p+1} or n_{p+1} may equal zero. (m_{p+1} either tends to ∞ or remains constant for all p sufficiently large.)

3. For every p, $\mathcal{V}_{p+1}$ shares no nodes in common with $\bigcup_{s=1}^{p-1} \mathcal{G}_s$; also, if $|p - m| > 1$, then $\mathcal{G}_p \cap \mathcal{G}_m$ is a finite set (possibly void) of $\mathcal{G}$-infinite nodes.

4. In each $\mathcal{G}_{p+1}$, there are m_{p+1} totally disjoint finite paths from the nodes in $\mathcal{V}_{p+1}$ to m_{p+1} of the m_{p+2} nodes in $\mathcal{V}_{p+2}$.

This ends the definition of the adjective "chainlike." (A graph consisting of a countable infinity of finite components is encompassed by this

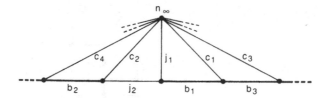

Figure 2.3. A three-times chainlike graph.

definition when $m_{p+1} = 0$ and $n_{p+1} = 0$ for all p. So too is a finite graph when we allow some of the $\mathcal{G}_p$ to be void; this we do in Section 2.4.)

If the m_p in item 2 remain bounded and are therefore equal to, say, $m < \infty$ for all p sufficiently large and if the total number n of $\mathcal{G}$-infinite nodes is finite, then $\mathcal{G}$ is said to be $(m + n)$-*times chainlike* or *finitely chainlike*. If, on the other hand, $m_p \to \infty$ or $n = \infty$, $\mathcal{G}$ is called *divergently chainlike*.

The chainlike structure is illustrated in Figure 2.2. Each finite subgraph is contained within the dotted lines labeled by the $\mathcal{G}_p$. The $\mathcal{G}$-finite nodes of $\mathcal{V}_{p+1}$ are represented by the crossed dots. The union of the paths indicated in item 4 is a collection of one-ended paths, which we call *spines*; these are indicated by the horizontal solid lines of Figure 2.2. The infinite nodes of $\mathcal{G}$ will also be called *spines*; we represent the latter in Figure 2.2 by the horizontal dot-dash lines and think of them as "one-ended paths of short circuits." Thus, a spine is either (i) a one-ended path, which starts at some node of, say, $\mathcal{V}_m$ and then passes through exactly one node of each set $\mathcal{V}_{m+1}, \mathcal{V}_{m+2}, \dots$, and possibly other nodes as well, or (ii) an infinite node, which meets an infinity of the $\mathcal{G}_p$ but not necessarily all the $\mathcal{G}_p$ after the first $\mathcal{G}_m$ that it meets. The set of spines is maximal in the sense that it contains all the $\mathcal{G}$-infinite nodes and that no other one-ended path exists in $\mathcal{G}$ that does not meet any spines. It follows from our definition that every finite component of $\mathcal{G}$ will be contained in a single $\mathcal{G}_p$ and will not meet any spine.

Here are two explicit examples of chainlike structures. Other examples are given in [165, Section 7].

Example 2.2-1. A three-times chainlike graph is shown in Figure 2.3. One of its spines is the infinite node n_∞. Its other two spines can be chosen as the two one-ended paths shown by the heavy lines. The branches labeled by j_k or by c_k are respectively "joints" and "chords," which will be defined later on. We can choose $\mathcal{G}_1$ to be the subgraph induced by j_1; $\mathcal{G}_2$ induced by j_2, b_1, c_1, c_2; $\mathcal{G}_3$ induced by b_2, b_3, c_3, c_4; and so on. ♣

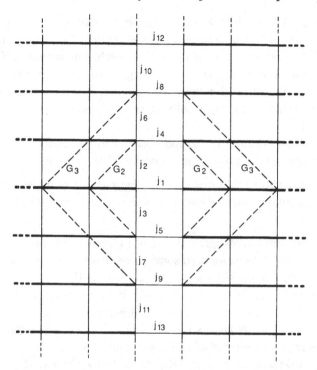

Figure 2.4. A divergently chainlike, locally finite graph.

Example 2.2-2. A divergently chainlike, locally finite graph is the two-dimensional rectangular grid shown in Figure 2.4. It has an infinity of spines, which can be chosen to be the one-ended paths indicated by the the heavy horizontal lines. Again the j_k denote "joints." The "chords" are the unlabeled vertical branches. We can choose $\mathcal{G}_1$ to be the joint j_1, $\mathcal{G}_2$ to be the subgraph induced by the branches within the dotted lines labeled G_2 including j_4 and j_5 but excluding j_1, $\mathcal{G}_3$ to be induced by the branches within G_3 including j_8 and j_9 but outside $\mathcal{G}_2$, and so forth. ♣

2.3 Halin's Result for Locally Finite Graphs

In this section we go through Halin's proof of the chainlike structure of any locally finite, countably infinite graph $\mathcal{H}$, which is in addition connected. These conditions on $\mathcal{H}$ will be understood throughout this section.

The assumption of connectedness entails no loss of generality. Indeed, if $\mathcal{H}$ has many infinite components, each of which is chainlike, we can arrange those components into a sequence $\mathcal{C}_1, \mathcal{C}_2, \mathcal{C}_3, \ldots$. Then, the first subgraph $\mathcal{H}_1$ of $\mathcal{H}$ can be taken to be the first subgraph of $\mathcal{C}_1$, the second subgraph of $\mathcal{H}_2$ of $\mathcal{H}$ can be taken to be the union of the second subgraph of $\mathcal{C}_1$ and the first subgraph of $\mathcal{C}_2$, and so forth. Furthermore, if $\mathcal{H}$ has finite components as well, they can be assigned to the $\mathcal{H}_p$ in a one-to-one fashion. All this will yield a chainlike structure for $\mathcal{H}$. It might also be worth mentioning that every locally finite, connected graph is countable [147, page 40].

Two standard results from graph theory will be needed. The first is König's lemma, and a proof of it is given in [147, page 40].

Lemma 2.3-1 (König's lemma): *For every node n of the locally finite, countably infinite, connected graph $\mathcal{H}$, there is a one-ended path starting at n.*

The second result is Menger's theorem [147, page 129] and it concerns finite graphs. To state it, we need some more definitions. Let n and m be two nodes in $\mathcal{H}$ and consider a collection of nm-paths. Those paths will be called *totally-disjoint-except-terminally* if, for every choice P_1 and P_2 of two of them, the nodes of P_1 are all different from the nodes of P_2 except for n and m. Now, assume that n and m are not adjacent; an *nm-separating set* is a set S of nodes not containing either n or m such that every nm-path meets S. Finally, let $\mathcal{V}$ and $\mathcal{W}$ be two disjoint sets of nodes in $\mathcal{H}$. A set S of nodes is said to *separate* $\mathcal{V}$ *from* $\mathcal{W}$ if every path connecting a node of $\mathcal{V}$ to a node of $\mathcal{W}$ meets S; in this case, we do not require that S be disjoint from $\mathcal{V}$ or $\mathcal{W}$.

Theorem 2.3-2 (Menger's Theorem): *Let n and m be two nonadjacent nodes of a finite connected graph $\mathcal{H}$. The maximum number of nm-paths that are totally-disjoint-except-terminally is equal to the minimum number of nodes among all the nm-separating sets.*

Actually, we shall apply a modified version of this theorem.

Corollary 2.3-3: *Let $\mathcal{V}$ and $\mathcal{W}$ be two disjoint sets of nodes in the finite connected graph $\mathcal{H}$. Assume that $|\mathcal{V}| \leq |\mathcal{W}|$ and that every node set S that separates $\mathcal{V}$ from $\mathcal{W}$ satisfies $|S| \geq |\mathcal{V}|$. Then, the maximum number of totally disjoint paths connecting nodes of $\mathcal{V}$ to nodes of $\mathcal{W}$ is $|\mathcal{V}|$. Consequently, there are $|\mathcal{V}|$ totally disjoint paths connecting all the nodes of $\mathcal{V}$ to some or all of the nodes of $\mathcal{W}$ such that only the end nodes of each path lie in $\mathcal{V}$ and $\mathcal{W}$ and each node of $\mathcal{V}$ is the end node of one and only one such path.*

Proof: Append to $\mathcal{H}$ a node n and $|\mathcal{V}|$ branches to make n adjacent to every node of $\mathcal{V}$. Also, append another node m and $|\mathcal{W}|$ branches to make m adjacent to every node of $\mathcal{W}$. Thus, n and m are not adjacent. Let $\mathcal{H}_1$ be the union of $\mathcal{H}$ and the appended branches. There is an obvious bijection from the nm-paths in $\mathcal{H}_1$ to the paths in $\mathcal{H}$ connecting nodes in $\mathcal{V}$ to nodes in $\mathcal{W}$. So, by hypothesis, the minimum number of nodes among all the nm-separating sets in $\mathcal{H}_1$ is $|\mathcal{V}|$. Now, invoke Menger's theorem to get the first conclusion. The second conclusion follows immediately. ♣

Here now is Halin's result.

Theorem 2.3-4: *Every locally finite, countably infinite, connected graph $\mathcal{H}$ is chainlike.*

Proof: Let $\mathcal{N}(\mathcal{M})$ denote the node set of any subgraph $\mathcal{M}$ of $\mathcal{H}$ (we allow $\mathcal{M} = \mathcal{H}$). Let $\mathcal{S}$ and $\mathcal{V}$ be finite subsets of $\mathcal{N}(\mathcal{H})$. We shall say that $\mathcal{S}$ *separates* $\mathcal{V}$ *from* ∞ if every one-ended path starting in $\mathcal{V}$ meets $\mathcal{S}$ – or, equivalently according to König's lemma, if every node of $\mathcal{V}$ that is not in $\mathcal{S}$ belongs to a finite component of $\mathcal{H} - \mathcal{S}$. Moreover, $\mathcal{S}$ is said to *separate a node n from* ∞ if $n \notin \mathcal{S}$ and if every one-ended path starting at n meets $\mathcal{S}$ – or, equivalently by König's lemma again, if n belongs to a finite component of $\mathcal{H} - \mathcal{S}$. Also, let $\mathcal{F}(\mathcal{V})$ denote the union of all the finite components of $\mathcal{H} - \mathcal{V}$. Since the removal of any node increases the number of components by no more than the degree of the node, it follows that $\mathcal{F}(\mathcal{V})$ is finite because $\mathcal{V}$ is finite and $\mathcal{H}$ is locally finite. $\mathcal{F}[\mathcal{V}]$ will denote the finite subgraph of $\mathcal{H}$ induced by the nodes of $\mathcal{F}(\mathcal{V})$ and $\mathcal{V}$.

Assume now that $\mathcal{S}$ does indeed separate $\mathcal{V}$ from ∞. Then, the finite set $\mathcal{S}'$ consisting of all nodes of $\mathcal{S}$ that are not in $\mathcal{F}(\mathcal{V})$ also separates $\mathcal{V}$ from ∞; moreover, $\mathcal{F}[\mathcal{S}']$ contains $\mathcal{F}[\mathcal{V}]$ as a subgraph. If the node set $\mathcal{R}$ separates $\mathcal{V}$ from $\mathcal{S}$, then $\mathcal{R}$ also separates $\mathcal{V}$ from ∞, and the set $\mathcal{R}'$ consisting of all nodes of $\mathcal{R}$ that are not in $\mathcal{F}(\mathcal{V})$ separates $\mathcal{V}$ from $\mathcal{S}'$; hence, $\mathcal{R}'$ separates $\mathcal{V}$ from ∞.

The rest of this proof presents an inductive construction of a chainlike structure for $\mathcal{H}$. Let n be any node of $\mathcal{H}$. There are finite node sets that separate n from ∞, for example the set of all nodes adjacent to n. Let $\mathcal{V}_2$ be a set of nodes separating n from ∞ and having the minimum cardinality for all such sets. Let $\mathcal{H}_1$ be the finite subgraph $\mathcal{F}[\mathcal{V}_2]$.

Let p be an integer no less than 2. Assume that the finite node sets $\mathcal{V}_2, \ldots, \mathcal{V}_p$ have been selected such that for every $i = 3, \ldots, p$ the following three conditions hold:

(i) $\mathcal{V}_i$ is disjoint from $\mathcal{V}_{i-1}$.

(ii) $\mathcal{V}_i$ separates $\mathcal{V}_{i-1}$ from ∞.

(iii) $\mathcal{V}_i$ has the minimum cardinality among all node sets satisfying
(i) and (ii).

We now indicate how a finite node set $\mathcal{V}_{p+1}$ can be chosen to satisfy
these three conditions when i is replaced by $p+1$. First of all, there is at
least one finite node set that satisfies (i) and (ii) for $i = p + 1$, namely,
the set of nodes adjacent to $\mathcal{V}_p$. Hence, there is a minimal one $\mathcal{V}_{p+1}$, and
that $\mathcal{V}_{p+1}$ will not have any node in common with $\mathcal{F}[\mathcal{V}_p]$.

Let $\mathcal{S}$ be a node set of minimum cardinality separating $\mathcal{V}_p$ from $\mathcal{V}_{p+1}$.
Hence, $|\mathcal{S}| \leq |\mathcal{V}_{p+1}|$, and $\mathcal{S}$ is finite. Moreover, $\mathcal{S} \cap \mathcal{F}(\mathcal{V}_p) = \emptyset$. Also, $\mathcal{S}$
separates $\mathcal{V}_p$ from ∞, and so $\mathcal{S}$ separates $\mathcal{V}_{p-1}$ from ∞ (or n from ∞ if
$p = 2$). Thus, $\mathcal{S} \cap \mathcal{V}_{p-1} = \emptyset$ because $\mathcal{V}_{p-1} \subset \mathcal{F}(\mathcal{V}_p)$. By the minimality
of $\mathcal{V}_p$, $|\mathcal{S}| \geq |\mathcal{V}_p|$. We can now invoke Corollary 2.3-3 to conclude that
there are $|\mathcal{V}_p|$ totally disjoint paths connecting all the nodes of $\mathcal{V}_p$ to
some or all of the nodes of $\mathcal{V}_{p+1}$ in accordance with the last conclusion
of that corollary.

$\mathcal{V}_p$ and the said paths are contained in $\mathcal{F}[\mathcal{V}_{p+1}]$, for otherwise $\mathcal{V}_{p+1}$
would not separate $\mathcal{V}_p$ from ∞. Let $\mathcal{H}_p$ be the finite subgraph of $\mathcal{H}$
induced by all the nodes of $\mathcal{F}[\mathcal{V}_{p+1}]$ that are not in $\mathcal{F}(\mathcal{V}_p)$.

Thus, by induction we can construct an infinite sequence of finite
node sets $\mathcal{V}_2, \mathcal{V}_3, \mathcal{V}_4, \ldots$ and an infinite sequence of finite subgraphs
$\mathcal{H}_1, \mathcal{H}_2, \mathcal{H}_3, \ldots$ such that $\mathcal{H}_{p+1} \cap (\bigcup_{i=1}^{p} \mathcal{H}_i) = \mathcal{V}_{p+1}$, $\mathcal{V}_{p+1} \subset \mathcal{H}_p$, $\mathcal{V}_{p+1} \cap$
$(\bigcup_{i=1}^{p-1} \mathcal{H}_i) = \emptyset$, and $|\mathcal{V}_p| \leq |\mathcal{V}_{p+1}|$. Furthermore, $\mathcal{H} = \bigcup_{p=1}^{\infty} \mathcal{H}_p$. Indeed,
since $\mathcal{H}$ is connected and since each $\mathcal{V}_{p+1}$ is disjoint from $\bigcup_{i=1}^{p-1} \mathcal{H}_i$, each
node of $\mathcal{H}$ will eventually be contained in some $\mathcal{F}[\mathcal{V}_{p+1}]$ and therefore
so too will every branch.

It is now routine to check that all the conditions for the chainlike
structure given in the preceding section are satisfied when it is assumed
that there are no infinite nodes. ♣

2.4 An Extension to Countably Infinite Graphs

Now, let $\mathcal{G}$ be a countably infinite graph with infinite nodes. As before,
there is no loss of generality in assuming that $\mathcal{G}$ is connected. We shall
establish a chainlike structure for $\mathcal{G}$ by extending that structure from
the locally finite subgraph $\mathcal{H}$ obtained from $\mathcal{G}$ by deleting all the infinite
nodes of $\mathcal{G}$.

In other words, the node set of $\mathcal{H}$ consists of all the finite nodes of $\mathcal{G}$,
and the branch set of $\mathcal{H}$ consists of all branches of $\mathcal{G}$ that are not incident
to infinite nodes. $\mathcal{H}$ is locally finite but need not be connected now. By
Theorem 2.3-4 and the second paragraph of the preceding section, $\mathcal{H}$ is

chainlike – in a trivial way if $\mathcal{H}$ is finite. In the following, we use the notation of the preceding section for the chainlike structure of $\mathcal{H}$. We can partition the nodes of $\mathcal{H}$ into two sets, the first consisting of those nodes that do not lie on one-ended paths and the second consisting of those that do. Let $\mathcal{K}$ and $\mathcal{M}$ be respectively the subgraphs of $\mathcal{H}$ induced by the nodes in the first and second sets. We allow either $\mathcal{K}$ or $\mathcal{M}$ to be void. By König's lemma, all the components of $\mathcal{K}$ are finite, whereas those of $\mathcal{M}$ are infinite.

Furthermore, $\mathcal{M}$, if it exists, must also be chainlike with the same sets $\mathcal{V}_p$ as those of $\mathcal{H}$. Indeed, by Condition 4 of Section 2.2, every node in any $\mathcal{V}_p$ lies on a one-ended path and hence must belong to $\mathcal{M}$. Also, each component of $\mathcal{K}$ must lie within a single $\mathcal{H}_p$ and possess no vertices in common with $\mathcal{V}_p$, with $\mathcal{V}_{p+1}$, or with the the totally disjoint finite paths described in Condition 4. (Moreover, by the finiteness of the $\mathcal{H}_p$, each $\mathcal{H}_p$ can contain no more than a finite number of the components of $\mathcal{K}$.) Thus, the removal of $\mathcal{K}$ from $\mathcal{H}$ will yield once again the same chainlike structure for the remaining graph $\mathcal{M}$.

Now, consider the union of all the finite paths mentioned in Condition 4. This union consists of m (or an infinity of) totally disjoint one-ended paths in $\mathcal{M}$ if $\mathcal{M}$ is m-times (or, respectively, divergently) chainlike. By a *spine* we mean either one of these one-ended paths or one of the infinite nodes of $\mathcal{G}$.

We now construct a chainlike structure for $\mathcal{G}$. The following procedure for doing so involves a number of steps in which certain nodes, branches, or components are designated. It can happen that some of these entities do not exist in $\mathcal{G}$. When this occurs, it is understood that the corresponding step is simply skipped.

Let $\mathcal{M}_p$ be that subgraph of $\mathcal{H}_p$ obtained by removing from $\mathcal{H}_p$ all components of $\mathcal{K}$ that are contained in $\mathcal{H}_p$. We also let q_k, where $k = 1, 2, \ldots$, denote the infinite nodes of $\mathcal{G}$, and we assign the indices of the q_k as follows. First of all, we arrange the infinite nodes that are not adjacent to any of the $\mathcal{M}_p$ into a sequence X. Then, we let $q_1, q_2, \ldots, q_a$ be the infinite nodes that are adjacent to $\mathcal{M}_1$ and let q_{a+1} be the first node in X; we also set $\mathcal{Q}_1 = \{q_1, \ldots, q_{a+1}\}$. Next, we let $q_{a+2}, q_{a+3}, \ldots, q_b$ be the infinite nodes that are adjacent to $\mathcal{M}_2$ but not adjacent to $\mathcal{M}_1$, and let q_{b+1} be the second node in X; also, $\mathcal{Q}_2 = \{q_{a+2}, \ldots, q_{b+1}\}$. Then, we let $q_{b+2}, q_{b+3}, \ldots, q_c$ be the infinite nodes that are adjacent to $\mathcal{M}_3$ but not adjacent to $\mathcal{M}_1 \cup \mathcal{M}_2$, and let q_{c+1} be the third node of X; also, $\mathcal{Q}_3 = \{q_{b+2}, \ldots, q_{c+1}\}$. We continue this numbering procedure until all infinite nodes are labeled; simultaneously, the set of all infinite nodes becomes partitioned into $\{\mathcal{Q}_s\}$. Note that, since each $\mathcal{M}_p$ is locally finite and has a finite number of nodes, it has only a finite number of adjacent infinite nodes. Consequently, this numbering system is feasible.

Observe that each component of $\mathcal{K}$ must be adjacent to at least one q_k but not to any nodes in $\mathcal{M}$ because $\mathcal{G}$ is connected and these components arise through the deletion of the q_k from $\mathcal{G}$. We label the components of $\mathcal{K}$ as $\mathcal{K}_{k,j}$ in the following way. $\mathcal{K}_{k,1}$, $\mathcal{K}_{k,2}$, $\mathcal{K}_{k,3}$, ... denote those components of $\mathcal{K}$ that are adjacent to q_k and are not adjacent to those q_i for which $i < k$. When $i > k$, $\mathcal{K}_{k,j}$ may be adjacent to q_i. For any fixed k, there may be a finite or infinite number of $\mathcal{K}_{k,j}$.

Finally, for each pair of infinite nodes q_i and q_j, we label the branches joining those two nodes by $b_{i,j,1}$, $b_{i,j,2}$, $b_{i,j,3}$, ...

Now, let $\mathcal{K}_1$ be the union of all $\mathcal{K}_{k,1}$ for which $q_k \in \mathcal{Q}_1$. $\mathcal{K}_1$ is a finite graph since each $\mathcal{K}_{k,1}$ is a finite graph and $\mathcal{Q}_1$ is a finite set. Let $\mathcal{P}_1$ be the union of all $b_{i,j,1}$ for which $q_i \in \mathcal{Q}_1$ and $q_j \in \mathcal{Q}_1$. Clearly, $\mathcal{P}_1$ is a finite graph. Let $\mathcal{R}_1$ be the union of $\mathcal{P}_1$, all branches joining $\mathcal{M}_1$ to $\mathcal{Q}_1$, and all branches joining $\mathcal{K}_1$ to any of the q_k (not necessarily in $\mathcal{Q}_1$). Since $\mathcal{M}_1$ and $\mathcal{K}_1$ are finite graphs, $\mathcal{R}_1$ is too. Finally, we let $\mathcal{G}_1$ be the finite subgraph of $\mathcal{G}$ induced by all the branches in $\mathcal{M}_1$, $\mathcal{K}_1$, and $\mathcal{R}_1$.

We inductively continue this process of constructing finite subgraphs $\mathcal{G}_p$ of $\mathcal{G}$ as follows. Let p be an integer greater than 1. Let $\mathcal{K}_p$ be the union of all $\mathcal{K}_{k,p-s+1}$ for which $q_k \in \mathcal{Q}_s$, where $s = 1, 2, \ldots, p$. As was true for $\mathcal{K}_1$, $\mathcal{K}_p$ is also a finite subgraph. Next, let $\mathcal{P}_p$ be the finite subgraph induced by all $b_{i,j,p-s+1}$ for which q_i and q_j both lie in $\bigcup_{r=1}^{s} \mathcal{Q}_r$ and at least one of them lies in $\mathcal{Q}_s$, where $s = 1, 2, \ldots, p$. Then, let $\mathcal{R}_p$ be the union of $\mathcal{P}_p$, all branches joining $\mathcal{M}_p$ to any of the q_k, and all branches joining $\mathcal{K}_p$ to any of the q_k. It follows from the finiteness of $\mathcal{P}_p$, $\mathcal{M}_p$, and $\mathcal{K}_p$ that $\mathcal{R}_p$ is a finite graph too. Finally, set $\mathcal{G}_p$ equal to the union of $\mathcal{M}_p$, $\mathcal{K}_p$, and $\mathcal{R}_p$. Thus, $\mathcal{G}_p$ is a finite subgraph of $\mathcal{G}$.

From the way we have constructed the $\mathcal{G}_p$ it can be seen that every branch of $\mathcal{G}$ lies in one and only one of the $\mathcal{G}_p$. That is, $\{\mathcal{G}_p\}_{p=1}^{\infty}$ is a partition of $\mathcal{G}$ into finite subgraphs. Moreover, $\mathcal{G}_p \cap \mathcal{G}_{p+1} = \mathcal{V}_{p+1} \cup \mathcal{W}_{p+1}$, where $\mathcal{V}_{p+1}$ is a finite set (possibly void) of $\mathcal{G}$-finite nodes in the chainlike structure of $\mathcal{H}$ and $\mathcal{W}_{p+1}$ is a finite set (possibly void) of $\mathcal{G}$-infinite nodes. Furthermore, $\mathcal{G}_p \cap \mathcal{G}_m$ is just a finite set (possibly void) of $\mathcal{G}$-infinite nodes if $|p - m| > 1$. Thus, the chainlike structure of $\mathcal{G}$ has been completely established.

We can summarize the discussions of this and the preceding section by

Theorem 2.4-1: *Every countably infinite graph is chainlike.*

2.5 Limbs

We continue to use the notation of the preceding section. Recall that a spine in $\mathcal{G}$ is defined to be any infinite node or any one of the one-ended paths arising from the union of the finite paths mentioned in Condition 4 of Section 2.2. Our next step is to construct in each $\mathcal{G}_p$ a spanning forest $\mathcal{F}_p$ such that each component of $\mathcal{F}_p$ meets one and only one spine and contains all the nodes and branches of that spine that lie in $\mathcal{G}_p$.

To this end, let $\mathcal{A}_1$ be the subgraph induced by all the branches in $\mathcal{G}_p$ that lie in spines. Choose any branch b_1 in $\mathcal{G}_p$ that is adjacent to $\mathcal{A}_1$. If b_1 has one node that is not in $\mathcal{A}_1$, set $\mathcal{A}_2 = \mathcal{A}_1 \cup b_1$; otherwise, set $\mathcal{A}_2 = \mathcal{A}_1$. Choose another branch b_2 in $\mathcal{G}_p$ that is adjacent to $\mathcal{A}_2$ and set $\mathcal{A}_3 = \mathcal{A}_2 \cup b_2$ if b_2 has one node that is not in $\mathcal{A}_2$; otherwise, set $\mathcal{A}_3 = \mathcal{A}_2$. Continue this procedure until all branches in $\mathcal{G}_p$ that are not in spines have been considered. This yields a spanning forest $\mathcal{F}_p$ in $\mathcal{G}_p$ with the desired properties.

We now set $\mathcal{F} = \bigcup_{p=1}^{\infty} \mathcal{F}_p$.

Lemma 2.5-1: $\mathcal{F}$ *is a spanning forest in* $\mathcal{G}$ *such that each component of* $\mathcal{F}$ *meets one and only one spine in* $\mathcal{G}$ *and contains all the nodes and branches of that spine.*

Proof: By virtue of the chainlike structure of $\mathcal{G}$, if $p \neq m$, a component of $\mathcal{F}_p$ and a component of $\mathcal{F}_m$ can meet only if they meet the same spine, and, if they do meet, their intersection lies in that spine and is either a $\mathcal{G}$-infinite node or, if $m = p + 1$, possibly a node in $\mathcal{V}_m$. It follows that $\mathcal{F}$ contains no loops and is therefore a forest. Moreover, $\mathcal{F}$ is spanning because it is spanning in each $\mathcal{G}_p$. The rest of the conclusion follows directly from the way each $\mathcal{F}_p$ was constructed in $\mathcal{G}_p$. ♣

We call each component of $\mathcal{F}$ a *limb*. The set of all limbs for a given spanning forest $\mathcal{F}$ constructed as indicated will be called a *full set of limbs*. Since $\mathcal{F}$ is spanning, every node of $\mathcal{G}$ belongs to a unique limb.

Lemma 2.5-2: *If* n_0 *is a node in* $\mathcal{G}$ *that is not in a spine, then there exists a unique finite path* P *contained in the limb* $\mathcal{L}$ *that contains* n_0 *such that* n_0 *is one end node of* P, *the other end node* n_1 *of* P *belongs to the spine of* $\mathcal{L}$, *and* P *lies entirely outside the spine of* $\mathcal{L}$ *except for* n_1. *If* n_0 *lies in a spine, then* $n_0 = n_1$ *and* P *is the trivial path consisting of* n_0 *alone.*

Proof: Choose any node n_2 in the spine S of $\mathcal{L}$. (When S is a $\mathcal{G}$-infinite node, we choose S itself.) Since $\mathcal{L}$ is a tree, there exists a unique finite path from n_0 to n_2 lying in $\mathcal{L}$. In tracing that path from n_0 to n_2, we find a first node n_1 that lies in S, and the traced path from n_0 to n_1 is the path P we seek. Again, P must be unique since $\mathcal{L}$ is a tree. ♣

Lemma 2.5-3: *The path P of Lemma 2.5-2 lies entirely within a single $\mathcal{G}_p$. When P is nontrivial, its nodes other than n_1 are not common to two or more of the $\mathcal{G}_p$.*

Proof: This follows from the facts that $\mathcal{L}$ is the union of its intersections with each $\mathcal{G}_p$ and two such intersections meet only at a single node in the spine of $\mathcal{L}$ if they meet at all. ♣

Lemma 2.5-4: (i) *If the spine of $\mathcal{L}$ is a one-ended path, then $\mathcal{L}$ is an infinite tree containing neither $\mathcal{G}$-infinite nodes nor endless paths. Moreover, given any node n_0 of $\mathcal{L}$, there exists a unique one-ended path starting at n_0 and contained entirely in $\bigcup_{s=p}^{\infty}(\mathcal{L} \cap \mathcal{G}_s)$, where $\mathcal{G}_p$ is the subnetwork containing n_0.*

(ii) *If the spine of $\mathcal{L}$ is a $\mathcal{G}$-infinite node, then $\mathcal{L}$ may be either a finite or infinite tree, but in either case it contains exactly one $\mathcal{G}$-infinite node, the spine itself, and does not contain any one-ended or endless path.*

Proof: (i) Since $\mathcal{L}$ is a component of the forest $\mathcal{F}$ and contains a one-ended path, it is an infinite tree. By Lemma 2.5-1, $\mathcal{L}$ contains exactly one spine. Since the $\mathcal{G}$-infinite nodes are all spines, $\mathcal{L}$ cannot contain a $\mathcal{G}$-infinite node. Furthermore, $\mathcal{L} = \bigcup_{p=1}^{\infty}(\mathcal{L} \cap \mathcal{G}_p)$. Each $\mathcal{L} \cap \mathcal{G}_p$ is a finite tree. Also, for $|p - m| > 1$, $\mathcal{L} \cap \mathcal{G}_p$ does not meet $\mathcal{L} \cap \mathcal{G}_m$, whereas $\mathcal{L} \cap \mathcal{G}_p$ meets $\mathcal{L} \cap \mathcal{G}_{p+1}$ at and only at the unique spine node of $\mathcal{L}$ residing in both $\mathcal{G}_p$ and $\mathcal{G}_{p+1}$. It follows that $\mathcal{L}$ cannot contain an endless path.

It also follows from this structure for $\mathcal{L}$ that there is exactly one one-ended path in $\bigcup_{s=p}^{\infty}(\mathcal{L} \cap \mathcal{G}_s)$ starting from a given node n_0 of $\mathcal{L} \cap \mathcal{G}_p$. That path lies entirely in $\mathcal{L}$'s spine if n_0 belongs to the spine. Otherwise, that path is the union of the n_0n_1-path of Lemma 2.5-2 and the one-ended path in the spine starting at n_1.

(ii) Now, assume that $\mathcal{L}$'s spine is the $\mathcal{G}$-infinite node q_k. As in (i), $\mathcal{L}$ is a tree and cannot contain any other $\mathcal{G}$-infinite node. $\mathcal{L} \cap \mathcal{G}_p$ is a finite tree for each p. For $p \neq m$, $\mathcal{L} \cap \mathcal{G}_p$ and $\mathcal{L} \cap \mathcal{G}_m$ meet at and only at q_k. Thus, $\mathcal{L} - q_k$ has only finite components. Consequently, $\mathcal{L}$ cannot contain any one-ended or endless path. Finally, in the process of constructing all the $\mathcal{F}_p$, either a finite or infinite number of branches may have been added to q_k to produce $\mathcal{L}$. ♣

Lemma 2.5-5: *If b is any branch of a limb $\mathcal{L}$, then $\mathcal{L} - b$ has exactly two components. One of them contains all of $\mathcal{L}$'s spine except possibly a finite portion of that spine. The other component is a finite tree all of whose nodes are $\mathcal{G}$-finite.*

Proof: Since $\mathcal{L}$ is a tree, the removal of b must yield exactly two components. If b is not a part of $\mathcal{L}$'s spine, all of that spine will appear in one of those components. The only other possibility is that the spine is a one-ended path and b is in that path. The removal of b will then split the spine into a finite path and a one-ended path, with the latter appearing in one of the aforementioned components. Finally, if the last conclusion were not true, either $\mathcal{L}$ would contain two $\mathcal{G}$-infinite nodes, or (by König's lemma) $\mathcal{L}$ would contain a $\mathcal{G}$-infinite node and a one-ended path, or $\mathcal{L}$ would contain an endless path, all of which are impossible according to Lemma 2.5-4. ♣

A branch of $\mathcal{G}$ that is not in the forest $\mathcal{F}$ will be called a *tie*. Whether or not a branch is a tie depends on the choice of $\mathcal{F}$. Let d be any tie. Two possibilities arise:

(i) Both nodes of d lie in the same limb $\mathcal{L}$: Since $\mathcal{L}$ is a tree, $\mathcal{L} \cup d$ contains exactly one loop, namely, the union of b and the unique path in $\mathcal{L}$ connecting the nodes of d. We refer to that loop as the *d-orb* or *tie orb*.

(ii) The two nodes of d lie in different limbs, say $\mathcal{L}_1$ and $\mathcal{L}_2$: In this case we define the *d-orb* or *tie orb* as the unique (finite or infinite) path $d \cup P_1 \cup P_2 \cup H_1 \cup H_2$, where P_1 and P_2 are the two finite paths in, respectively, $\mathcal{L}_1$ and $\mathcal{L}_2$ as specified in Lemma 2.5-2. Also, if the spine of $\mathcal{L}_1$ is a $\mathcal{G}$-infinite node, H_1 is that node; if the spine of $\mathcal{L}_1$ is a one-ended path, H_1 is the one-ended path lying in the spine and starting at the node where P_1 ends. H_2 lies in the spine of $\mathcal{L}_2$ and is defined similarly, Thus, we see that this d-orb lies in $\mathcal{L}_1 \cup d \cup \mathcal{L}_2$.

Lemma 2.5-6: *Let $\mathcal{G}_p$ be the subgraph that contains a given tie d. Then, the d-orb is contained entirely within $\bigcup_{s=p}^{\infty} \mathcal{G}_s$. Moreover, the d-orb is contained entirely within $\mathcal{G}_p$ alone under either one of the following conditions:*

(i) *Both nodes of d lie in the same limb.*
(ii) *The two nodes of d lie in different limbs, both of which have spines that are $\mathcal{G}$-infinite nodes.*

Proof: Since $\{\mathcal{G}_p\}$ is a partition of $\mathcal{G}$, d must lie in a single $\mathcal{G}_p$. For any limb $\mathcal{L}$, $\mathcal{L} \cap \mathcal{G}_p$ is a tree. Hence, under Condition (i), the d-orb will lie in $\mathcal{G}_p$. On the other hand, if d's nodes lie in different limbs, the d-orb is $d \cup P_1 \cup P_2 \cup H_1 \cup H_2$. By Lemma 2.5-3, P_1 and P_2 also lie in $\mathcal{G}_p$. When the spines of both limbs are $\mathcal{G}$-infinite nodes, H_1 and H_2 are those nodes, which implies that the d-orb lies in $\mathcal{G}_p$. The only other possibility is that one or both of H_1 and H_2 are one-ended paths, but in either case H_1

and H_2 will lie in $\bigcup_{s=p}^{\infty} \mathcal{G}$ according to Lemma 2.5-4(i). Therefore, so too will the d-orb. ♣

Lemma 2.5-7: *Let b be any branch in $\mathcal{F}$. Then, there are only a finite number of tie orbs that contain b.*

Proof: Let $\mathcal{L}$ be the limb that contains b. Let $\mathcal{H}$ be that finite component of $\mathcal{L} - b$ whose nodes are all $\mathcal{G}$-finite (Lemma 2.5-5). Let d be any tie. The following results follow directly from the definition of the d-orb. If d is not adjacent to $\mathcal{H}$, then the d-orb does not contain b. If one node of d is in $\mathcal{H}$ and the other node of d is not in $\mathcal{H}$, then the d-orb contains b. Finally, if both nodes of d lie in $\mathcal{H}$, then the d-orb does not contain b. Thus, the only tie orbs that contain b are those whose ties have one and only one node in $\mathcal{H}$. Since $\mathcal{H}$ is finite and all its nodes are $\mathcal{G}$-finite, there can be only a finite number of such ties. ♣

2.6 Current Regimes Satisfying Kirchhoff's Current Law

We now assign electrical parameters to the graph $\mathcal{G}$ to obtain an electrical network $\mathbf{N}$ in accordance with Section 1.4. We take it that each branch is in its Thevenin equivalent form shown in Figure 1.1(b). For the sake of a more general phasor analysis, we assume that the branch voltage v_j and branch current i_j for the jth branch are complex numbers. Also, we denote the given branch voltage source by e_j (rather than E_j), another complex number. To adhere to convention, we replace r_j by the symbol z_{jj}, a complex number denoting the branch's *self-impedance*. We also allow coupling between branches, that is, dependent sources. In particular, for $k \neq j$ a current i_k in the kth branch may produce a voltage $z_{jk}i_k$ in the jth branch; here, z_{jk} is the *transimpedance*. Thus, we may have

$$v_j = -e_j + \sum_k z_{jk}i_k \tag{2.5}$$

where the summation is over all branch indices including $k = j$. By some conditions that will be imposed later on, for each j all but a finite number of the z_{jk} will be zero.

$\mathbf{N}$'s graph is $\mathcal{G}$, which we continue to assume throughout this chapter is a connected, countably infinite graph. For the sake of brevity, we simply transfer all the graph-theoretic terminology from $\mathcal{G}$ to $\mathbf{N}$. For example, we may speak of "subnetworks" instead of "subgraphs" and of "$\mathbf{N}$-finite" nodes instead of "$\mathcal{G}$-finite" nodes.

Every branch of $\mathbf{N}$ has an orientation. Thus, if branch b and node n are incident, then b may be said to be either *incident away from* or *incident toward n.* In addition, if d is a tie, we assign to the d-orb that orientation which agrees with the orientation of d (i.e., while tracing the d-orb in the direction of the d-orb's orientation, we pass through d in the direction of d's orientation). Finally, throughout this chapter we shall impose Kirchhoff's current law only at the finite nodes of **N**, not at its infinite nodes.

Lemma 2.6-1: *The specification of the currents in all the ties and the imposition of Kirchhoff's current law at only the finite nodes of **N** uniquely determine the current in each branch of the corresponding forest $\mathcal{F}$.*

Proof: We assume that $\mathcal{F}$ has at least one branch. Then, by Lemma 2.5-5, $\mathcal{F}$ must have at least one end node that is **N**-finite. After numbering all the branches in $\mathcal{F}$ in any fashion, we let b_1, b_2, b_3,... denote those branches. Among all the end branches of $\mathcal{F}$, choose the one, say b_{k_1}, having the least index k_1 and let n_1 be its **N**-finite end node. Thus, all other branches in **N** that are incident to n_1 are ties; they are finite in number and have specified currents. Kirchhoff's current law therefore uniquely determines the current in b_{k_1}.

Inductively, assume that the currents have already been determined in the branches $b_{k_1},\ldots,b_{k_m}$ of $\mathcal{F}$. Among all the end branches of $\mathcal{F} - b_{k_1} - \cdots - b_{k_m}$, choose that end branch $b_{k_{m+1}}$ having the lowest index and let n_2 be its **N**-finite end node. All other branches in **N** that are incident to n_2 will be finite in number and their currents will be known. So, Kirchhoff's current law determines the current in $b_{k_{m+1}}$ uniquely.

Note that at no step will $\mathcal{F} - b_{k_1} - \cdots - b_{k_m}$ have a component that does not contain at least part of a spine. In the event that $\mathcal{F}$ contains a finite limb $\mathcal{L}$ and all but one, say b_{k_i}, of the branches of $\mathcal{L}$ have been treated, it follows that b_{k_i} will have one **N**-finite node n_r and one **N**-infinite node q_s. The procedure then applies Kirchhoff's current law to n_r to determine the current in b_{k_i}. No contradiction can arise at q_r because Kirchhoff's current law is not applied to the **N**-infinite nodes.

Let b_k be any branch of $\mathcal{F}$ and let $\mathcal{L}$ be the limb that contains b_k. This procedure will eventually assign a current to b_k. Indeed, by Lemma 2.5-5, one of the two components, say, $\mathcal{H}$ of $\mathcal{L} - b_k$ is a finite tree all of whose nodes are **N**-finite. Thus, the procedure will eventually assign a current to every branch of $\mathcal{H}$ and then to b_k as well. Moreover, the current in b_k will depend only on the ties adjacent to $\mathcal{H}$ and therefore will be independent of the way the branches in $\mathcal{F}$ were numbered. That is, the current in b_k is uniquely determined by the tie currents. This completes the proof. ♣

Under the notation defined in Lemma 2.5-7 and its proof, b lies in a given d-orb if and only if one but not both of the nodes of d lies in $\mathcal{H}$. Now, assume that the current in d is i and the currents in all other ties are zero. Then, a repetition of the proof of Lemma 2.6-1 shows that Kirchhoff's current law requires that every branch not in the d-orb have zero current and that every branch b in the d-orb have the current $\pm i$; here again, the plus (minus) sign is used if b's orientation agrees (disagrees) with the d-orb's orientation. We say that the tie d *induces* the current $\pm i$ (or zero) in a branch if that branch is (is not) in d's orb. (Thus, d induces i in itself.)

Now, assume that arbitrary currents are assigned to all the ties in $\mathbf{N}$. By virtue of Lemma 2.5-7, only a finite number of ties induce nonzero currents in any given branch. Therefore, we may apply superposition to conclude that the currents induced in all branches of $\mathbf{N}$ by all the ties are finite and satisfy Kirchhoff's current law at all finite nodes. In view of the uniqueness assertion of Lemma 2.6-1, we can conclude with the following.

Lemma 2.6-2: *Let there be given a current regime in $\mathbf{N}$ such that Kirchhoff's current law is satisfied at finite nodes. Then, for any choice of a full set of limbs, the current in any branch is equal to the finite sum of the currents induced in that branch by the ties.*

2.7 Joints and Chords

We now return to $\mathcal{G}$ and a graph-theoretic discussion. In order to make use of Kirchhoff's voltage law, which by definition applies so far only to finite loops, we construct a spanning tree in $\mathcal{G}$ by adding certain ties to the chosen forest $\mathcal{F}$.

In particular, append to $\mathcal{F}_1$ as many ties in $\mathcal{G}_1$ as possible without forming any loops in the resulting subnetwork. Continue the procedure considering in turn $\mathcal{F}_1$, $\mathcal{F}_2$, $\mathcal{F}_3,\ldots$ as follows. Let $j_1,\ldots,j_m$ be the ties that have been added to $\mathcal{F}_1, \mathcal{F}_2,\ldots,\mathcal{F}_{p-1}$, where $p \geq 2$. Then, add to $\mathcal{F}_p$ as many ties in $\mathcal{G}_p$ as possible without forming any loops in the union of those ties in $\mathcal{G}_p$ with $\mathcal{F}_p \cup \cdots \cup \mathcal{F}_{p-1} \cup j_1 \cup \cdots \cup j_m$. After completing this procedure, let $\mathcal{J}$ be the set of all added ties and let $\mathcal{J}'$ be the subgraph of $\mathcal{G}$ induced by those ties. The members of $\mathcal{J}$ are called *joints*, and $\mathcal{J}$ is called a *full set of joints*. (See Examples 2.2-1 and 2.2-2 in this regard.)

Actually, the full sets of joints for a finitely chainlike infinite graph satisfy the base axioms of a matroid [133].

Lemma 2.7-1: $\mathcal{F} \cup \mathcal{J}'$ *is a spanning tree in $\mathcal{G}$.*

Proof: Since $\mathcal{F}$ is spanning in $\mathcal{G}$, so too is $\mathcal{F} \cup \mathcal{J}'$. Also, $\mathcal{F} \cup \mathcal{J}'$ will not contain any loops because no loops were allowed during the process of constructing $\mathcal{J}$. Finally, suppose that $\mathcal{F} \cup \mathcal{J}'$ is not connected. Let n_1 and n_2 be two nodes appearing in different components of $\mathcal{F} \cup \mathcal{J}'$. Since $\mathcal{G}$ is connected, there exists a path P in $\mathcal{G}$ joining n_1 and n_2. In tracing P we find at least one tie, say, d joining two different components of $\mathcal{F} \cup \mathcal{J}'$. But this is a contradiction; for, in the process of constructing $\mathcal{J}$, d would have been chosen as a joint, thereby connecting those components. ♣

Those ties that are not joints are called *chords*. (Examples 2.2-1 and 2.2-2 illustrate chords.) Set $\mathcal{T} = \mathcal{F} \cup \mathcal{J}'$. Since $\mathcal{T}$ is a spanning tree, each chord a generates in conjunction with $\mathcal{T}$ a unique loop, which we call either the $a \cup \mathcal{T}$ *loop* or the *chord-tree loop*. It is the union of a with the unique path in $\mathcal{T}$ connecting the nodes of a. Assume that a lies in $\mathcal{G}_p$. If both nodes of a lie in the same limb of $\mathcal{F}$, then the $a \cup \mathcal{T}$ loop is identical with the a-orb and lies entirely within $\mathcal{G}_p$. However, if the two nodes of a lie in different limbs, say $\mathcal{L}_1$ and $\mathcal{L}_2$, then the $a \cup \mathcal{T}$ loop is different from the a-orb (since the latter is now a path). Moreover, $a \cup \mathcal{T}$ lies in $\bigcup_{s=1}^{p} \mathcal{G}_s$. Indeed, if there were no path in $\mathcal{T} \cap (\bigcup_{s=1}^{p} \mathcal{G}_s)$ joining the nodes of a, then a would have been chosen as a joint, in contradiction to the assumption that a is a chord. Thus, such a path does exist, and its union with a yields the $a \cup \mathcal{T}$ loop lying in $\bigcup_{s=1}^{p} \mathcal{G}_p$. This result coupled with Lemma 2.5-6 yields the following.

Lemma 2.7-2: *Let $\mathcal{G}$ be any connected countable graph. Choose a chain-like partition $\{\mathcal{G}_s\}_{s=1}^{\infty}$ for $\mathcal{G}$ and then the spanning forest $\mathcal{F}$ as stated in Section 2.5. Also, choose a full set $\mathcal{J}$ of joints as stated in this section. Set $\mathcal{T} = \mathcal{F} \cup \mathcal{J}'$. Let a be any chord in $\mathcal{G}_p$. Then, the a-orb lies in $\bigcup_{s=p}^{\infty} \mathcal{G}_s$ and the $a \cup \mathcal{T}$ loop lies in $\bigcup_{s=1}^{p} \mathcal{G}_s$.*

2.8 The Equations for a Limb Analysis

A limb analysis uses Kirchhoff's voltage law to generate equations for unknown chord currents, given a particular choice of the set of limbs and joints. Kirchhoff's current law is automatically satisfied just by the way those equations are set up. In general, there will be an infinity of chords. As a result, we will be working with the space $\mathcal{C}^{\infty}$ of all infinite (vertical) vectors of the form $\mathbf{x} = [x_1, x_2, x_3, \ldots]^T$ where the components x_k are complex numbers. (The superscript T denotes matrix transpose.) No restriction is placed on the growth of the x_k as $k \to \infty$. Multiplication by a complex number and addition are defined componentwise, and this makes $\mathcal{C}^{\infty}$ a linear space.

Now, consider an infinite matrix of the form

$$Z = \begin{bmatrix} Z_{11} & Z_{12} & Z_{13} & \cdots \\ Z_{11} & Z_{12} & Z_{13} & \cdots \\ Z_{11} & Z_{12} & Z_{13} & \cdots \\ \vdots & \vdots & \vdots & \end{bmatrix}$$

where each Z_{jk} is a complex number. Z defines a mapping $\mathbf{x} \mapsto Z\mathbf{x}$ of C^∞ into C^∞ by means of the matrix product $Z\mathbf{x}$ if and only if every row of Z has no more than a finite number of nonzero entries. In this case Z is said to be *row-finite*, and the mapping $\mathbf{x} \mapsto Z\mathbf{x}$ is linear.

Assume in addition that Z has the partitioned form

$$Z = \begin{bmatrix} Z_1 & & & \\ \hline W_2 & Z_2 & & \quad 0 \\ \hline & W_3 & Z_3 & \\ \hline & & & \end{bmatrix} \tag{2.6}$$

where each main-diagonal block Z_p is a square finite $k_p \times k_p$ matrix and all entries to the right of these Z_p are zero. If every Z_p is nonsingular, then the equation $Z\mathbf{x} = \mathbf{y}$, where $\mathbf{y}$ is a given vector in C^∞, has a unique solution $\mathbf{x} \in C^\infty$. It can be obtained by first solving

$$Z_1[x_1, \ldots, x_{k_1}]^T = [y_1, \ldots, y_{k_1}]^T$$

for the first k_1 components of $\mathbf{x}$. Then, the next k_2 components of $\mathbf{x}$ can be determined by solving

$$Z_2[x_{k_1+1}, \ldots, x_{k_1+k_2}]^T = [y_{k_1+1}, \ldots y_{k_1+k_2}]^T - W_2[x_1, \ldots, x_{k_1}]^T.$$

Continuing in this way, we can determine all the components of $\mathbf{x}$. When Z has the form of (2.6), wherein each Z_p is nonsingular, we say that Z is *invertible in blocks*.

We now turn to the equations generated by a limb analysis. As was indicated in Example 1.6-3, Kirchhoff's voltage and current laws, coupled with (2.5), are not in general enough to force a unique current flow in **N**. This is reflected in the fact that the customary mesh analysis of finite networks fails in general for infinite networks; for one thing, given a fundamental system of mesh currents with respect to some spanning tree, it can happen that an infinity of such currents flow through a single branch, which can lead in turn to divergent series in the analysis.

On the other hand, because of Lemmas 2.6-2 and 2.7-2, we can apply another kind of analysis, which we call *limb analysis*, as follows: We continue to impose the same conditions on the electrical network **N** as those specified at the beginning of Section 2.6. However, a limb analysis of **N** requires an additional condition on the transimpedance z_{jk}: Assume that, if b_j is a branch in $\mathcal{G}_p$, then $z_{jk} \neq 0$ only if $b_k \in \mathcal{G}_s$ where $s \leq p$. This means that voltages are induced in b_j only by a finite number of branch currents, namely, the currents on some or all of the branches in $\bigcup_{s=1}^{p} \mathcal{G}_s$. Next, number all the joints consecutively using the positive integers. Let $\mathbf{j} = [j_1, j_2, \ldots]^T$ be the (finite or infinite) vector of all joint currents where j_k is the current in the kth joint. Furthermore, consecutively number the chords $1, 2, 3, \ldots$, starting first with the chords in $\mathcal{G}_1$, then proceeding to the chords in $\mathcal{G}_2$, then to the chords in $\mathcal{G}_3$, and so forth. Let $\mathbf{c} = [c_1, c_2, \ldots]^T$ be the vector of chord currents where c_k is the current in the kth chord.

If Kirchhoff's current law is satisfied, the current in each limb branch can be written as the finite sum of the currents induced in that limb branch by the chords and joints (see Lemma 2.6-2). Moreover, if Kirchhoff's voltage law is satisfied, we can write a sequence of Kirchhoff's voltage law equations, one for each chord-tree loop, in the order of the chord indices; in doing so, each chord-tree loop is assigned the orientation that agrees with its chord's orientation. Thus, in summing the branch voltages around a chord-tree loop, a plus (minus) sign is attached to a branch voltage if that branch's orientation agrees (disagrees) with the orientation of the chord-tree loop. Upon transposing the terms involving joint currents to the right-hand side, we obtain the matrix equation

$$Z\mathbf{c} = \mathbf{f}, \qquad (2.7)$$

where $Z = [Z_{jk}]$ is a matrix, whose entries are linear combinations of the self-impedances and transimpedances in **N**. Also, $\mathbf{f}$ is a vector whose entries are linear combinations of the branch voltage sources and the joint currents where the coefficients of the joint currents are self-impedances and transimpedances. We assume that the voltage sources and impedances are given, and we assign the values of the joint currents arbitrarily. Our assumption in Section 2.6 about coupling between branches in conjunction with Lemma 2.7-2 shows that Z has the partitioned form of (2.6). If each Z_p therein is nonsingular, then Z is invertible in blocks. This allows us to solve for $\mathbf{c}$, after which we can compute all the branch currents by using Lemma 2.6-2.

Before discussing conditions on the branch impedances which ensure that Z is invertible in blocks, let us note how this limb analysis avoids the aforementioned pitfall that renders the customary mesh analysis inoperative for infinite networks. First of all, it identifies a set of branches, namely, the joints to which one is free to assign currents arbitrarily, leading thereby to unique currents in the remaining branches. Moreover, by Lemma 2.6-2 only a finite number of chord and joint currents flow through any branch. (Contrast this to the application of mesh analysis to infinite networks wherein an infinity of mesh currents will in general pass through a given tree branch.) This in turn allows us to apply Kirchhoff's voltage law around the chord-tree loops to get network equations represented by (2.7) wherein Z is row-finite. Actually, our numbering procedure, Lemma 2.7-2, and our hypothesis on coupling between branches forces Z to have the partitioned form of (2.6).

2.9 Chord Dominance

The kth equation in the expansion of (2.7) corresponds to Kirchhoff's voltage law written for the $a_k \cup T$ loop, where a_k is the kth chord. But the only chord that induces a nonzero current in a_k is a_k itself, and the only chord contained in the $a_k \cup T$ loop is again a_k. These facts imply that the self-impedance z_{a_k} of a_k appears only as an added term in the kth main-diagonal entry Z_{kk} of Z and nowhere else. That is,

$$Z_{kk} = z_{a_k} + z^{kk},$$

where z^{kk} is independent of z_{a_k}; also, Z_{ms} is independent of z_{a_k} if either one or both of m and s are not equal to k. Therefore, by varying z_{a_k}, we vary only Z_{kk} and no other entry of Z. In particular, if the absolute values $|z_{a_k}|$ of all the chord self-impedances are chosen sufficiently large, then all the blocks Z_p in (2.6) will become dominated by their diagonal elements and thereby nonsingular [96, page 32].

An explicit condition of this nature can be obtained if we examine how the various branch impedances appear in the entries of the block Z_p in (2.6). We have

$$Z_{kk} = z_{a_k} + \sum_j \pm z_j^{kk}$$

and, for $s \neq k$,

$$Z_{ks} = \sum_j \pm z_j^{ks}.$$

Here both summations have a finite number of terms. Also, $\sum_j \pm z_j^{kk}$ contains the self-impedances of all the branches other than a_k in the intersection of the $a_k \cup T$ loop with the a_k-orb as well as those trans-impedances that couple currents in the branches of the a_k-orb to voltages in branches of the $a_k \cup T$ loop. Finally, $\sum_j \pm z_j^{ks}$, where $k \neq s$, contains the self-impedances of all branches in the intersection of the $a_k \cup T$ loop with the a_s-orb as well as those transimpedances that couple currents in the branches of the a_s-orb to voltages in branches of the $a_k \cup T$ loop. The plus (minus) sign in front of z_j^{kk} or z_j^{ks} is used if a positive current in chord a_k or, respectively, chord a_s produces via this impedance a positive (negative) voltage in the $a_k \cup T$ loop when z_j^{kk} or z_j^{ks} is taken to be one ohm.

Now, assume that the chord a_k lies in $\mathcal{G}_p$ amd let $l, l+1, \ldots, m$ be the indices of all the chords in $\mathcal{G}_p$. Thus, $l \leq k \leq m$. If

$$|z_{a_k}| > |\sum_j \pm z_j^{kk}| + \sum_{s=l, s \neq k}^{m} |\sum_j \pm z_j^{ks}|, \qquad (2.8)$$

then

$$|Z_{kk}| \geq |z_{a_k}| - |\sum_j \pm z_j^{kk}| > \sum_{s=l, s\neq k}^{m} |\sum_j \pm z_j^{ks}| = \sum_{s=l, s\neq k}^{m} |Z_{ks}|.$$

Hence, if (2.8) holds for every chord a_k in **N**, then each Z_p in (2.6) will truly be dominated along its rows by its diagonal elements. Similarly, if

$$|z_{a_k}| > |\sum_j \pm z_j^{kk}| + \sum_{s=l, s \neq k}^{m} |\sum_j \pm z_j^{sk}|, \qquad (2.9)$$

then, as above,

$$|Z_{kk}| > \sum_{s=l, s\neq k}^{m} |Z_{sk}|.$$

So, if (2.9) holds for every chord a_k in **N**, then each Z_p will be dominated along its columns by its diagonal elements. Finally, if either (2.8) holds for every chord in **N** or (2.9) holds for every chord in **N**, we say that **N** is *chord-dominant with respect to T and $\mathcal{J}$*, where as always T is the chosen tree $\mathcal{F} \cup \mathcal{J}'$ and $\mathcal{J}$ is the chosen full set of joints. In this case, Z is invertible in blocks.

2.10 Limb Analysis, Summarized

Theorem 2.10-1: *Let* $\mathbf{N}$ *be a connected, countably infinite network. Then, there exists in* $\mathbf{N}$ *a spanning forest* $\mathcal{F}$*, a full set* $\mathcal{J}$ *of joints, and a chainlike partition* $\{\mathcal{G}_s\}_{s=1}^{\infty}$ *of the graph* $\mathcal{G}$ *of* $\mathbf{N}$ *into finite subgraphs* $\mathcal{G}_s$ *such that, for* $\mathcal{T} = \mathcal{F} \cup \mathcal{J}'$ *and for each chord* a *in* $\mathcal{G}_p$*, the corresponding* $a \cup \mathcal{T}$ *loop lies in* $\bigcup_{s=1}^{p} \mathcal{G}_s$ *and the corresponding a-orb lies in* $\bigcup_{s=p}^{\infty} \mathcal{G}_s$*. Assume that the electrical parameters of* $\mathbf{N}$ *satisfy the conditions stated in the first paragraph of Section 2.6, and also assume that coupling between branches is such that a current in branch* b_k *of* $\mathcal{G}_s$ *produces a nonzero voltage in branch* b_j *of* $\mathcal{G}_p$ *only if* $s \le p$*. Arbitrarily assign values to all the joint currents. Number the chords as stated in Section 2.8. Upon writing Kirchhoff's voltage law around each chordtree loop and invoking Kirchhoff's current law to express each branch current as the finite sum of the chord and joint currents induced in that branch (Lemma 2.6-2), we obtain a system of equations that has the matrix form (2.7), where* $\mathbf{c}$ *is the unkown vector of chord currents and* $\mathbf{f}$ *is a known vector depending on the branch voltage sources, the self-impedances and transimpedances, and the joint currents; moreover,* Z *has the partitioned form of (2.6).* Z *is invertible on* $\mathcal{C}^{\infty}$ *whenever each main-diagonal block* Z_p *in* Z*'s partitioned form is nonsingular. A sufficient condition for this to be so is that* $\mathbf{N}$ *be chord-dominant with respect to* $\mathcal{T}$ *and* $\mathcal{J}$*. When* Z *is invertible on* $\mathcal{C}^{\infty}$*,* $\mathbf{c}$ *will be uniquely determined, and, according to Lemma 2.6-2, so too will be all the branch currents. Moreover, when* Z *is invertible on* $\mathcal{C}^{\infty}$*, any set of branch currents that satisfy Kirchhoff's current law at finite nodes and Kirchhoff's voltage law (around finite loops) will correspond in this way to a particular choice of joint currents.*

Proof: Everything has been established by our foregoing arguments except for the last sentence. Under any given set of branch currents, the joint currents will be specified. Because Kirchhoff's current and voltage laws are satisfied as stated, (2.7) holds. The invertibility of Z implies that the chord currents, as determined by (2.7), must coincide with the given chord currents. The limb-branch currents, as determined by Lemma 2.6-2, must also coincide with the given ones by virtue of the uniqueness assertion of Lemma 2.6-1. ♣

Finally, let us note in passing that the analysis of [165, Section 6] can now be applied to determine the dimension dim $\mathcal{H}$ of the linear space $\mathcal{H}$ of all homogeneous current flows in $\mathbf{N}$. By a *homogeneous current flow* we mean a vector of all the branch currents in $\mathbf{N}$, where Kirchhoff's

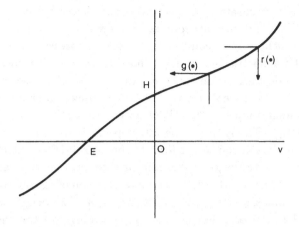

Figure 2.5. A voltage-current characteristic for a nonlinear branch.

current law at finite nodes and Kirchhoff's voltage law (around finite loops) are satisfied as well as (2.5) with all $e_j = 0$. When **N** satisfies the hypothesis of Theorem 2.10-1, we have dim $\mathcal{H} = |\mathcal{J}|$, where $|\mathcal{J}|$ is the cardinality of the full set of joints. Since **N** is connected, $|\mathcal{J}| = x - 1$, where x is the cardinality of the set of spines in **N**.

2.11 Nonlinear Networks

A limb analysis can be extended to nonlinear networks in various ways not only for resistive networks but also for nonlinear RLC networks as well [167],[168]. In the next two sections we shall present two such extensions for resistive networks having no coupling between branches. (Later on, we will indicate how such coupling can be taken into account.) This section is devoted to a reformulation of Kirchhoff's current and voltage equations into a form suitable for those extensions.

We now assume that each branch b is characterized by a voltage-current curve as is indicated in Figure 2.5. The correspondence (v, i) given by this curve determines a *voltage-controlled conductance* $g(\cdot)$ if $v \mapsto g(v) = i$ is a function (i.e., a single-valued mapping) for all $v \in R^1$. In this case, the branch itself is said to be *voltage-controlled*. Similarly, that correspondence determines a *current-controlled resistance* $r(\cdot)$ if $i \mapsto r(i) = v$ is a function for all $i \in R^1$, and the branch too is called *current-controlled*. We do not require that $g(0)$ or $r(0)$ be zero. Moreover, if g or r is a bijection of R^1 onto R^1, then the branch is both voltage-controlled and current-controlled. Also, every constant voltage source is trivially a current-controlled resistor, and every constant current source is trivially a voltage-controlled conductor. These are the only sources we allow.

Linear branches are encompassed by these representations. In this case, the characteristic is a straight line. Its slope with respect to the v-axis is its constant conductance g', and $g(v) = g'v$. Also, $r' = 1/g'$ is its constant resistance with $r(i) = r'i$. Its v-axis intercept E is the constant voltage source in the Thevenin equivalent form of the branch (Figure 1.1(b)). Finally, its i-axis intercept H is the constant current source in the Norton equivalent form of the branch (Figure 1.1(c)).

We return now to the general nonlinear case. Assume that, among all the possible choices for the chainlike representation of the network along with a set of limbs, joints, and chords, there is at least one choice under which all the limb branches are current-controlled, all the chords are voltage-controlled, and all joints are either voltage-controlled or current-controlled. Upon selecting that choice, we may use Kirchhoff's current law to write an equation for the current i_k in each limb branch b_k of some fixed subgraph $\mathcal{G}_p$ as follows:

$$i_k - \sum_{k,q} \pm g_q(v_q) = f_k + \sum_{k,t} \pm j_t. \qquad (2.10)$$

Here, v_q is the voltage on the chord b_q in $\mathcal{G}_p$, and $g_q(\cdot)$ is the conductance of b_q. The summation $\sum_{k,q}$ is over all chords c_q in $\mathcal{G}_p$ that induce a nonzero current $g_q(v_q)$ in b_k; the plus (minus) sign is chosen for $g_q(v_q)$ if the induced current's orientation agrees (disagrees) with b_k's orientation. f_k is the sum of all currents induced in b_k by all the joints and chords in $\bigcup_{s=1}^{p-1} \mathcal{G}_s$ and is measured with respect to b_k's orientation. f_k is zero if b_k is not in a spine. If b_k is in a spine, f_k equals (plus or minus) the current entering $\mathcal{G}_p$ at the node of that spine in $\mathcal{V}_p$. j_t is the current in the tth joint of $\mathcal{G}_p$. The summation $\sum_{k,t}$ is over all joints j_t in $\mathcal{G}_p$ that induce nonzero currents in b_k, the plus (minus) sign being used with j_t if the induced current's orientation agrees (disagrees) with b_k's orientation.

Kirchhoff's voltage law generates a similar set of equations. Continue to keep $\mathcal{G}_p$ fixed. Let b_q be any chord in $\mathcal{G}_p$. The union of the chosen limbs and joints is the spanning tree $\mathcal{T}$, and $b_q \cup \mathcal{T}$ denotes the chord-tree loop for b_q. An equation for the voltage v_q on b_q is

$$v_q - \sum_{q,k} \pm r_k(i_k) = a_q + \sum_{q,t} \pm w_t. \qquad (2.11)$$

Here, i_k is the current in the limb branch b_k in $\mathcal{G}_p$, and $r_k(\cdot)$ is the resistance of b_k. The summation $\sum_{q,k}$ is over all limb branches b_k in $\mathcal{G}_p$ that lie in the $b_q \cup \mathcal{T}$ loop; the plus (minus) sign is chosen for $r_k(i_k)$ if the orientation of b_q disagrees (agrees) with b_k's orientation with respect to a tracing of the $b_q \cup T$ loop. a_q is the voltage across that portion of the $b_q \cup T$ loop that lies in $\bigcup_{s=1}^{p-1} \mathcal{G}_s$ and is measured with respect to a tracing of that loop that disagrees with b_q's orientation. a_q equals zero if that loop lies entirely within $\mathcal{G}_p$. w_t is the voltage on the tth joint of $\mathcal{G}_p$. The summation $\sum_{q,t}$ is over all joints j_t in $\mathcal{G}_p$ lying in the $b_q \cup T$ loop; the plus (minus) sign is used with w_t if b_q's orientation disagrees (agrees) with j_t's orientation with respect to a tracing of the $b_q \cup T$ loop.

Eventually our argument will rely on a recursive determination of the limb-branch currents and chord voltages in the subgraphs $\mathcal{G}_1, \mathcal{G}_2, \ldots$ considered sequentially. Thus, if we assume that all such currents and voltages in $\bigcup_{s=1}^{p-1} \mathcal{G}_s$ are known, then the right-hand sides of (2.10) and (2.11) are also known. Upon collecting all the equations (2.10) and (2.11) for the limb-branch currents i_k and the chord voltages v_q in $\mathcal{G}_p$, we get a nonlinear equation in matrix form: $x - W(x) = y$ or

$$[I - W(\cdot)]x = y \qquad (2.12)$$

Upon properly renumbering the branches of $\mathcal{G}_p$, we can give these matrices the following forms. For n limb branches and m chords in $\mathcal{G}_p$, x is the unknown $(n+m) \times 1$ vector

$$x = [i_1, \ldots, i_n, v_{n+1}, \ldots, v_{n+m}]^T.$$

y is a known $(n+m) \times 1$ vector whose elements are the right-hand sides of (2.10) and (2.11). I is the $(n+m) \times (n+m)$ identity matrix. Finally,

$$W(\cdot) = \begin{bmatrix} 0 & A(\cdot) \\ B(\cdot) & 0 \end{bmatrix}$$

where $A(\cdot)$ is an $n \times m$ matrix and $B(\cdot)$ is an $m \times n$ matrix. For $k = 1, \ldots, n$ and $q = n+1, \ldots, n+m$, the kth row of $W(\cdot)$ has either the mapping $\pm g_q(\cdot)$ or 0 in the qth column, and the qth row of $W(\cdot)$ has either the mapping $\pm r_k(\cdot)$ or 0 in the kth column. In fact, the vector $W(x)$ has $\sum_{k,q} \pm g_q(v_q)$ in the kth position and $\sum_{q,k} \pm r_k(i_k)$ in the qth position.

2.12 A Contraction Mapping Result

One version of the contraction-mapping principle for k-dimensional real Euclidean space R^k is given by the next lemma. We shall make use of it in establishing an existence and uniqueness theorem for the voltage-current regime in our network. Let $\| \cdot \|$ denote the Euclidean norm for R^k, and let I denote the identity mapping on R^k. A mapping F of R^k

into R^k is said to be a *contraction mapping* if for all u and v in R^k we have $||F(u+v) - F(u)|| \leq \alpha||v||$, where $0 < \alpha < 1$.

Lemma 2.12-1: *If F is a contraction mapping of R^k into R^k, then $I - F$ is a bijection of R^k onto R^k.*

Proof: Since F is a contraction mapping, it has a unique fixed point x_0, that is, $F(x_0) = x_0$ [94, page 314]. Now, let $y \in R^k$ be fixed. We wish to show that there is a unique $z \in R^k$ such that $z - F(z) = y$. Set $F_y(\cdot) = y + F(\cdot)$. It follows immediately that $F_y(\cdot)$ is also a contraction mapping; indeed, $||F_y(u+v) - F_y(u)|| = ||F(u+v) - F(u)|| \leq \alpha||v||$. Therefore, $F_y(\cdot)$ has a unique fixed point, say, z. Whence, $z - F(z) = y$. ♣

The Schwarz inequality implies that, for any set of k real numbers $a_1, \ldots, a_k$, we have $(\sum_{i=1}^{k} a_i)^2 \leq K \sum_{i=1}^{k} a_i^2$. Hence, for any $(n+m) \times 1$ vector $\mathbf{h} = [h_1, \ldots, h_{n+m}]^T$ of real numbers h_i, we may write the following for the operator W defined in the preceding section.

$$||W(x+h) - W(x)||^2$$

$$= \sum_{s=1}^{n} \left\{ \sum_{s,q} \pm[g_q(v_q + h_q) - g_q(v_q)] \right\}^2$$

$$+ \sum_{s=n+1}^{n+m} \left\{ \sum_{s,k} \pm[r_k(i_k + h_k) - r_k(i_k)] \right\}^2$$

$$\leq \sum_{q=n+1}^{n+m} M_q[g_q(v_q + h_q) - g_q(v_q)]^2$$

$$+ \sum_{k=1}^{n} M_k[r_k(i_k + h_k) - r_k(i_k)]^2$$

where M_q (or M_k) is a number determined as follows. Consider all the rows of $W(\cdot)$ in which $g_q(\cdot)$ (or, respectively, $r_k(\cdot)$) appears. Let $M_{q,\nu}$ (or $M_{k,\nu}$) be the number of terms in the νth such row. Then, $M_q = \sum_\nu M_{q,\nu}$ (or $M_k = \sum_\nu M_{k,\nu}$).

A way to determine M_q that avoids setting up the matrix $W(\cdot)$ is as follows. Let P_q be the orb for the chord in $\mathcal{G}_p$ whose conductance is $g_q(\cdot)$. For any limb-branch b_k lying in both $\mathcal{G}_p$ and P_q, count the number of chords in $\mathcal{G}_p$ whose orbs pass through b_k. Add together all such numbers obtained by letting b_k traverse all the limb-branches that lie in both $\mathcal{G}_p$ and P_q. This gives the M_q for $g_q(\cdot)$.

To determine M_k in an alternative way, let b_k be the limb-branch whose resistance is $r_k(\cdot)$ and let d_q be any chord in $\mathcal{G}_p$ whose $d_q \cup T$

loop contains b_k. Count the number of limb-branches in $\mathcal{G}_p$ that lie in the $d_q \cup \mathcal{T}$ loop. Add together all such numbers obtained by letting d_q traverse all chords in $\mathcal{G}_p$ whose chord-$\mathcal{T}$ loops contain b_k. This gives the M_k for $r_k(\cdot)$.

Now, assume that, for all values of their arguments, the functions g_q and r_k satisfy the Lipschitz conditions:

$$[g_q(v_q + h_q) - g_q(v_q)]^2 \leq \gamma_q h_q^2, \quad q = n+1, \ldots, n+m \tag{2.13}$$

and

$$[r_k(i_k + h_k) - r_k(i_k)]^2 \leq \gamma_k h_k^2, \quad k = 1, \ldots, n \tag{2.14}$$

where γ_q and γ_k are real numbers such that

$$M_q \gamma_q < 1, \quad M_k \gamma_k < 1. \tag{2.15}$$

(These Lipschitz conditions imply that g_q and r_k are continuous functions.) It follows from (2.13), (2.14), and (2.15) that

$$\|W(x+h) - W(x)\|^2 \leq \alpha^2 \sum_{s=1}^{n+m} h_s^2 = \alpha^2 \|h\|^2 \tag{2.16}$$

where $0 < \alpha < 1$. Thus, W is a contraction mapping. We may therefore invoke Lemma 2.12-1 to conclude that (2.12) has a unique solution x for each vector y.

In other words, if we choose arbitrarily the voltages on the voltage-controlled joints and the currents in the current-controlled joints, then a knowledge of the limb-branch currents and the chord voltages in $\bigcup_{s=1}^{p-1} \mathcal{G}_p$ allows us to compute those quantities in $\mathcal{G}_p$ as well. Thus, a recursive computation uniquely determines the currents and voltages in all of the network. We have hereby established the following.

Theorem 2.12-2: *Assume that a (possibly nonlinear) resistive network* **N** *has a connected, countably infinite graph $\mathcal{G}$ and the following properties:*

(i) *There is no coupling between branches.*

(ii) *Among the possible choices of the chainlike partition $\{\mathcal{G}_p\}_{p=1}^{\infty}$ of $\mathcal{G}$ along with a spanning forest $\mathcal{F}$ and a full set of joints, there is one for which every limb branch is a current-controlled resistance and every chord is a voltage-controlled conductance.*

(iii) *For that choice and for every $\mathcal{G}_p$, the conductance $g_q(\cdot)$ for each chord in $\mathcal{G}_p$ satisfies (2.13) and (2.15) and the resistance $r_k(\cdot)$ for each limb-branch in $\mathcal{G}_p$ satisfies (2.14) and (2.15), where the M_q and M_k are defined above.*

Then, any assignment of voltage-current pairs to all the joints uniquely determines under Kirchhoff's current law at finite nodes and Kirchhoff's voltage law (for finite loops) the currents and voltages in all branches of **N***. They can be computed recursively by solving in turn an equation of the form (2.12) for each of the finite subgraphs* $\mathcal{G}_1, \mathcal{G}_2, \ldots$.

We note in passing that the assumption of no coupling between branches can be relaxed. For example, given any limb-branch b_k in $\mathcal{G}_p$, if a voltage on any chord of $\mathcal{G}_p$ (or on any branch of $\bigcup_{s=1}^{p-1} \mathcal{G}_p$) induces by means of a transconductance a current in any other chord of $\mathcal{G}_p$ whose orb passes through b_k, then an additional unknown (respectively, known) term appears in (2.10). Similarly, given any chord b_q in $\mathcal{G}_p$, if a current in any limb of $\mathcal{G}_p$ (or in any branch of $\bigcup_{s=1}^{p-1} \mathcal{G}_p$) induces by means of a transresistance a voltage in some branch in $\mathcal{G}_p$ of the $b_q \cup \mathcal{T}$ loop other than b_q, then another unknown (respectively, known) term appears in (2.11). Upon appropriately restricting such coupling within $\mathcal{G}_p$ for these additional unknown terms, the inequality (2.16) can again be satisfied, and therefore Theorem 2.12-2 can be extended. However, coupling that induces a current or voltage in a branch of $\mathcal{G}_p$ from a voltage or current in a branch of $\mathcal{G}_n$, where $n > p$, cannot be encompassed by our analysis.

It is also worth mentioning that Theorem 2.12-2 extends directly to the case where the currents and voltages are elements of a real or complex Banach space B and the resistances and conductances are nonlinear operators on B.

2.13 A More General Fixed Point Theorem

We can derive a more generally applicable result than Theorem 2.12-2 by using Brouwer's fixed-point theorem instead of the contraction-mapping principle – at the expense, however, of a weakened conclusion, namely, the uniqueness of the voltage-current regime may be lost. Brouwer's theorem may be stated in the following way [48, page 468]. Let D denote a closed spherical region in R^{n+m}: $D = \{x \in R^{n+m} : \|x\| \leq \rho > 0\}$.

Lemma 2.13-1 (Brouwer's theorem): *Any continuous mapping* $F(\cdot)$ *of D into D has at least one fixed point $x_0 \in D$, i.e., $F(x_0) = x_0$.*

We can generalize Theorem 2.12-2 by dropping the Lipschitz conditions (2.13) and (2.14) and using instead the boundedness conditions, given by (2.19) and (2.20) below, along with some continuity assumptions. For the conductance of each chord b_q we write

$$g_q(v) = g_q(0) + \hat{g}_q(v) = H_q + \hat{g}_q(v). \qquad (2.17)$$

Also, for the resistance of each limb-branch b_k we write

$$r_k(i) = r_k(0) + \hat{r}_k(i) = E_k + \hat{r}_k(i). \tag{2.18}$$

Thus, each chord b_q is represented in its Norton form by a pure current source H_q in parallel with a conductance $\hat{g}_q(\cdot)$ whose characteristic is assumed to be continuous and passes through the origin, and each limb-branch b_k is represented in its Thevenin form by a pure voltage source E_k in series with a resistance $\hat{r}_k(\cdot)$ whose characteristic is also assumed to be continuous and passes through the origin. Moreover, we require that for each chord b_q

$$|\hat{g}_q(v)| \leq Q_q|v| \tag{2.19}$$

and for each limb-branch b_k

$$|\hat{r}_k(i)| \leq Q_k|i|. \tag{2.20}$$

We may rewrite equations (2.10) and (2.11) as follows.

$$i_k - \sum_{k,q} \pm\hat{g}_q(v_q) = f_k + \sum_{k,t} \pm j_t + \sum_{k,q} \pm H_q \tag{2.21}$$

$$v_q - \sum_{q,k} \pm\hat{r}_k(i_k) = a_q + \sum_{q,t} \pm w_t + \sum_{q,k} \pm E_k \tag{2.22}$$

Here, all the symbols have exactly the same meanings as before, except for our altered forms of the chord conductances and limb-branch resistances. As before, the right-hand sides of (2.21) and (2.22) are known. Furthermore, let M_q and M_k be defined exactly as in the preceding section and assume that

$$M_q Q_q^2 < 1, \quad M_k Q_k^2 < 1. \tag{2.23}$$

Next, upon numbering all the chords and limb-branches in $\mathcal{G}_p$ as in Section 2.12, we obtain (2.12) once again except that g_q is replaced by $\hat{g}_q$, r_k by $\hat{r}_k$, and y is altered by the addition or subtraction of the H_q and E_k. We can employ Schwarz's inequality as before to obtain

$$\|W(x)\|^2 = \sum_{s=1}^{n} \left[\sum_{s,q} \pm\hat{g}_q(v_q) \right]^2 + \sum_{s=n+1}^{n+m} \left[\sum_{s,k} \pm\hat{r}_k(i_k) \right]^2$$

$$\leq \sum_{q=n+1}^{n+m} M_q|\hat{g}_q(v_q)|^2 + \sum_{k=1}^{n} M_k|\hat{r}_k(i_k)|^2.$$

By (2.19) and (2.20),

$$||W(x)||^2 \le \sum_{q=n+1}^{n+m} M_q Q_q^2 v_q^2 + \sum_{k=1}^{n} M_k Q_k^2 i_k^2.$$

In view of (2.23), we therefore have $||W(x)|| \le \delta||x||$ for some $\delta < 1$.

In the present case with regard to (2.12), y is the known vector in R_{n+m} obtained by gathering the right-hand sides of (2.21) and (2.22) in accordance with our numbering system $k = 1, \ldots, n$ for the limb-branches and $q = n+1, \ldots, n+m$ for the chords in $\mathcal{G}_p$. Let D represent the spherical region:

$$D = \left\{ x \in R^{n+m} : ||x|| \le \frac{||y||}{1-\delta} \right\}.$$

Also, let $z = y/(1-\delta)$. Then, for every $x \in D$,

$$||y + W(x)|| \le ||y|| + \delta||x|| \le (1-\delta)||z|| + \delta||z|| = ||z||.$$

Thus, $y + W(\cdot)$ maps D into D. We have assumed that all chord conductances and all limb-branch resistances are continuous functions on R_1. Consequently, $y + W(\cdot)$ is continuous on D. So, by Lemma 2.13-1, there exists at least one $x_0 \in D$ such that $y + W(x_0) = x_0$; that is, (2.12) has a solution x_0.

Upon applying this argument inductively to $\mathcal{G}_1, \mathcal{G}_2, \mathcal{G}_3, \ldots$, we obtain

Theorem 2.13-2: *Assume that the hypothesis of Theorem 2.12-2 holds except for condition (iii), which is replaced by the following: (iii') For every $\mathcal{G}_p$, each chord conductance $g_q(\cdot)$ in $\mathcal{G}_p$ is a continuous function and its Norton form (2.17) satisfies (2.19), and each limb-branch resistance $r_k(\cdot)$ in $\mathcal{G}_p$ is a continuous function and its Thevenin form (2.18) satisfies (2.20); in addition, (2.23) is satisfied.*

Then, the conclusion of Theorem 2.12-2 holds again except that the determined voltage-current regime may not be unique.

Here too, coupling between branches can be encompassed as indicated at the end of the last section.

Chapter 3

Finite-power Regimes:
The Linear Case

In 1971 Harley Flanders [51] opened a door by showing how a unique, finite-power, voltage-current regime could be shown to exist in an infinite resistive network whose graph need not have a regular pattern. To be sure, infinite networks had been examined at least intermittently from the earliest days of circuit theory, but those prior works were restricted to simple networks of various sorts, such as ladders and grids. For example, infinite uniform ladder networks were analyzed in [31], [73], and [139], works that appeared 70 to 80 years ago. Early examinations of uniform grids and the discrete harmonic operators they generate can be found in [35], [44], [45], [47], [52], [54], [65], [84], [126], [135], [143].

Flanders' theorem, an exposition of which starts this chapter, is restricted to locally finite networks with a finite number of sources. Another tacit assumption in his theory is that only open-circuits appear at the infinite extremities of the network. The removal of these restrictions, other extensions, and a variety of ramifications [158], [159], [163], [177] comprise the rest of this chapter. However, the assumptions that the network consists only of linear resistors and independent sources and is in a finite-power regime are maintained throughout this chapter.

Actually, finite-power theories for nonlinear networks are now available, and they apply just as well to linear networks as special cases. One is due to Dolezal [40], [41], and the other to DeMichelle and Soardi [37].

However, since the nonlinear case is such an important generalization, we shall postpone these theories to the next chapter.

As was noted in Examples 1.6-3 and 1.6-4 and examined in Chapter 2, Kirchhoff's laws and Ohm's law are enough to establish the existence of a voltage-current regime in an infinite network, but are not enough for its uniqueness. One objective of this chapter is to find additional conditions that yield uniqueness. This is resolved by specifying in effect the "connections at infinity" along with the finite-power requirement. However, it is a fact that there are many networks for which the finite-power condition is all that is needed beyond Kirchhoff's and Ohm's laws in order to achieve uniqueness. For instance, this is so for the network of Figure 1.6 when every resistor is 1 Ω. In fact, we might say for some network that "infinity is imperceptible" (otherwise, "infinity is perceptible") if its voltage-current regime is independent of (dependent upon) what the connections at infinity might be. How to characterize networks, for which "infinity is imperceptible," is examined in Chapters 6 and 7, but only for cascades and grids. Moreover, Sections 3.10 and 3.11 of this chapter present some related results. The general problem is a difficult one. It is related to the transience or recurrence of random walks on infinite graphs and to the parabolicity or hyperbolicity of infinite networks; in this regard, see the references cited in Section 8.3.

3.1 Flanders' Theorems

Flanders' seminal paper [51] rigorously established for the first time the existence of a voltage-current regime in a locally finite, resistive network having a finite number of sources. He showed that the regime is unique among all regimes whose power dissipations are finite and whose current vectors are the limits in a certain sense of linear combinations of loop currents. The last condition asserts in effect that there are no connections to the network at infinity. In this exposition we will simplify Flanders' original development by assuming that all branch resistances are positive; Flanders allowed nonnegative branch resistances so long as every loop had a positive resistance. Flanders also allowed both current sources and voltage sources, but we will assume for now that only voltage sources are present. Both of these simplifying restrictions will be removed later on when we introduce pure-source branches.

Conditions 3.1-1: *Let* **N** *be a locally finite, countably infinite network, every branch b_j of which is in the Thevenin form with a positive resistance*

r_j and a voltage source e_j (as in Figure 1.1(b) but with E replaced by e_j). Assume that all but a finite number of the e_j are zero.

As always, we take it that all the branches are indexed by $j = 1, 2, 3, \ldots$ and are assigned orientations. Now, any assignment of real numbers x_j to the branches of $\mathbf{N}$ defines a vector $\mathbf{x} = (x_1, x_2, x_3, \ldots)$; the set of branches for which $x_j \neq 0$ is called the *support of* $\mathbf{x}$. Thus, the vector $\mathbf{e} = (e_1, e_2, e_3, \ldots)$ has a finite support, according to Conditions 3.1-1. Also, as always, the symbol $\sum$ will denote a summation over all branch indices j unless something else is explicitly indicated.

To proceed, let $\mathcal{I}$ denote the set of all branch-current vectors $\mathbf{i} = (i_1, i_2, \ldots)$ whose power dissipations are finite, i.e., $\sum i_j^2 r_j < \infty$. The members of $\mathcal{I}$ need not satisfy Kirchhoff's current law. The linear operations are defined componentwise on the current vectors in $\mathcal{I}$, and this makes $\mathcal{I}$ a linear space. $\mathcal{I}$ is equipped with the inner product

$$(\mathbf{i}, \mathbf{s}) = \sum r_j i_j s_j \tag{3.1}$$

where $\mathbf{i}, \mathbf{s} \in \mathcal{I}$; $\| \cdot \|$ will now denote the corresponding norm.

Lemma 3.1-2: *Convergence in $\mathcal{I}$ implies componentwise convergence.*

Proof: Let $\mathbf{i}_k = (i_{k1}, i_{k2}, \ldots)$ converge to $\mathbf{i} = (i_1, i_2, \ldots)$ in $\mathcal{I}$ as $k \to \infty$. Thus,

$$\| \mathbf{i}_k - \mathbf{i} \|^2 = \sum r_j (i_{kj} - i_j)^2 \to 0.$$

Since every $r_j > 0$, we must have that $i_{kj} \to i_j$ for each j. ♣

Lemma 3.1-3: *$\mathcal{I}$ is a Hilbert space*

Proof: We have to show that $\mathcal{I}$ is complete. This can be done in a standard way [89, page 21]. Let $\{\mathbf{i}_k\}_{k=1}^{\infty}$ be a Cauchy sequence in $\mathcal{I}$ and set $\mathbf{i}_k = (i_{k1}, i_{k2}, \ldots)$. Thus, given any $\epsilon > 0$ there exists a positive integer n_ϵ such that, for all $k, m > n_\epsilon$,

$$\| \mathbf{i}_k - \mathbf{i}_m \|^2 = \sum r_j (i_{kj} - i_{mj})^2 < \epsilon. \tag{3.2}$$

Since every $r_j > 0$, it follows that, for each fixed j, $\{i_{kj}\}_{k=1}^{\infty}$ is a Cauchy sequence in R^1. Hence, there exists an $i_j \in R^1$ such that $i_{kj} \to i_j$ as $k \to \infty$. Set $\mathbf{i} = (i_1, i_2, \ldots)$.

It follows from (3.2) that, for each positive integer p,

$$\sum_{j=1}^{p} r_j (i_{kj} - i_{mj})^2 < \epsilon$$

whenever $k, m > n_\epsilon$. We may let $m \to \infty$ to obtain

$$\sum_{j=1}^{p} r_j (i_{kj} - i_j)^2 \le \epsilon.$$

We then let $p \to \infty$ to get

$$\sum r_j (i_{kj} - i_j)^2 \le \epsilon \qquad (3.3)$$

whenever $k > n_\epsilon$. Hence, $\mathbf{i}_k - \mathbf{i} \in \mathcal{I}$. Since $\mathbf{i}_k \in \mathcal{I}$, $\mathbf{i} \in \mathcal{I}$ too. Finally, (3.3) also indicates that $\mathbf{i}_k \to \mathbf{i}$ in $\mathcal{I}$. ♣

Next, we assign an orientation to every loop in $\mathbf{N}$. A *loop current of value* $x \in R^1$ is a current vector $\mathbf{i} = (i_1, i_2, \dots)$ such that, for some loop L, $i_j = 0$ if branch $b_j \notin L$ and $i_j = \pm x$ if $b_j \in L$, where the plus (minus) sign is used if the orientation of b_j agrees (disagrees) with the orientation of L. Thus, $\mathbf{i}$ satisfies Kirchhoff's current law. Since every loop is by definition finite, $\mathbf{i}$ has a finite support. This implies in turn that its power dissipation is finite, and therefore $\mathbf{i} \in \mathcal{I}$.

We let $\mathcal{K}^0$ denote the span of all loop currents and let $\mathcal{K}$ be the closure of $\mathcal{K}^0$ in $\mathcal{I}$. Thus, $\mathcal{K}$ is a closed subspace of $\mathcal{I}$ and is a Hilbert space by itself with the inner product (3.1). Every member of $\mathcal{K}^0$ has a finite support, but this is not in general true for the members of $\mathcal{K}$.

Let $\mathbf{u} = (u_1, u_2, \dots)$ be any voltage vector. We define $\mathbf{u}$ as a functional on at least some $\mathbf{i} \in \mathcal{I}$ through the pairing

$$\langle \mathbf{u}, \mathbf{i} \rangle = \sum u_j i_j \qquad (3.4)$$

whenever the summation converges. In the particular case where $\mathbf{u}$ is the vector $\mathbf{e} = (e_1, e_2, \dots)$ of all branch voltage sources, the summation will exist for all $\mathbf{i} \in \mathcal{I}$ because $\mathbf{e}$ has a finite support. Moreover, $\mathbf{e}$ is clearly a linear functional on $\mathcal{I}$.

Lemma 3.1-4: $\mathbf{e}$ *is a continuous linear functional on* $\mathcal{I}$.

Proof: Let $\mathbf{i}_k \to 0$ in $\mathcal{I}$ as $k \to \infty$. By Lemma 3.1-2, $e_j i_{kj} \to 0$ for every fixed j. Since $e_j \ne 0$ only for a finite number of j, $\langle \mathbf{e}, \mathbf{i}_k \rangle = \sum e_j i_{kj} \to 0$ as well. ♣

R will denote the operator that assigns to every branch-current vector $\mathbf{i} = (i_1, i_2, \dots)$ the vector $(r_1 i_1, r_2 i_2, \dots)$ consisting of the voltages across the branch resistances. We call R the *resistance operator*.

The principal theorem of this section is a consequence of Riesz's representation theorem, which can be stated as follows [94, page 393].

Lemma 3.1-5 (Riesz's representation theorem): *If f is a continuous linear functional on a Hilbert space* $\mathcal{X}$, *then there exists a unique* $y \in \mathcal{X}$ *such that* $f(x) = (x, y)$ *for every* $x \in \mathcal{X}$.

Here finally is Flanders' theorem, which launched the mathematical theory of infinite networks with arbitrary graphs.

Theorem 3.1-6: *Given a network $\mathcal{N}$ that satisfies Conditions 3.1-1, there exists a unique $\mathbf{i} \in \mathcal{K}$ such that*

$$\langle \mathbf{e} - R\mathbf{i}, \mathbf{s} \rangle = 0 \qquad (3.5)$$

for every $\mathbf{s} \in \mathcal{K}$. Moreover, $\mathbf{i}$ is the limit of a sequence in $\mathcal{K}^0$ and is in fact uniquely determined by (3.5) even when $\mathbf{s}$ is restricted to $\mathcal{K}^0$.

Proof: Since $\mathcal{K}$ is a subspace of $\mathcal{I}$ with the same inner product, it follows from Lemma 3.1-4 that $\mathbf{e}$ is a continuous linear functional on $\mathcal{K}$. By Lemma 3.1-5, since $\mathcal{K}$ is a Hilbert space by itself, there exists a unique $\mathbf{i} \in \mathcal{K}$ such that $\langle \mathbf{e}, \mathbf{s} \rangle = (\mathbf{s}, \mathbf{i})$ for all $\mathbf{s} \in \mathcal{K}$. But $(\mathbf{s}, \mathbf{i}) = \sum r_j i_j s_j = \langle R\mathbf{i}, \mathbf{s} \rangle$. Thus, we have (3.5). The second statement follows from the fact that $\mathcal{K}^0$ is dense in $\mathcal{K}$. ♣

Equation (3.5) is an extension to infinite networks of Tellegen's equation. We shall examine a number of its implications, such as the satisfaction of Kirchhoff's laws, in later sections after we generalize the theorem in several ways.

Flanders also showed that the regime dictated by Theorem 3.1-6 can be approximated by the regimes in a sequence of expanding finite subnetworks of $\mathbf{N}$. To be precise, let $\{J_k\}_{k=1}^{\infty}$ be a sequence of subsets of the index set J for the branches of $\mathbf{N}$ such that $J_k \subset J_{k+1}$ for every k and $J = \bigcup_{k=1}^{\infty} J_k$. Let $\mathbf{N}_k$ be the subnetwork induced by all the branches b_j with $j \in J_k$. In words of customary engineering phraseology, $\mathbf{N}_k$ is obtained by open-circuiting every branch b_j with $j \notin J_k$ and removing every node that is not incident to a branch b_j with $j \in J_k$. In this case, we shall say that $\mathbf{N}_k$ is an *opened subnetwork of* $\mathbf{N}$ and that $\{\mathbf{N}_k\}_{k=1}^{\infty}$ is an *expanding sequence* of opened subnetworks of $\mathbf{N}$. Note that, since $J = \bigcup_{k=1}^{\infty} J_k$, every branch of $\mathbf{N}$ is eventually in some $\mathbf{N}_k$. Finally, we call $\mathbf{N}_k$ *finite* if J_k is a finite set. It is a fact that Theorem 3.1-6 is true even when $\mathbf{N}$ is a finite network; every argument in its derivation continues to hold – often in a trivial way.

Lemma 3.1-7: *Let $\mathbf{N}$ and $\mathbf{i}$ be as asserted in Theorem 3.1-6. Choose any $\epsilon > 0$ and let $\mathbf{x} \in \mathcal{K}^0$ be such that $\|\mathbf{x} - \mathbf{i}\| \leq \epsilon$. Let $\mathbf{N}_k$ be any finite opened subnetwork which contains the support of both $\mathbf{x}$ and $\mathbf{e}$. Let $\mathbf{i}_k$ be the unique current vector dictated by Theorem 3.1-6 when applied to $\mathbf{N}_k$. Then, $\|\mathbf{i}_k - \mathbf{i}\| \leq 2\epsilon$.*

Proof: Let $\mathbf{s}$ be any member of $\mathcal{K}$ whose support is in $\mathbf{N}_k$. Then, by

(3.5), $(\mathbf{s}, \mathbf{i}_k) = \langle R\mathbf{i}_k, \mathbf{s} \rangle = \langle \mathbf{e}, \mathbf{s} \rangle = \langle R\mathbf{i}, \mathbf{s} \rangle = (\mathbf{s}, \mathbf{i})$. Hence, $|(\mathbf{s}, \mathbf{x} - \mathbf{i}_k)| = |(\mathbf{s}, \mathbf{x}) - (\mathbf{s}, \mathbf{i}_k)| = |(\mathbf{s}, \mathbf{x}) - (\mathbf{s}, \mathbf{i})| = |(\mathbf{s}, \mathbf{x} - \mathbf{i})| \leq ||\mathbf{s}|| \, ||\mathbf{x} - \mathbf{i}|| \leq ||\mathbf{s}||\epsilon$, according to the Schwarz inequality and the hypothesis. We may set $\mathbf{s} = \mathbf{x} - \mathbf{i}_k$ to get $||\mathbf{x} - \mathbf{i}_k||^2 \leq \epsilon ||\mathbf{x} - \mathbf{i}_k||$. Thus, $||\mathbf{x} - \mathbf{i}_k|| \leq \epsilon$. By the triangle inequality for norms, $||\mathbf{i}_k - \mathbf{i}|| \leq ||\mathbf{i}_k - \mathbf{x}|| + ||\mathbf{x} - \mathbf{i}|| \leq 2\epsilon$. ♣

Flanders' approximation theorem is the following.

Theorem 3.1-8: *Let $\{\mathbf{N}_k\}_{k=1}^{\infty}$ be an expanding sequence of finite opened subnetworks of $\mathbf{N}$ and assume that the support of $\mathbf{e}$ is contained in $\mathbf{N}_1$. Let $\mathbf{i}$ (or $\mathbf{i}_k$) be the unique vector dictated by Theorem 3.1-6 when applied to $\mathbf{N}$ (respectively, to $\mathbf{N}_k$). Then, $\mathbf{i}_k \to \mathbf{i}$ in $\mathcal{I}$, that is, $||\mathbf{i}_k - \mathbf{i}|| \to 0$ as $k \to \infty$.*

Note: By Lemma 3.1-2, the conclusion implies that $i_{kj} \to i_j$ in every branch b_j.

Proof: Choose any $\epsilon > 0$. According to Theorem 3.1-6, $\mathbf{i} \in \mathcal{K}$. Hence, there is an $\mathbf{x} \in \mathcal{K}_0$ such that $||\mathbf{i} - \mathbf{x}|| \leq \epsilon$. Since $\{\mathbf{N}_k\}$ is an expanding sequence and $\mathbf{x}$ has a finite support, there is an integer k_0 such that $\mathbf{N}_{k_0}$ contains the support of $\mathbf{x}$. So, for every $k \geq k_0$, $||\mathbf{i}_k - \mathbf{i}|| \leq 2\epsilon$ according to Lemma 3.1-7. ♣

3.2 Connections at Infinity: 1-Graphs

As was indicated in Example 1.6-4, the idea, "connections to a network at its infinite extremities," can have a meaning. Flanders' theory does not encompass this possibility, but it is a fruitful avenue of investigation. It can explain why a network may have a variety of different finite-power regimes – depending on what the connections at infinity happen to be.

In this section we present a precise definition of the "infinite extremities" of a countably infinite graph. This idea was first promulgated in [163] and developed further in [177]. However, the terminology we now employ will not conform with that used in those two works because of still another generalization we will make in Chapter 5, namely, our theory of transfinite networks. The present terminology is chosen to facilitate that generalization. Since there are so many new definitions, we will display the principal ones in separate paragraphs to make them easier to find.

Our definitions will build on the concept of a 0-graph, which was discussed in Section 1.3. Let $\mathcal{G}^0$ be a countably infinite 0-graph. "Shorts at infinity" will be called "1-nodes" to distinguish them from the 0-nodes defined in Section 1.3. Thus, a *node* may now be either a 0-node or a

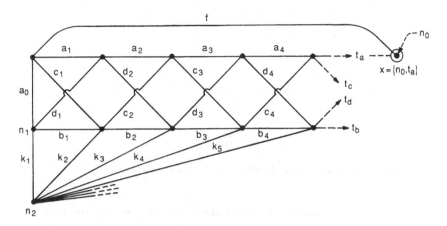

Figure 3.1. A countably infinite 1-graph. t_a denotes the 0-tip having as a representative the one-ended path consisting of the branches a_m. The branch f is incident to the node n_0, which is embraced by the 1-node $x = \{n_0, t_a\}$.

1-node. To define 1-nodes, we first need to make precise the idea of an "infinite extremity," which we shall henceforth refer to as a "0-tip."

Two one-ended paths in $\mathcal{G}^0$ are said to be *equivalent* if they differ only for a finite number of 0-nodes and branches. For example, in Figure 3.1 the two one-ended paths, specified by the branch sequences $\{a_1, a_2, a_3, a_4, a_5, \ldots\}$ and $\{c_2, a_3, a_4, a_5, \ldots\}$, are equivalent. This defines, in fact, an equivalence relationship, and it partitions the set of all one-ended paths in $\mathcal{G}^0$ into equivalence classes.

Definition of a 0-tip: Each such equivalence class will be called a *0-tip*.

(0-tips were called "pathlike extremities" in [177].) Each one-ended path is called a *representative* of the 0-tip in which it appears. Any subgraph that contains that representative is said to *have* that 0-tip.

Example 3.2-1. Although $\mathcal{G}^0$ is countable, it may have an uncountable set of 0-tips. This is so, for example, for the infinite binary tree of Figure 3.2. Although each 0-tip therein has an infinity of representatives, it can be identified with its unique representative that starts at the 0-node n_0. The cardinality of the set of all one-ended 0-paths starting at n_0 is **c**, the cardinality of the continuum, and hence there is a continuum of 0-tips. ♣

Example 3.2-2. As another example, consider the infinite 0-graph of Figure 3.3. Here too we can restrict our attention to those representatives of 0-tips that start at the 0-node n_0, but now each 0-tip has an

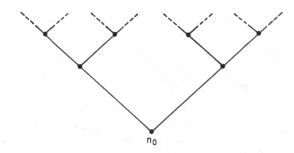

Figure 3.2. The infinite binary tree. It has a continuum of 0-tips.

infinity of such representatives. Every one of those representatives can be designated by a sequence of branches of the form $\{z_1, z_2, z_3, \ldots\}$, where each z_i can be chosen to be either a_i or b_i. There is a continuum of such representatives. Moreover, each 0-tip can be identified as an equivalence class of such sequences, where two sequences are taken to be equivalent if they differ by no more than a finite number of the z_i. Given any such sequence S we can count all the others in its equivalence class; indeed, count the one that differs from S only in z_1, then count the two additional ones that differ from S at most only in z_1 and z_2, then count the four additional ones that differ from S at most only in z_1, z_2, and z_3, and so forth. Thus, the cardinality of the set of representatives for any 0-tip is $\aleph_0$, the cardinal number of a denumerably infinite set. Consequently, the cardinality of the set of 0-tips is $\mathbf{c} \div \aleph_0 = \mathbf{c}$ [118, page 299]. ♣

We turn now to the construction of a 1-node. Note that the 0-graph $\mathcal{G}^0$ may not possess any 0-tips because it may not contain any one-ended paths. If, however, $\mathcal{G}^0$ does possess 0-tips, partition the set T^0 of all 0-tips into subsets T_τ^0; thus, $T^0 = \cup T_\tau^0$, where τ denotes the indices of the partition. Each T_τ^0 may be either finite, denumerable, or uncountably infinite, but it is not void. Furthermore, for each τ let $\mathcal{N}_\tau^0$ be either the void set or a singleton whose element is a 0-node; we require that $\mathcal{N}_{\tau_1}^0 \cap \mathcal{N}_{\tau_2}^0 = \emptyset$ if $\tau_1 \neq \tau_2$.

Definition of a 1-node: For each τ the set $T_\tau^0 \cup \mathcal{N}_\tau^0$ is called a *1-node.*

In the event that T_τ^0 is a countable set, the corresponding 1-node x^1 may be written out as

$$x^1 = \{x_0^0, t_1^0, t_2^0, t_3^0, \ldots\}$$

where the t_m^0 are 0-tips and x_0^0 is a 0-node, which may not be present. Our definition ensures that every 1-node contains at least one 0-tip and every

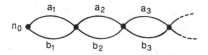

Figure 3.3. A 1-times chainlike infinite graph having a continuum of 0-tips.

0-tip appears in one and only one 1-node. Also, every 1-node contains at most one 0-node and perhaps none at all, and no 0-node appears in more than one 1-node; in fact, a particular 0-node may not be a member of any 1-node.

A physical interpretation of a 1-node is that of a short circuit connected to its elements. This may allow the flow of current along paths out to infinity (i.e., along representatives of a 0-tip), through a short circuit at infinity, and then along other infinite paths. Alternatively, a current may jump along a short circuit from a 0-node out to infinity and then continue along infinite paths.

Definition of an exceptional element: If $\mathcal{N}_\tau^0$ is not void, the 0-node in $\mathcal{N}_\tau^0$ is called the *exceptional element* of the 1-node $\mathcal{N}_\tau^0 \cup \mathcal{T}_\tau^0$.

Definition of "to embrace" for 0-nodes and 1-nodes: We say that a 0-node *embraces* itself and also all its elementary tips, but nothing else. Similarly, a 1-node x^1 is said to *embrace* itself, all its elements, and all the elementary tips in its exceptional element if it has one. We take it that x^1 does *not* embrace the representatives of its 0-tips, nor any other entity.

Definition of an ordinary node: An *ordinary* node is a 0-node (finite or infinite) that is not embraced by any 1-node; otherwise, it is called an *embraced* 0-node.

We will use some natural vocabulary in discussing these concepts; in order to be precise, let us explain that vocabulary. Assume that x_0^0 is the exceptional element of a 1-node x^1. A branch that is incident to x_0^0 is said to *meet* x^1. A 0-path P^0 that contains x_0^0 is said to *meet* x^1 at x_0^0, and, if x_0^0 is a terminal node of P^0, P^0 is said to *terminate at* x^1 *with* x_0^0. Similarly, a one-ended or endless 0-path that contains a representative of a 0-tip t^0 in x^1 is said to *meet* x^1 *with* t^0.

Example 3.2-3. An example of a 1-node is indicated in Figure 3.1. Let t_a be the 0-tip having the one-ended path P_a, specified by the branch sequence $\{a_1, a_2, a_3, \ldots\}$, as one of its representatives. Also, f denotes the branch with n_0 as one of its 0-nodes. We can stipulate an electrical short between n_0 and the "infinite extremity" of P_a by specifying $x = \{n_0, t_a\}$ as a 1-node. Moreover, we say that P_a meets x with t_a. ♣

Definition of a 1-graph: A *1-graph* is a triplet $\mathcal{G}^1 = (\mathcal{B}, \mathcal{N}^0, \mathcal{N}^1)$, where $\mathcal{B}$ and $\mathcal{N}^0$ denote the branch set and 0-node set of a 0-graph and $\mathcal{N}^1$ is a specified set of 1-nodes constructed from the 0-tips and 0-nodes of that 0-graph. It is required that $\mathcal{N}^1$ be nonvoid; otherwise, the 1-graph is taken to be nonexistent.

Note again that the possibility of a void $\mathcal{N}^1$ exists because there may not be any one-ended paths. Also, to identify $\mathcal{N}^1$, we need merely specify the 1-nodes that are not singletons. As with 0-graphs, we can uniquely associate an ordinary graph as defined in Section 1.2 with a 1-graph by invoking the incidences between branches and their 0-nodes, but now the shorts at infinity are lost in this transition from a 1-graph to its ordinary graph.

Example 3.2-4.   With regard to Figure 3.1 again, if we also stipulate that all 1-nodes other than x are singletons, we will have completely specified that diagram as a 1-graph. Alternatively, we could specify another 1-graph by taking $\{n_0, t_a\}$ and $\{t_b, t_c, t_d\}$ as the only nonsingleton 1-nodes, where t_b, t_c, and t_d are the 0-tips with the representative paths $\{b_1, b_2, \ldots\}$, $\{c_1, c_2, \ldots\}$, and $\{d_1, d_2, \ldots\}$ respectively. Still another 1-graph is obtained when we declare that all the 0-tips and n_0 as well are embraced by a single 1-node; in this case, the network is "shorted everywhere at infinity," and that short also includes n_0. ♣

Let $\mathcal{G}^1 = (\mathcal{B}, \mathcal{N}^0, \mathcal{N}^1)$ be a 1-graph. Given a subset $\mathcal{B}_*$ of $\mathcal{B}$, we wish to define the "reduction of $\mathcal{G}^1$ induced by $\mathcal{B}_*$". Let $\mathcal{T}_*$ be the union of the branches in $\mathcal{B}_*$, as before. Starting with only the elementary tips in $\mathcal{T}_*$, we retrace the construction of $\mathcal{G}^1$ and strike out all void sets as they arise in the process. More specifically, let $(\mathcal{B}_*, \mathcal{N}_*^0)$ be the reduced 0-graph obtained from $(\mathcal{B}, \mathcal{N}^0)$ as stated in Section 1.3. Corresponding to every 0-tip of $(\mathcal{B}_*, \mathcal{N}_*^0)$ there is a unique 0-tip of $(\mathcal{B}, \mathcal{N}^0)$ containing the representatives of the first 0-tip. However, there may be 0-tips in $(\mathcal{B}, \mathcal{N}^0)$ which do not exist as 0-tips in $(\mathcal{B}_*, \mathcal{N}_*^0)$ because $(\mathcal{B}_*, \mathcal{N}_*^0)$ does not contain any representatives of those 0-tips. Now, let x^1 be any 1-node for the 1-graph $\mathcal{G}^1 = (\mathcal{B}, \mathcal{N}^0, \mathcal{N}^1)$. Remove every 0-tip in x^1 that does not exist as a 0-tip in $(\mathcal{B}_*, \mathcal{N}_*^0)$. The resulting set x_*^1 is called a *reduced 1-node* (*induced by* $\mathcal{B}_*$) if it possesses at least one 0-tip – and even if no 0-tips were removed. Let $\mathcal{N}_*^1$ be the set of all reduced 1-nodes. If $\mathcal{N}_*^1$ is not void, let $\mathcal{G}_*$ be the 1-graph $(\mathcal{B}_*, \mathcal{N}_*^0, \mathcal{N}_*^1)$; if $\mathcal{N}_*^1$ is void, let $\mathcal{G}_*$ be the 0-graph $(\mathcal{B}_*, \mathcal{N}_*^0)$.

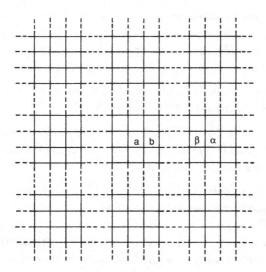

Figure 3.4. A 1-graph consisting of an infinity of infinite square-grid 0-graphs connected at their adjacent 0-tips by 1-nodes. For example, let t_1 and t_2 be the 0-tips with representatives specified by the branch sequences $\{a, b, \ldots\}$ and $\{\alpha, \beta, \ldots\}$ respectively; $\{t_1, t_2\}$ is one of the 1-nodes.

Definition of a reduced 1-graph: $\mathcal{G}_*$ is called the *reduction of* $\mathcal{G}^1$ *with respect to* $\mathcal{B}_*$ or the *reduced graph induced by* $\mathcal{B}_*$.

With 1-nodes in hand we have implicitly generalized the idea of connectedness. Recall the definition of a path given in Section 1.2. Such a path will be called a 0-*path* to distinguish it from the idea of a "1-path," which will be introduced shortly. Similarly, a loop as defined in Section 1.2 will also be called a 0-*loop*. We shall now say that a 0-graph or a 1-graph $\mathcal{G}$ is 0-*connected* or *finitely connected* if for every pair of 0-nodes n_a, n_b there exists in $\mathcal{G}$ a finite 0-path terminating at n_a and n_b. Thus, the 1-graph of Figure 3.1 is truly 0-connected.

Example 3.2-5. On the other hand, consider the 1-graph of Figure 3.4. Each section of solid lines represents an infinite two-dimensional square grid. Two adjacent square grids are taken to be connected at their adjacent 0-tips by 1-nodes. More specifically, the central square grid has a 0-tip t_1 with the horizontal representative $\{a, b, \ldots\}$ and the square grid directly to its right has a 0-tip t_2 with the horizontal representative $\{\alpha, \beta, \ldots\}$, where a, b, α, and β denote branches. We take $\{t_1, t_2\}$ to be a 1-node. In the same way every pair of adjacent 0-tips arising from horizontal or vertical representatives are shorted together by 1-nodes.

Although there are 1-node connections between all adjacent square grids, this 1-graph is not 0-connected, for there is no finite 0-path between nodes that lie in different square grids. However, by generalizing the idea of a path, we can say that $\mathcal{G}$ is connected in a wider sense. ♣

We need some more definitions. Let x be either a 0-node or a 1-node, and similarly for y. Also, let P^0 be a 0-path.

Definition of "totally disjoint" for 0-nodes and 1-nodes: x and y are said to be *totally disjoint* if they do not embrace a common element. Also, x and P^0 are called *totally disjoint* if P^0 does not meet x.

Definition of "terminally incident": x and P^0 are said to be *terminally incident* if P^0 meets x either with a terminal node or with a 0-tip. In this case, x and P^0 are said to be *terminally incident but otherwise totally disjoint* if they do not meet at any other node or with another 0-tip of P^0; thus, P^0 meets x with only one (not both) of its ends or tips.

Now consider the alternating sequence

$$\{\ldots, x_m^1, P_m^0, x_{m+1}^1, P_{m+1}^0, \ldots\} \tag{3.6}$$

where each P_m^0 is a nontrivial 0-path and each x_m^1 is a 1-node except possibly when (3.6) terminates on the left and/or on the right. In the latter case, each terminal element is either a 0-node or a 1-node.

Definition of a 1-path: Such a sequence is called a 1-*path* if in addition it satisfies

Conditions 3.2-6:

(i)   *The indices m are restricted to the integers, and they number the elements of (3.6) consecutively as indicated.*

(ii) *Each P_m^0 is terminally incident to the two nodes immediately preceding and succeeding it in (3.6) but is otherwise totally disjoint from those nodes.*

(iii) *Every two elements in (3.6) that are not adjacent therein are totally disjoint.*

Condition (i) may appear to be superfluous since a sequence can, by definition, be numbered as stated. However, later on in Section 5.4, we will discuss other kinds of paths whose elements cannot be so numbered – as, for example, when the indices m extend into the transfinite ordinals; because of this, we explicitly state (i) to emphasize the distinction.

A 1-path is called *nontrivial* if it has at least three elements. A 1-path is called either *finite*, or *one-ended*, or *endless* if its sequence (3.6) is respectively finite, or one-way infinite, or two-way infinite.

Definition of a 1-loop: A *1-loop* is a finite 1-path except for the following requirement: One of the two terminal elements embraces the other.

Let P_1^1 and P_2^1 each denote either a 1-path or a 1-loop.

Definition of "totally disjoint" for 1-paths and 1-loops: P_1^1 and P_2^1 are said to be *totally disjoint* if every node or 0-path in P_1^1 is totally disjoint from every node and every 0-path in P_2^1.

Example 3.2-7. For an example, refer to Figure 3.1 once again. With $x = \{n_0, t_a\}$ as indicated, there is a 1-loop $\{x, P^0, x\}$ having the branches $f, a_1, a_2, a_3, \ldots$ in P^0; there are also many other 1-loops meeting x with n_0 and t_a. On the other hand, if $y = \{t_b, t_c\}$ is declared to be another 1-node, no new 1-loop is generated because no representative of t_b is totally disjoint from any representative of t_c. ♣

Definition of "to embrace" for 0-paths and 1-paths: A 0-path is said to *embrace* itself, its elements, and also the elementary tips in its 0-nodes and branches, but not any other entity. Similarly, a 1-path is said to *embrace* itself, its elements, and all the elements embraced by its elements, but not any other entity.

Definition of "1-connected": Let x_a be a 0-node or a 1-node, and similarly for x_b. x_a and x_b are said to be *1-connected* if there exists a finite 1-path with x_a and x_b as its terminal elements. (This meaning for "1-connected" is different from the customary one in conventional graph theory [147, page 131].) Two branches are said to be *1-connected* if their 0-nodes are 1-connected. A 1-graph is said to be *1-connected* if every two nodes of ranks 0 or 1 are 1-connected.

It follows directly from our definitions that, if two nodes x_a and x_b are 0-connected, then they are also 1-connected. Indeed, let P^0 be a finite 0-path with x_a and x_b as its terminal nodes. Then, $\{x_a, P^0, x_b\}$ is a finite 1-path.

Definition of a 1-section: A *1-section* of a 1-graph $\mathcal{G}$ is a reduction of $\mathcal{G}$ induced by a maximal set of branches that are pairwise 1-connected.

For instance, the graph $\mathcal{G}^1$ of Figure 3.4 is 1-connected but not 0-connected and is a 1-section in itself. On the other hand, each square-grid section is a 0-section of $\mathcal{G}^1$. A 0-path cannot pass from one 0-section to another 0-section; a 1-path can do so – but only if it passes through a 1-node.

3.3 Existence and Uniqueness

We can now extend Flanders' theory by allowing various shorts at infinity. In Chapter 5 we will present a further generalization wherein infinite collections of infinite networks are connected at their extremities in a hierarchical fashion to allow nodes of still higher ranks. For the present, we will only allow 0-nodes and 1-nodes. Moreover, we shall allow an infinity of sources so long as the total power available from those sources is finite. It will now be shown that, when these shorts and sources are appropriately specified, the network will have one and only one finite-power regime whose current vector lies in a Hilbert space constructed out of 0-loop and 1-loop currents.

By assigning electrical parameters to the branches of a 0-graph or a 1-graph, as stated in Section 1.4, we obtain a 0-*network* or a 1-*network* respectively.

Conditions 3.3-1: *Let* $\mathbf{N}$ *be a 0-network or a 1-network. Assume that every branch* b_j *of* $\mathbf{N}$ *is in the Thevenin form with a positive resistance* r_j *and a voltage source* e_j. *Assume, moreover, that* $\sum e_j^2 g_j < \infty$, *where* $g_j = 1/r_j$.

The quantity $\sum e_j^2 g_j$ will be called the *total isolated power* of the network. It is the power delivered by all the sources when a short is placed across every branch having a source. As will be shown below, the total power absorbed by all the resistors will be equal to the total power delivered by all the sources and will be no larger than $\sum e_j^2 g_j$. Thus, the assumption that $\sum e_j^2 g_j < \infty$ ensures that the voltage-current regime of the network will be of finite power.

As in Section 3.1, $\mathcal{I}$ denotes the Hilbert space of all branch-current vectors $\mathbf{i}$ of finite power dissipation and is equipped with the inner product (3.1). Thus, for each $\mathbf{i} \in \mathcal{I}$, $\|\mathbf{i}\|^2 = \sum i_j^2 r_j < \infty$.

Let L be a 0-loop or a 1-loop and assign to it an orientation. Then, a *loop current of value* $i \in R^1$ is a branch-current vector $\mathbf{i}$ for which $i_j = 0$ if branch $b_j \notin L$ and $i_j = \pm i$ if $b_j \in L$, where the plus (minus) sign is used if the orientations of b_j and L agree (disagree). In this case, we say that i *flows around* L. Recall that any 0-loop is automatically a 1-loop. We shall say that a 1-loop is *proper* if it is not a 0-loop. This is equivalent to saying that the proper 1-loop embraces at least one 0-path having a 0-tip. We also say that a 1-loop current is *proper* if its 1-loop is proper and that it *meets* a node if its 1-loop does so. If L is a 0-loop, then $\mathbf{i} \in \mathcal{I}$; however, when L is a proper 1-loop, $\mathbf{i} \in \mathcal{I}$ if and only if $\sum_{(L)} r_j < \infty$, where $\sum_{(L)}$ is a summation over the branch indices

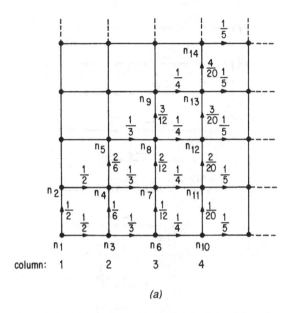

(a)

Figure 3.5(a). A 1-basic current described in Example 3.3-2.

for L. When $\sum_{(L)} r_j < \infty$, L will be called *perceptible*. A loop current will satisfy Kirchhoff's current law at every ordinary 0-node, but not necessarily at an embraced 0-node because a tracing of the loop may jump from that 0-node to a 0-tip.

We now define a *basic current* as a certain kind of branch-current vector having either rank 0 or rank 1. A *0-basic current* is simply a 0-loop current. A *1-basic current* is a branch-current vector $\mathbf{i}$ of the form $\mathbf{i} = \sum \mathbf{i}_m$, where

   (i)   the set of summands $\mathbf{i}_m$ is countable,

   (ii)   each $\mathbf{i}_m$ is a proper 1-loop current,

   (iii) the support of $\mathbf{i}$ is contained in finitely many 1-sections, and

   (iv) only finitely many of the $\mathbf{i}_m$ meet any ordinary 0-node.

Actually, with appropriate choices of the $\mathbf{i}_m$, a 1-basic current may become one or many (even infinitely many) 0-loop currents. Also, Condition (iv) implies that a 1-basic current satisfies Kirchhoff's current law at every ordinary 0-node. Furthermore, it is possible for a 1-basic current to be a member of $\mathcal{I}$ without any of its summands $\mathbf{i}_m$ being in $\mathcal{I}$; to see this, consider

Example 3.3-2. A 1-basic current $\mathbf{i}$ in a quarter-plane square grid is illustrated in Figure 3.5(a). The vertical columns (i.e., paths) of that

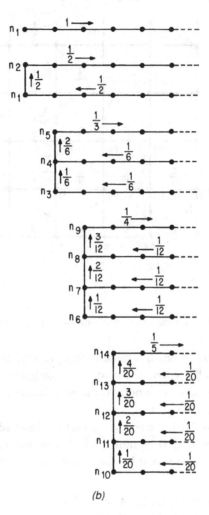

(b)

Figure 3.5(b). Its decomposition into finite sums of 1-loop currents.

grid are numbered by the indices $\nu = 1, 2, 3, \ldots$ as shown. Assume that the resistance of every vertical branch in the νth column and of every horizontal branch between columns ν and $\nu+1$ is $1/\nu$ Ω. Assume that the 0-node n_1 and all the 0-tips having horizontal paths as representatives comprise a 1-node and that all other 1-nodes are singletons. The fractions in Figure 3.5(a) represent branch currents; an unlabeled branch carries no current.

To see that this current regime $\mathbf{i}$ is a 1-basic current, decompose it into the infinity of current regimes suggested by Figure 3.5(b). Each of the latter is a finite superposition of 1-loop currents. Clearly, the infinite superposition of all of these 1-loop currents satisfies the requirements for a 1-basic current. By summing the powers dissipated in all the branches by $\mathbf{i}$, we see that $\mathbf{i} \in \mathcal{I}$. On the other hand, none of the 1-loop currents comprising $\mathbf{i}$ is in $\mathcal{I}$ because no corresponding 1-loop is perceptible. ♣

For a given 0-network, $\mathcal{K}^0$ will denote the span of all 0-loop currents. For a given 1-network, $\mathcal{K}^0$ will denote the span of all 0-basic currents and all 1-basic currents that are members of $\mathcal{I}$. In both cases, $\mathcal{K}$ will denote the closure of $\mathcal{K}^0$ in $\mathcal{I}$. Thus, $\mathcal{K} \subset \mathcal{I}$. In fact, $\mathcal{K}$ is a Hilbert space by itself when it is equipped with the inner product of $\mathcal{I}$. By Lemma 3.1-2, convergence in $\mathcal{K}$ implies componentwise convergence. Actually, enlarged 1-nodes may effectively be introduced when taking the closure of $\mathcal{K}^0$.

Example 3.3-3. Refer to the 1-graph of Figure 3.6. Assume that the branch resistance values decay so rapidly as one proceeds to the right that $\sum r_j < \infty$. Figure 3.6(a) shows a 1-loop current consisting of a flow along the branches a_m, through a 1-node $c = \{n, t_a\}$, and back through a single return branch f; n is an embraced 0-node, to which the return branch is incident, and t_a is the 0-tip having as a representative the 0-path of the a_m branches. Since this 1-loop is perceptible, its loop current is a member of $\mathcal{K}^0$. Also, all 0-loop currents – and in particular those indicated in Figure 3.6(b) – are members of $\mathcal{K}^0$ too. Furthermore, we assume that the 0-tip t_b corresponding to the b_m branches comprises a singleton 1-node. However, if all the loop currents suggested in parts (a) and (b) of Figure 3.6 have 1-ampere values, their superposition will be the 1-loop current indicated in Figure 3.6(c) and will be a member of $\mathcal{K}$ too because it is the limit in $\mathcal{I}$ of a sequence in $\mathcal{K}^0$. Thus, an enlarged 1-node $d = \{n, t_a, t_b\}$ has effectively been introduced by taking the closure of $\mathcal{K}^0$, even though d was not declared to be a 1-node for this 1-network. In fact, by appropriately altering the directions of the loop currents of Figure 3.6(b), we see that every 0-tip is in effect in an enlarged 1-node. ♣

Consider now the voltage sources. Any branch voltage-source vector $\mathbf{e}$ defines a functional on $\mathcal{K}$ according to the pairing

$$\langle \mathbf{e}, \mathbf{i} \rangle = \sum e_j i_j, \quad \mathbf{i} \in \mathcal{K} \tag{3.7}$$

so long as $\sum e_j i_j$ converges.

Lemma 3.3-4: *Under Conditions 3.3-1, $\mathbf{e}$ is a continuous linear functional on $\mathcal{K}$ according to (3.7).*

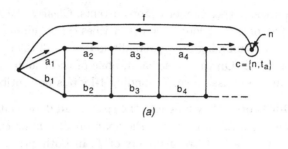

(a)

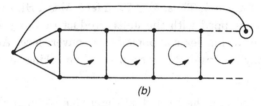

(b)

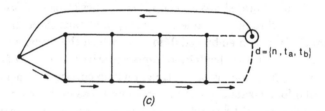

(c)

Figure 3.6. The effective introduction of an expanded 1-node when taking the closure of $\mathcal{K}^0$.

Proof: We first show that $\sum e_j i_j$ converges absolutely. By Schwarz's inequality,

$$\sum |e_j i_j| = \sum |g_j^{1/2} e_j r_j^{1/2} i_j| \leq \left[\sum g_j e_j^2 \sum r_j i_j^2\right]^{1/2}. \tag{3.8}$$

The right-hand side is finite by virtue of Conditions 3.3-1 and the fact that $\mathbf{i} \in \mathcal{K}$.

Since absolutely convergent series can be rearranged, the functional defined by (3.7) is linear. Moreover, it is continuous because, according to (3.8) and the norm of $\mathcal{K}$,

$$|\langle \mathbf{e}, \mathbf{i} \rangle| \leq \sum |e_j i_j| \leq \left[\sum g_j e_j^2\right]^{1/2} \|\mathbf{i}\|. \quad \clubsuit$$

Here is a fundamental existence and uniqueness theorem for the voltage-current regime of a 1-network. Its proof is virtually the same as that of Theorem 3.1-6; just replace Lemma 3.1-4 by Lemma 3.3-4.

Theorem 3.3-5: *Under Conditions 3.3-1, there exists a unique* $i \in \mathcal{K}$ *such that*

$$\langle e - Ri, s \rangle = 0 \tag{3.9}$$

for every $s \in \mathcal{K}$. *This equation implies the uniqueness of* i *in* $\mathcal{K}$ *even when* s *is restricted to* $\mathcal{K}^0$.

Equation (3.9) is known in the electrical engineering literature as Tellegen's equation. It, rather than Kirchhoff's laws, is the governing equation that determines the voltage-current regime for our 1-network **N**. Actually, the uniqueness of that regime arises from the conjunction of the finite-total-isolated-power condition ($\sum e_j^2 g_j < \infty$), the restriction of the allowable branch-current vectors to $\mathcal{K}$, and Tellegen's equation (3.9). Nonetheless, as we shall see in the next section, Kirchhoff's laws do hold in certain circumstances, even though they have been relegated to a secondary role in this theory. Also, Ohm's law has been imposed by virtue of the term Ri in (3.9).

Corollary 3.3-6: *Under Conditions 3.3-1, the total power* $\sum i_j^2 r_j$ *dissipated in all the resistors equals the total power* $\sum e_j i_j$ *supplied by all the voltage sources and is no larger than the finite total isolated power* $\sum e_j^2 g_j$ *available from all the voltage sources.*

Proof: Set $s = i$ in (3.9) to get

$$\sum i_j^2 r_j = \langle Ri, i \rangle = \langle e, i \rangle = \sum e_j i_j .$$

Upon combining this with (3.8), we obtain $\sum i_j^2 r_j \leq \sum e_j^2 g_j$. ♣

A different interpretation of Corollary 3.3-6 is worth mentioning. In a finite-dimensional space, the linearity of an operator implies its continuity, but this is not in general true in an infinite-dimensional space. However, continuity does hold under Theorem 3.3-5 in the sense that a small change in the voltage-source vector **e** produces no more than a small change in the current vector **i**, or for that matter in the voltage vector $v = e - Ri$. In other words the network is "stable." To explicate what we mean by small, we may use the norm of $\mathcal{I}$ as a measure of size for **i** and may use $(\sum e_j^2 g_j)^{1/2}$ as a measure of size for **e**. Then the power inequality of Corollary 3.3-6 may be rewritten to assert that a small change in **e** implies a small change in **i** as follows.

$$\|i_1 - i_2\| \leq \left[\sum (e_{1j} - e_{2j})^2 g_j \right]^{1/2} .$$

Here, i_1 and i_2 are the current vectors produced by e_1 and e_2 in accordance with Theorem 3.3-5.

Another ramification of Theorem 3.3-5 is the *reciprocity principle*, which for finite networks asserts that, if a voltage source e in branch b_j produces a current i in branch b_k of an otherwise sourceless network, then e in b_k produces i in b_j. This principle continues to hold for 1-networks even when b_j and b_k are infinitely distant. Indeed, a more general result that encompasses the reciprocity principle as a special case, is the following.

Corollary 3.3-7: *Let* $\mathbf{e} = (e_1, e_2, \dots)$ *and* $\mathbf{f} = (f_1, f_2, \dots)$ *be two voltage source vectors for a network* $\mathbf{N}$ *satisfying Conditions 3.3-1. In accordance with Theorem 3.3-5, let* $\mathbf{i} = (i_1, i_2, \dots)$ *be the current vector produced by* $\mathbf{e}$ *and let* $\mathbf{x} = (x_1, x_2, \dots)$ *be the current vector produced by* $\mathbf{f}$. *Then,* $\sum e_j x_j = \sum f_j i_j$.

Proof: Invoke (3.9) with $\mathbf{s} = \mathbf{x}$ when the voltage-source vector is $\mathbf{e}$ and with $\mathbf{s} = \mathbf{i}$ when the voltage-source vector is $\mathbf{f}$. This yields

$$\langle \mathbf{e}, \mathbf{x} \rangle = \langle R\mathbf{i}, \mathbf{x} \rangle = \sum r_j i_j x_j = \langle R\mathbf{x}, \mathbf{i} \rangle = \langle \mathbf{f}, \mathbf{i} \rangle . \quad \clubsuit$$

Example 3.3-8. Let us show with a simple network that the current regime truly does depend in general on what the shorts at infinity are. Consider the infinite series connection of Figure 1.5 but make the following changes: The source branch is in its Thevenin form of a 1-V voltage source in series with a 1 Ω resistor; the resistance values to the right of the source branch are consecutively 2^{-k} Ω, $k = 1, 2, 3, \dots$, and similarly for the resistance values to the left. Thus, the sum of all the resistances is 3 Ω. Consider two cases: In the first one, there is an open circuit at infinity, that is, the two 0-tips are declared to be in separate singleton 1-nodes. Therefore, $\mathcal{K}^0 = \mathcal{K} = \{0\}$. It follows that the unique current regime dictated by Theorem 3.3-5 has 0 A in every branch. In the second case, let there be a short circuit at infinity, that is, the two 0-tips are members of the same 1-node. Now, $\mathcal{K}^0 = \mathcal{K}$ is the one-dimensional space consisting of current flows around the one and only 1-loop. In this case, Theorem 3.3-5 dictates that every branch current be 1/3 A. $\clubsuit$

3.4 The Validity of Kirchhoff's Laws

Under the regime dictated by Theorem 3.3-5, Kirchhoff's current law will hold at certain nodes and Kirchhoff's voltage law will hold around certain loops. Consider first Kirchhoff's current law, which (in accordance with Section 1.5) states that

$$\sum_{(n)} \pm i_j = 0 \qquad (3.10)$$

where n is a 0-node, $\sum_{(n)}$ is a summation over the indices j for the branches b_j incident to n, i_j is the corresponding branch current, and the plus (minus) sign is used if b_j is incident toward (away from) n. The index of an incident self-loop appears twice and its current term appears once with a plus sign and once with a minus sign. The 0-node n is called *restraining* if the sum of the conductances of all the branches incident to n is finite, that is, if $\sum_{(n)} g_j < \infty$; a self-loop contributes twice its conductance to the summation. A finite node is restraining, but an infinite node may or may not be restraining. Now, recall that an *ordinary node* is a 0-node that is not embraced by a 1-node.

Theorem 3.4-1: *If n is an ordinary restraining 0-node, then, under the voltage-current regime dictated by Theorem 3.3-5, Kirchhoff's current law (3.10) is satisfied at n; moreover, the series on the left-hand side of (3.10) converges absolutely when n is an infinite node.*

Proof: Let $\mathbf{i} \in \mathcal{K}$. Then,

$$\sum_{(n)} |i_j| = \sum_{(n)} r_j^{1/2} |i_j| g_j^{1/2} \leq \left(\sum_{(n)} r_j i_j^2 \sum_{(n)} g_j \right)^{1/2} \leq \|\mathbf{i}\| \left(\sum_{(n)} g_j \right)^{1/2} .$$

Since n is restraining, the right-hand side is finite, which establishes the asserted absolute convergence.

Next, as was noted in the preceding section, every basic current satisfies (3.10) at n. Consequently, so too does every member of $\mathcal{K}^0$ since each such member is a (finite) linear combination of basic currents. Since $\mathcal{K}^0$ is dense in $\mathcal{K}$, we can choose a sequence $\{\mathbf{i}_k\}_{k=0}^{\infty}$ in $\mathcal{K}^0$ which converges in $\mathcal{K}$ to the unique $\mathbf{i} \in \mathcal{K}$ specified in Theorem 3.3-5. Thus, with i_{kj} (or i_j) denoting the jth component of $\mathbf{i}_k$ (or $\mathbf{i}$), we may write $\sum_{(n)} \pm i_{kj} = 0$ and

$$\left| \sum_{(n)} \pm i_j \right| = \left| \sum_{(n)} \pm i_j - \sum_{(n)} \pm i_{kj} \right| \leq \sum_{(n)} |i_j - i_{kj}| = \sum_{(n)} r^{1/2} |i_j - i_{kj}| g_j^{1/2}$$

$$\leq \left(\sum_{(n)} r_j (i_j - i_{kj})^2 \sum_{(n)} g_j \right)^{1/2} \leq \|\mathbf{i} - \mathbf{i}_k\| \left(\sum_{(n)} g_j \right)^{1/2} \to 0$$

as $k \to \infty$. Thus, $\mathbf{i}$ satisfies Kirchhoff's current law at n. ♣

It is worth mentioning that, even if the 0-node n is not restraining, Kirchhoff's current law may still be satisfied at n; indeed, other branches not incident to n may be so configured that (3.10) is forced into fulfillment at n.

Example 3.4-2.   An example of a voltage-current regime dictated by Theorem 3.3-5 that does not satisfy Kirchhoff's current law at an infinite node is provided by the network discussed in Example 1.6-1. See Figure 1.4(a); number its branches by $j = 0, 1, 2, \ldots$, proceeding from left to right, and orient the source branch upward and all other branches downward. The nodes n_1 and n_2 are ordinary but not restraining. This simple network is a 0-network because there are no 0-tips. Therefore $\mathcal{K}^0$ is the span of 0-loop currents only. Thus, according to the second conclusion of Theorem 3.3-5, the current vector is that unique member of $\mathcal{K}$ for which Kirchhoff's voltage law is satisfied around every loop. Now, the current vector $\mathbf{i} = (1, 0, 0, 0, \ldots)$ is the limit in $\mathcal{I}$ of the current vectors

$$\mathbf{i}_m = (i_{m0}, i_{m1}, \ldots, i_{m,m}) \in \mathcal{K}^0$$

where $i_{m0} = m/(m+1)$ and $i_{mj} = 1/(m+1)$ for $j = 1, \ldots, m$. Indeed, $\|\mathbf{i} - \mathbf{i}_m\|^2 = \sum_{j=0}^{m} 1/(m+1)^2 = 1/(m+1) \to 0$ as $m \to \infty$. Hence, $\mathbf{i} \in \mathcal{K}$. Clearly, Kirchhoff's voltage law is satisfied for this $\mathbf{i}$. So, $\mathbf{i}$ is the unique current solution dictated by Theorem 3.3-5. However, $\mathbf{i}$ does not satisfy Kirchhoff's current law at nodes n_1 and n_2.   ♣

Consider now Kirchhoff's voltage law. This too was stated in Section 1.5, but now we generalize that statement by allowing L to be either a 0-loop or a 1-loop. Thus,

$$\sum_{(L)} \pm v_j = 0 \tag{3.11}$$

where $\sum_{(L)}$ is now a summation over the indices j for all the branches b_j in a given oriented 0-loop or 1-loop L, v_j is the corresponding branch voltage, and the plus (minus) sign is used if the orientations of b_j and L agree (disagree). As was defined in Section 3.3, L is called *perceptible* if $\sum_{(L)} r_j < \infty$. Obviously, every 0-loop is perceptible since it contains only a finite number of branches, but a 1-loop may or may not be perceptible.

Theorem 3.4-3: *If L is a 0-loop or a perceptible 1-loop, then, under the voltage-current regime dictated by Theorem 3.3-5, Kirchhoff's voltage law (3.11) holds for L, and the series on the left-hand side of (3.11) converges absolutely when L is a perceptible 1-loop.*

Proof: Let $\mathbf{s}$ be the loop current corresponding to a flow of 1 A around L. Since $v_j = r_j i_j - e_j$ for each branch, the substitution of $\mathbf{s}$ into (3.9) yields (3.11).

Let us now show that the left-hand side of (3.11) converges absolutely when L is a perceptible 1-loop. We may write

$$\sum_{(L)} |e_j| = \sum_{(L)} r^{1/2} |e_j| g_j^{1/2} \leq \left(\sum_{(L)} r_j \sum_{(L)} e_j^2 g_j \right)^{1/2} .$$

By the perceptibility of L and Conditions 3.3-1, the right-hand side is finite. Similarly,

$$\sum_{(L)} |r_j i_j| = \sum_{(L)} r_j^{1/2} |i_j| r_j^{1/2} \leq \left(\sum_{(L)} r_j i_j^2 \sum_{(L)} r_j \right)^{1/2} .$$

Since $\mathbf{i} \in \mathcal{K}$, the last right-hand side is finite too. Since $v_j = r_j i_j - e_j$, we are done. ♣

Here too, Kirchhoff's voltage law may be satisfied around a nonperceptible 1-loop if the rest of the network is appropriately configured.

Although Kirchhoff's laws do not hold in general, they do play a more dominant role when all the nodes are restraining nodes and there are no shorts at infinity; in particular, Kirchhoff's and Ohm's laws coupled with a finite power requirement suffice to establish a unique voltage-current regime, namely, the one asserted by Theorem 3.3.5. A precise statement of this will be given in Theorem 3.5-6.

3.5 A Dual Analysis

For finite networks mesh and nodal analysis are "dual" in the circuit-theory sense, that is, currents and voltages exchange their roles with currents playing the fundamental role in mesh analysis and voltages doing the same in nodal analysis. The question naturally arises as to whether Theorem 3.3-5 has a dual in a similar sense. The answer is yes. To show this, we first present in this section an existence-and-uniqueness theorem based on voltage vectors.

Conditions 3.5-1: *Let* $\mathbf{N}$ *be a 0-network or a 1-network. Assume that every branch* b_j *of* $\mathbf{N}$ *is in the Norton form with a positive conductance* g_j *and a current source* h_j *(as in Figure 1.1(c) but with H replaced by* h_j*). Assume, moreover, that* $\sum h_j^2 r_j < \infty$, *where* $r_j = 1/g_j$.

The condition $\sum h_j^2 r_j < \infty$ is just a rearranged form of the finite-total-isolated-power condition $\sum e_j^2 g_j < \infty$ of Conditions 3.3-1. Indeed, a Norton-Thevenin conversion of each branch has it that $r_j = 1/g_j$ and $e_j = -h_j r_j$, from which we get $\sum e_j^2 g_j = \sum h_j^2 r_j$.

Given such a branch current-source vector $\mathbf{h} = (h_1, h_2, \ldots)$, we search for a solution for the branch-voltage vector $\mathbf{v}$ satisfying another generalized form of Tellegen's equation [see (3.12) below]. That vector is also required to be of finite power. This is ensured by requiring the voltage vector to be a member of the space $\mathcal{V}$, which is defined as the space of all $\mathbf{v} = (v_1, v_2, \ldots)$ such that $\sum v_j^2 g_j < \infty$ and $\langle \mathbf{v}, \mathbf{s} \rangle = \sum v_j s_j = 0$ for all $\mathbf{s} \in \mathcal{K}^0$. This last condition ensures in addition that Kirchhoff's voltage law is satisfied around every 0-loop and every perceptible 1-loop but is actually a stronger condition because it holds for all basic currents $\mathbf{s}$ in $\mathcal{I}$. The linear operations are defined on $\mathcal{V}$ componentwise. We assign to $\mathcal{V}$ the inner product $(\mathbf{v}, \mathbf{w}) = \sum g_j v_j w_j$ for all $\mathbf{v}, \mathbf{w} \in \mathcal{V}$. $\| \cdot \|$ denotes the corresponding norm. The proof of Lemma 3.1-3 shows that $\mathcal{V}$ is complete and therefore a Hilbert space. The *conductance operator* G assigns to each $\mathbf{v} \in \mathcal{V}$ the current vector $(g_1 v_1, g_2 v_2, \ldots)$.

With these definitions in hand, we may repeat the proofs of Lemma 3.3-4, Theorem 3.1-6, and Corollary 3.3-6, interchanging voltages and currents as well as resistances and conductances, to obtain a result dual to Theorem 3.3-5 and Corollary 3.3-6.

Theorem 3.5-2: *Under Conditions 3.5-1 there exists a unique $\mathbf{v} \in \mathcal{V}$ such that*

$$\langle \mathbf{w}, \mathbf{h} - G\mathbf{v} \rangle = 0 \tag{3.12}$$

for every $\mathbf{w} \in \mathcal{V}$. Moreover, the total power $\sum v_j^2 g_j$ dissipated in all the conductances equals the total power $\sum v_j h_j$ supplied by all the current sources and is no larger than the finite total isolated power $\sum h_j^2 r_j$ available from all the current sources.

We now prove in two steps that Theorems 3.3-5 and 3.5-2 are equivalent in the sense that they yield the same voltage-current regime when the branches of Conditions 3.3-1 are related to the branches of Conditions 3.5-1 through a Thevenin–Norton conversion.

Theorem 3.5-3: *Let $\mathbf{i}$ be dictated by Theorem 3.3-5 and set $\mathbf{v} = R\mathbf{i} - \mathbf{e}$. Then, $\mathbf{v}$ is the unique member of $\mathcal{V}$ specified by Theorem 3.5-2.*

Proof: We have already noted that $\sum e_j^2 g_j = \sum h_j^2 r_j$; so, Conditions 3.3-1 are satisfied if and only if Conditions 3.5-1 are satisfied after a Thevenin–Norton conversion is made. We now show that $\mathbf{v} = R\mathbf{i} - \mathbf{e}$ is a member of $\mathcal{V}$. Indeed, for every $\mathbf{s} \in \mathcal{K}^0$, $\langle \mathbf{v}, \mathbf{s} \rangle = \langle R\mathbf{i} - \mathbf{e}, \mathbf{s} \rangle$, and this is equal to zero according to (3.9). Moreover, we can invoke Minkowski's inequality and the relation $r_j g_j = 1$ to write

$$\left(\sum v_j^2 g_j\right)^{1/2} = \left(\sum (r_j i_j - e_j)^2 g_j\right)^{1/2} = \left(\sum (i_j r_j^{1/2} - e_j g_j^{1/2})^2\right)^{1/2}$$

$$\leq \left(\sum i_j^2 r_j\right)^{1/2} + \left(\sum e_j^2 g_j\right)^{1/2} .$$

The first sum on the right-hand side is finite because $\mathbf{i} \in \mathcal{K}$ according to Theorem 3.3-5. So too is the second sum by Conditions 3.3-1. Thus, $\sum v_j^2 g_j < \infty$. So truly, $\mathbf{v} \in \mathcal{V}$.

Finally, we have to show that (3.12) is satisfied. For any $\mathbf{w} \in \mathcal{V}$,

$$\langle \mathbf{w}, G\mathbf{v} \rangle = \langle \mathbf{w}, G(R\mathbf{i} - \mathbf{e}) \rangle = \langle \mathbf{w}, \mathbf{i} - G\mathbf{e} \rangle = \langle \mathbf{w}, \mathbf{i} + \mathbf{h} \rangle$$

since $-G\mathbf{e} = \mathbf{h}$ according to the Thevenin-Norton conversion. Now, $\langle \mathbf{w}, \mathbf{i} \rangle = 0$ because of the following facts: $\mathbf{i} \in \mathcal{K}$; $\mathcal{K}^0$ is dense in $\mathcal{K}$; $\langle \mathbf{w}, \mathbf{s} \rangle = 0$ for every $\mathbf{s} \in \mathcal{K}^0$ by the definition of $\mathcal{V}$; $\mathbf{w}$ defines a continuous linear functional on $\mathcal{K}$ because, for $\mathbf{s} \in \mathcal{K}$,

$$|\langle \mathbf{w}, \mathbf{s} \rangle| = \left|\sum w_j g_j^{1/2} s_j r_j^{1/2}\right| \leq \left(\sum w_j^2 g_j \sum s_j^2 r_j\right)^{1/2} = \|\mathbf{w}\| \|\mathbf{s}\| .$$

Thus, $\langle \mathbf{w}, G\mathbf{v} \rangle = \langle \mathbf{w}, \mathbf{h} \rangle$, which is (3.12). ♣

Theorem 3.5-4: *Let $\mathbf{v}$ be dictated by Theorem 3.5-2 and set $\mathbf{i} = G\mathbf{v} - \mathbf{h}$. Then, $\mathbf{i}$ is the member of $\mathcal{K}$ specified by Theorem 3.3-5.*

Proof: We start by noting again that Conditions 3.3-1 and 3.5-1 are equivalent. We must show that $\mathbf{i} \in \mathcal{K}$. By using Minkowski's inequality and $i_j = g_j v_j - h_j$ as in the preceding proof, we obtain

$$\left(\sum i_j^2 r_j\right)^{1/2} \leq \left(\sum v_j^2 g_j\right)^{1/2} + \left(\sum h_j^2 r_j\right)^{1/2} .$$

The right-hand side is finite because $\mathbf{v} \in \mathcal{V}$ and Conditions 3.5-1 are assumed. Thus, $\mathbf{i} \in \mathcal{I}$.

To complete the proof that $\mathbf{i} \in \mathcal{K}$, we make use of Hilbert's coordinate space l_{2r} of one-way infinite vectors of quadratically summable real numbers. Let $\mathcal{K}_*$ be the space of all vectors $(r_1^{1/2} s_1, r_2^{1/2} s_2, \ldots)$, where $\mathbf{s} = (s_1, s_2, \ldots) \in \mathcal{K}$, and let $\mathcal{V}_*$ be the space of all vectors $(g_1^{1/2} v_1, g_2^{1/2} v_2, \ldots)$, where $\mathbf{v} = (v_1, v_2, \ldots) \in \mathcal{V}$. Then, $\mathcal{K}_*$ and $\mathcal{V}_*$ are both subspaces of l_{2r}. For any $\mathbf{v}_* \in \mathcal{V}_*$ and any $\mathbf{s}_* \in \mathcal{K}_*$, we have $(\mathbf{v}_*, \mathbf{s}_*) = \sum g_j^{1/2} v_j r_j^{1/2} s_j = \sum v_j s_j$. By definition, $\mathcal{V}$ is the set of all voltage vectors $\mathbf{v}$ of finite power for which $\sum v_j s_j = 0$ for all $\mathbf{s} \in \mathcal{K}^0$, and thereby for all $\mathbf{s} \in \mathcal{K}$ as in the last part of the preceding proof. This shows that $\mathcal{V}_*$ and $\mathcal{K}_*$ are orthogonal complements of each other in l_{2r}; that is, $\mathcal{K}_* = \mathcal{V}_*^\perp$. Let $\mathbf{w}$ be any member of $\mathcal{V}$ and let $\mathbf{w}_*$ be the corresponding member of $\mathcal{V}_*$. We may write

$$\langle \mathbf{w}, \mathbf{i} \rangle = \sum g_j^{1/2} w_j r_j^{1/2} i_j = (\mathbf{w}_*, \mathbf{i}_*).$$

On the other hand, by (3.12)

$$\langle \mathbf{w}, \mathbf{i} \rangle = \langle \mathbf{w}, G\mathbf{v} - \mathbf{h} \rangle = 0.$$

Hence, $\mathbf{i}_* \in \mathcal{V}_*^\perp = \mathcal{K}_*$. Therefore, $\mathbf{i} \in \mathcal{K}$.

Finally, we show that (3.9) is satisfied. Let $\mathbf{s}$ be any member of $\mathcal{K}$. Then,

$$\langle R\mathbf{i}, \mathbf{s} \rangle = \langle R(G\mathbf{v} - \mathbf{h}), \mathbf{s} \rangle = \langle \mathbf{v} - R\mathbf{h}, \mathbf{s} \rangle = \langle \mathbf{v} + \mathbf{e}, \mathbf{s} \rangle.$$

Since $\mathbf{v} \in \mathcal{V}$, $\langle \mathbf{v}, \mathbf{s} \rangle = 0$. Thus, $\langle R\mathbf{i}, \mathbf{s} \rangle = \langle \mathbf{e}, \mathbf{s} \rangle$, which is (3.9). ♣

We now examine a special class of networks whose voltage-current regimes are uniquely determined just by Kirchhoff's and Ohm's laws and the finite-power requirement.

Conditions 3.5-5: N *is a 0-connected 0-network or a 0-connected 1-network all of whose 1-nodes are singletons. There is at most one non-restraining node n_0. Every branch of* N *is in the Norton form. Finally,* $\sum h_j^2 r_j < \infty$.

We are requiring in effect that N be open-circuited everywhere at infinity – in addition to being 0-connected. As a result, there are no 1-loops, and therefore $\mathcal{K}^0$ is just the span of all 0-loop currents.

Furthermore, all 0-nodes are now ordinary. The single nonrestraining node n_0, if it exists, will be considered to be the *ground node* at zero electric potential, with respect to which all other node potentials are measured. This can be done because any other node is reachable from n_0 through a finite 0-path. In the absence of the nonrestraining node, we fix the ground node n_0 as one of the restraining nodes. The node potential W_k at any other node n_k is $\sum_{(P)} \pm v_j$, where the v_j are the branch voltages along a 0-path P traced from n_k to n_0, the plus (minus) sign being chosen if the branch orientation agrees (disagrees) with the tracing. The satisfaction of Kirchhoff's voltage law around 0-loops will insure that the node voltages are uniquely determined with respect to the fixed n_0.

Theorem 3.5-6: *Under Conditions 3.5-5, there exists a unique voltage-current regime $\{\mathbf{v}, \mathbf{i}\}$ such that Kirchhoff's voltage law is satisfied around all 0-loops, Kirchhoff's current law is satisfied at all restraining nodes, and the total dissipated power $\sum v_j^2 g_j$ is finite. The voltage vector $\mathbf{v}$ is identical to that given by Theorem 3.5-2.*

Proof: Note that Conditions 3.5-5 imply the fulfillment of Conditions 3.5-1. $\mathcal{V}$ is now the space of all voltage vectors $\mathbf{v}$ of finite power ($\sum v_j^2 g_j < \infty$) satisfying Kirchhoff's voltage law around 0-loops.

Let $\{\mathbf{v}, \mathbf{i}\}$ be any voltage-current regime satisfying the conditions stated in this theorem. Hence, $\mathbf{v} \in \mathcal{V}$. We shall now show that the series

$$\langle \mathbf{w}, \mathbf{h} - G\mathbf{v} \rangle = \sum w_j(h_j - g_j v_j)$$

is equal to zero for every $\mathbf{w} \in \mathcal{V}$. This will satisfy the requirement (3.12) of Theorem 3.5-2. The series converges absolutely by virtue of Schwarz's inequality. Consequently, we may rearrange and sum terms in any fashion. With n_0 denoting the ground node as before, let n_k, where $k = 1, 2, \ldots$, denote all the other nodes. Consider any branch b_j incident at some chosen n_k. Given any $\mathbf{w} \in \mathcal{V}$, we have $w_j = \pm(W_k - W_q)$, where W_k and W_q are the voltages at the nodes n_k and n_q of b_j and the plus (minus) sign is chosen if b_j is oriented away from (toward) n_k. In particular, $W_0 = 0$. Now, gather all the terms in the series that contain W_k. We get

$$W_k \sum_{(n_k)} \pm(h_j - g_j v_j).$$

This last sum equals zero because Kirchhoff's current law is satisfied at n_k. Since this is so for every n_k other than n_0 and since $W_0 = 0$, the above series also equals zero. Hence, $\{\mathbf{v}, \mathbf{i}\}$ is the voltage-current regime dictated by Theorem 3.5-2. The uniqueness assertion of that theorem completes the proof. ♣

In the event that all the resistance values are uniformly bounded above and also uniformly bounded away from zero, we may alter Theorem 3.5-6 as follows. As before, l_{2r} denotes Hilbert's real coordinate space.

Corollary 3.5-7: *Under Conditions 3.5-5, if there exist two constants α and β with $0 < \alpha < \beta < \infty$ such that $\alpha \leq r_j \leq \beta$ for all j, then Theorem 3.5-6 continues to hold when the conditions $\sum h_j^2 r_j < \infty$ and $\sum v_j^2 g_j < \infty$ are replaced by $\mathbf{h} \in l_{2r}$ and $\mathbf{v} \in l_{2r}$.*

Proof: The condition $\sum h_j^2 g_j < \infty$ will be satisfied if and only if $\mathbf{h} \in l_{2r}$, and similarly for $\mathbf{v}$. ♣

3.6 Transferring Pure Sources

The theory developed so far can be extended to networks having pure-source branches such as those shown in Figure 1.2. Moreover, such sources can be connected to the network at infinity. The latter will

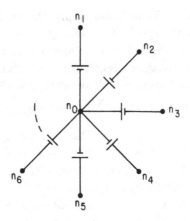

Figure 3.7. A star network of pure voltage sources.

occur when a node of a pure-source branch is embraced by a 1-node. A theory for networks with pure sources is the objective of this and the next section.

Let **N** be a 0-network or a 1-network and let **R**, **E**, and **H** be the subnetworks of **N** induced by the resistive branches (Figure 1.1(a)), the pure voltage-source branches (Figure 1.2(a)), and the pure current-source branches (Figure 1.2(b)) respectively. Also, let **R** ∪ **E** denote the subnetwork of **N** induced by the resistive branches and the pure voltage-source branches taken together. Here, by "subnetwork" we mean the network whose 0-graph or 1-graph is the reduction of the 0-graph or 1-graph of **N** induced by the said branches. Some restrictions have to be placed on **R**, **E**, and **H** in order to avoid impossible situations such as two unequal pure voltage sources connected in parallel or two unequal pure current sources connected in series.

Consider **E**. We allow **E** to have many (possibly an infinity of) components. We require

Conditions 3.6-1: E *is a 0-network. Each component of* **E** *is a finite or infinite star network, every end node of which is ordinary.*

A star network is illustrated in Figure 3.7, a special case being a single branch. Every node of each star network is a (finite or infinite) 0-node of **N**, but only the central node n_0 is allowed to be (but need not be) embraced by some 1-node of **N**. The other nodes $n_1, n_2, \ldots$ of the star network are required to be ordinary (i.e., nonembraced). Thus, a pure voltage source can "reach to infinity" with only one of its nodes. Actually, we could also allow a component of **E** to be any 0-network, so long as

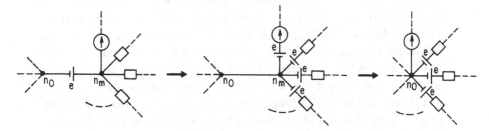

Figure 3.8. The transferring of a pure voltage source e into adjacent branches. The boxes represent resistive branches. At the third and final stage, the node n_m has been absorbed into the central node n_0, and the voltage source in series with the pure current source has been removed by virtue of Norton's theorem.

Kirchhoff's voltage law is satisfied on each of its 0-loops and there is no more than one embraced node. However, each such component could then be replaced by an equivalent star network.

As for **H**, we assume

Conditions 3.6-2: **H** *is a 0-network or a 1-network. Every node of* **H** *is embraced by a 0-node or a 1-node of* **R** $\cup$ **E**, *and, for every branch b of* **H**, *there is a 0-path or a 1-path in* **R** $\cup$ **E** *connecting the nodes of b.*

Thus, a pure current source may be "out at infinity" with both of its nodes. Moreover, **H** has no internal nodes, that is, a node all of whose incident branches are pure current sources. Actually, we could also allow any number of internal finite nodes in **H** so long as Kirchhoff's current law is satisfied at those nodes, but any such node n could then be eliminated by connecting other current sources to the nodes adjacent to n.

In order to use Hilbert-space techniques, we have to insure that the sources – including the sources at infinity – do not inject an infinite amount of power into the network **N**. To do so, we will first transfer all the pure sources into resistive branches following a standard technique [107, pages 131–132] and then will impose a certain finite-power condition on the resulting branch sources.

Transferring pure voltage sources: The technique for transferring pure voltage sources is illustrated in Figure 3.8. The left-hand part of Figure 3.8 indicates one branch of a star component of **E** and its adjacent branches. n_0 is the central 0-node and n_m is an ordinary end node of the star network. The other branches incident to n_m are either resistive branches or pure current sources. The polarity of e must not be altered when transferring e into the other branches incident to n_m. The transference of e into those branches makes the $n_0 n_m$-branch a short, as shown in the middle part of Figure 3.8. That short is then eliminated by combining n_0 and n_m into a single node, as is indicated in the right-hand part of Figure 3.8. If a transferred e appears in series with a current source, it can be shorted out according to Norton's theorem.

This process of transferring voltage sources will convert a star component $\mathbf{C}$ of $\mathbf{E}$ into a single node. Moreover, every resistive branch or pure current source connected between two nodes of $\mathbf{C}$ will become a self-loop. Note that the algebraic sum of the branch voltage-source values along any loop in $\mathbf{R} \cup \mathbf{E}$ that passes through n_m and/or n_0 remains the same before and after the transference.

The purpose of transferring e through, say, n_m is to obtain a finite value for $\sum e_k^2 g_k$, where $e_k = e$ and the summation is over all resistive branches incident to n_m; $\sum e^2 g_k$ will be finite if and only if n_m is a restraining node. In the event it is not, one can transfer some or all of the transferred voltage sources e still further along a subnetwork emanating from n_m. It may be possible to find some distribution of the sources e throughout only resistive branches such that $\sum e_k^2 g_k = e^2 \sum g_k < \infty$, where now g_k denotes the conductances of the branches into which e has been transferred. More generally, one may resort to transferring successively only a portion of each e_k in order to achieve a finite $\sum \hat{e}_k^2 g_k < \infty$ where now $|\hat{e}_k| \leq e$.

Example 3.6-3. Consider the network of Figure 3.9, where originally a 1-V pure voltage source appeared in the $n_0 n_m$ branch. All the other branches were purely resistive; their resistances are as follows: Let $\nu = 1, 2, 3, \ldots$ be indices for the rows of the square grid, as shown. The resistance of every horizontal branch in row ν is ν Ω. The resistance of every vertical branch between rows ν and $\nu+1$ is ν Ω too. The resistance of every diagonal branch between node n_m and the first node of row ν is also ν Ω. The sequence of pure-voltage-source transferences is as follows: The original 1-V source is first transferred through n_m into all the diagonal branches and of course into the $n_m n_1$ branch as well to obtain a 1-V source in each of those branches and a short for the $n_0 n_m$ branch. Next, a 1/2-V source is transferred through n_2 to obtain 1/2-V sources for the $n_1 n_2$ and $n_2 n_3$ branches, as well as for the $n_2 n_4$ branch. Next, a 2/3-V source is transferred through n_4 to obtain the 1/6-V source shown for the $n_2 n_4$ branch and 2/3-V sources for the $n_4 n_5$ and $n_4 n_6$ branches. Fractionating and transferring the sources in this way infinitely often, we get the set of transferred voltage sources in the infinite wedge-shaped

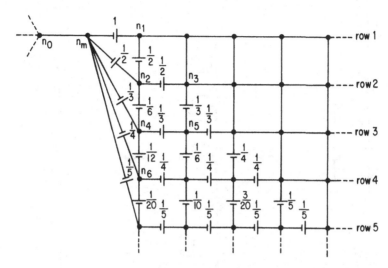

Figure 3.9. The result of repeated and fractionated transferences of a 1-V voltage source, originally between the 0-nodes n_0 and n_m, to obtain an infinite distribution of voltage sources in a wedge-shaped infinite subnetwork.

subnetwork indicated in Figure 3.9. A summation over all branches with these transferred voltage sources yields $\sum \hat{e}_k^2 g_k < \infty$. ♣

Some care must be taken when transferring a voltage source infinitely often, for it is possible to eliminate sources from all branches in this way. To avoid this paradoxical situation, we restrict ourselves to a "proper" transference of voltage sources. As is indicated in Figure 3.8, let the pure voltage source e oriented from n_0 to n_m be transferred first through n_m and then (possibly fractionated and) transferred through other nodes to obtain a distribution of voltage sources $\hat{e}_j$ in resistive branches. Let P be any one-ended 0-path starting at the node coalesced from n_0 and n_m. Let $\sum_P \pm \hat{e}_j$ be the algebraic sum of the $\hat{e}_j$ in branches of P oriented from the coalesced node toward infinity. If $e = \sum_P \pm \hat{e}_j$ for every choice of P, then e is said to be *properly transferred*. It may be noted that the transference illustrated in Figure 3.9 is proper.

Transferring pure current sources: By Conditions 3.6-2, for any pure current source h there is a 0-path or a 1-path P_* in $\mathbf{R} \cup \mathbf{E}$ connecting the nodes of h. Therefore, h can be transferred into the branches of P_*, as is illustrated in Figure 3.10 for the case of a one-ended P_* where one node n_0 of h is the end node of P_* and the other node of h is embraced by a 1-node containing the 0-tip t of P_*. However, if a branch

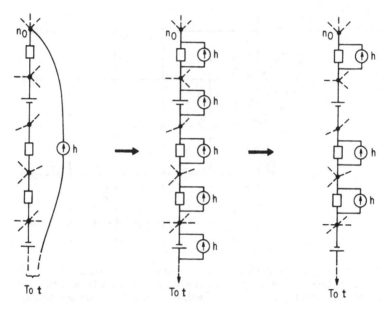

Figure 3.10. The transferring of a pure current source h into a one-ended path in $\mathbf{R} \cup \mathbf{E}$ for the case where one node n_0 of h is ordinary and other node of h is embraced by a 1-node containing the 0-tip t. The boxes represent resistive branches. At the last stage, the current sources across the pure voltage sources have been removed by virtue of Thevenin's theorem.

in P_* was originally a pure voltage source, its transferred parallel h may be discarded according to Thevenin's theorem. Note that the algebraic sum of all the current-source values at any node between two resistive branches in P_* is not altered by this process.

Our purpose is to find a path P_* for which $\sum h^2 r_k < \infty$. As with a loop, we say that a 0-path or a 1-path is *perceptible* if the sum of its branch resistances is finite. Thus, we seek a perceptible path in $\mathbf{R} \cup \mathbf{E}$ connecting the nodes of h. However, if none is available, it may be possible to ensure the finiteness of the power dissipated in the resistances by the flow of h through $\mathbf{R} \cup \mathbf{E}$ by spreading portions of h along an infinity of paths all of which meet the same 1-node as does h. The result will be a distribution of branch current sources, which by themselves satisfy Kirchhoff's current law at all nodes except those of h.

Here too, we must restrict the way current-source transferences are repeated infinitely often so as not to lose any current sources. Let n_1 and n_2 be the nodes of an original current source h oriented from n_1 to

n_2. Let C be a cutset that separates n_1 and n_2, that is, a minimal set of branches such that any 0-path or 1-path that embraces n_1 and n_2 also embraces a branch of C. We can orient C according to the orientation from n_1 to n_2. Now, let $\hat{h}_j$ be the current sources after transference. Let $\sum_C \pm \hat{h}_j$ be the algebraic sum of all transferred current sources in branches of C oriented according to C. If $h = \sum_C \pm \hat{h}_j$ for every choice of C, h is said to be *properly transferred*.

Example 3.6-4. An example of such a spreading of h is illustrated in Figure 3.5(a) for a quarter-plane resistive square grid. We take it that a pure current source feeds a 1-A current from the 0-node n_1 into a 0-node n_2 embraced by a 1-node containing the 0-tips of all the one-ended horizontal paths. By successively fractionating and transferring h infinitely often, we can obtain the infinite distribution of branch current sources $\hat{h}_j$ suggested in Figure 3.5(a). With branch resistance values as stated in Example 3.3-2, it follows that $\sum \hat{h}_k^2 r_k < \infty$, as desired. It can be seen that this transference is proper. ♣

Now, assume that a transferring of all the pure sources into the resistive branches has already occurred, and let $\mathbf{N}_c$ denote the resulting network. We refer to $\mathbf{N}_c$ as the *converted network*. All the branches of $\mathbf{N}_c$ are resistive as in Figure 3.11(a) and have the same resistance values as they do in $\mathbf{N}$. The 0-graph or 1-graph of $\mathbf{N}_c$ can be obtained from that of $\mathbf{R}$ by combining into a single 0-node each maximal set of 0-nodes in $\mathbf{R}$ that are common to a single star component of $\mathbf{E}$.

Let I_j be an index set for all the pure voltage sources e_η that get transferred into a given resistive branch b_j. Similarly, let J_j be an index set for all the pure current sources h_μ that get transferred into b_j. Set $g_j = 1/r_j$. We now define the *total isolated power* P of the network $\mathbf{N}_c$ to be

$$P = \sum_j \left[g_j^{1/2} \left(e_j + \sum_{\eta \in I_j} \pm e_\eta \right) - r_j^{1/2} \left(h_j + \sum_{\mu \in J_j} \pm h_\mu \right) \right]^2 \quad (3.13)$$

where the summation on j is over all branches in $\mathbf{N}_c$ and the plus or minus signs are chosen according to whether or not the polarities of e_η and h_μ conform with the orientations indicated in Figure 3.11(a). The significance of this is that the squared term under the summation on j is the power dissipated in r_j when the jth branch in $\mathbf{N}_c$ is converted into a Thevenin equivalent branch and then shorted. Note that P depends in general on the way the pure sources are transferred into the branches

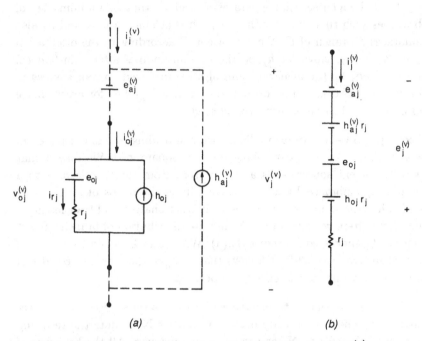

Figure 3.11. (a) A typical branch in $\mathbf{N}_c$ with resistance r_j. $h_{aj}^{(\nu)}$ and $e_{aj}^{(\nu)}$ are the sums of transferred current sources and transferred voltage sources respectively for the νth way of converting $\mathbf{N}$ to $\mathbf{N}_c$, and h_{oj} and e_{oj} are the original branch sources. $i_j^{(\nu)}$ is that branch's current in $\mathbf{N}_c$ for the νth way, and $i_{oj}^{(\nu)}$ is that branch's corresponding current in $\mathbf{N}$. (b) The Thevenin's equivalent branch.

and may become either finite or infinite in certain cases. So, we have to state our *finite-total-isolated-power condition* for $\mathbf{N}_c$ as follows.

Condition 3.6-5: *There exists at least one way of properly transferring all the pure sources of* $\mathbf{N}$ *into resistive branches such that the four series*

$$\sum_j g_j e_j^2, \quad \sum_j g_j \left(\sum_{\eta \in I_j} \pm e_\eta \right)^2$$

$$\sum_j r_j h_j^2, \quad \sum_j r_j \left(\sum_{\mu \in J_j} \pm h_\mu \right)^2$$

are all finite for the resulting converted network $\mathbf{N}_c$.

Minkowski's inequality coupled with Condition 3.6-5 insures that P is finite. It can be shown [177, Section IX] that the following four (more easily checked) conditions are sufficient for the fulfillment of Condition 3.6-5: (i) There are only a finite number of pure sources in $\mathbf{N}$. (ii) Every end node of every star component of $\mathbf{E}$ is restraining. (iii) There is a perceptible path in $\mathbf{R} \cup \mathbf{E}$ connecting the nodes of each pure current source. (iv) The branch sources (other than the pure sources before conversion) satisfy $\sum e_j^2 g_j < \infty$ and $\sum h_j^2 r_j < \infty$.

3.7 Networks with Pure Sources

Our method for studying a 1-network $\mathbf{N}$ with pure sources is to transfer properly all the pure sources into resistive branches to get a converted network $\mathbf{N}_c$ in accordance with Condition 3.6-5. Theorem 3.3-5 then asserts the existence of a unique voltage-current regime in $\mathbf{N}_c$, which can in turn be uniquely related to a voltage-current regime in the original network $\mathbf{N}$. The question is: Does the regime in $\mathbf{N}$ depend on the way the pure sources are transferred?

Note that, after all pure sources have been properly transferred, every branch in $\mathbf{N}_c$ will be resistive and will consist of its original parameters (shown with the solid lines in Figure 3.11(a) and the subscript o) along with the transferred sources $e_{aj} = \sum_{\eta \in I_j} \pm e_\eta$ and $h_{aj} = \sum_{\mu \in J_j} \pm h_\mu$ (shown with the dotted lines in Figure 3.11(a) and the subscript a). Subsequently, we shall show that the original branch voltage v_{oj} and the original branch current i_{oj} will remain unchanged as one proceeds from $\mathbf{N}$ to $\mathbf{N}_c$, even though the total branch voltage $v_{oj} - e_{aj}$ and the total branch current $i_{oj} - h_{aj}$ in $\mathbf{N}_c$ will depend in general on the way the pure sources are transferred. It is in this way that the unique regime for $\mathbf{N}$ is established.

In order to invoke Theorem 3.3-5, we convert every branch in $\mathbf{N}_c$ into its Thevenin equivalent, as shown in Figure 3.11(b), where $e_j = e_{aj} + e_{oj} - (h_{aj} + h_{oj})r_j$, $v_j = v_{oj} - e_{aj}$, and $i_j = i_{oj} - h_{aj}$. In $\mathbf{N}_c$ there will be no branches for those indices j corresponding to the pure sources in $\mathbf{N}$.

We now wish to transfer the conclusion of Theorem 3.3-5, as applied to $\mathbf{N}_c$, to the original network $\mathbf{N}$. Note, first of all, that one gets from $\mathbf{R} \cup \mathbf{E}$ to $\mathbf{N}_c$ simply by shorting out star components of $\mathbf{E}$. Now, by virtue of Kirchhoff's current law, there is a bijective correspondence from all the basic currents in $\mathbf{R} \cup \mathbf{E}$ to all the basic currents in $\mathbf{N}_c$ because each basic current in $\mathbf{R} \cup \mathbf{E}$ is a superposition of loop currents with the property that only a finite number of them flow through any ordinary node. It is this last property that allows the imposition of Kirchhoff's current law at the end nodes of the star components of E even when such an end node is not restraining.

Throughout this section, voltage or current vectors in $\mathbf{N}$ are designated by subscript asterisks to distinguish them from the corresponding entities in $\mathbf{N}_c$. Thus, $\mathbf{s}_*$ and $\mathbf{s}$ may denote respectively a basic current in $\mathbf{N}$ and its uniquely corresponding basic current in $\mathbf{N}_c$. Let $\mathbf{e}_* = (e_{*1}, e_{*2}, \ldots)$ be the vector of voltage sources in $\mathbf{N}$. The superscript ν will index the different ways of transferring pure sources to satisfy Condition 3.6-5. For the νth such way, $\mathbf{N}_c^{(\nu)}$ denotes the resulting converted network. The spaces $\mathcal{K}^0$ and $\mathcal{K}$ for $\mathbf{N}_c^{(\nu)}$ do not change as ν changes because all the $\mathbf{N}_c^{(\nu)}$ have the same 1-graph or perhaps 0-graph and the same branch resistances. The only things that change as ν changes are the branch source values. For $\mathbf{N}_c^{(\nu)}$, $\mathbf{e}_o = (e_{o1}, e_{o2}, \ldots)$ and $\mathbf{e}_a^{(\nu)} = (e_{a1}^{(\nu)}, e_{a2}^{(\nu)}, \ldots)$ are respectively the vector of original voltage sources and the vector of transferred voltage sources for the resistive branches alone, the terms for the pure source branches being absent. (See Figure 3.11(a) again.)

For any basic current $\mathbf{s}_*$ in $\mathbf{R} \cup \mathbf{E}$ and its corresponding basic current $\mathbf{s}$ in $\mathbf{N}_c^{(\nu)}$, we have

$$\langle \mathbf{e}_*, \mathbf{s}_* \rangle = \langle \mathbf{e}_a^{(\nu)} + \mathbf{e}_o, \mathbf{s} \rangle. \tag{3.14}$$

To see this, choose for $\mathbf{s}_*$ a loop-current superposition $\mathbf{s}_* = \sum_\mu \mathbf{f}_{*\mu}$ in $\mathbf{R} \cup \mathbf{E}$ in accordance with the definition of a basic current and let ϕ_μ be the common absolute value of the nonzero components of $\mathbf{f}_{*\mu}$. Next, let ϵ be a single pure voltage source in $\mathbf{R} \cup \mathbf{E}$. The contribution of ϵ to the left-hand side of (3.14) is $\sum_k \pm \epsilon \phi_{\mu_k}$, where $\{\mu_k\}$ is the finite set of indices for the loop currents $\mathbf{f}_{*\mu}$ passing through ϵ. Transfer ϵ through a single ordinary node n_1 but do not as yet coalesce the original branch for ϵ into a single node. Denote the new network by $\mathbf{N}_1$. In $\mathbf{N}_1$, ϵ appears in every branch incident to n_1 except for the branch in which it originally appeared (and which is now a short circuit acting as a branch). Let $\mathbf{f}_\mu$ be the loop current in $\mathbf{N}_1$ corresponding to $\mathbf{f}_{*\mu}$. It follows that the contribution of ϵ as a functional on $\sum \mathbf{f}_\mu$ in $\mathbf{N}_1$ is again $\sum_k \pm \epsilon \phi_{\mu_k}$. In addition, a pair of values $+\epsilon\phi$ and $-\epsilon\phi$ will appear on the right-hand side of (3.14) for every loop current $\mathbf{f}$ that passes through a pair of branches incident to n_1 but not through the original branch of ϵ. These pairs contribute nothing to the right-hand side of (3.14). Moreover, for the given basic current $\mathbf{s}_*$ there are only a finite number of component loop currents passing through n_1. So truly, (3.14) holds for $\mathbf{R} \cup \mathbf{E}$ replaced by $\mathbf{N}_1$ when one transfers ϵ only through n_1. Finally, after all the voltage sources have been transferred out of a star component of $\mathbf{E}$, we may coalesce that star component into a single node. This may change $\mathbf{s}$ but not the value of the right-hand side of (3.14). In the same way, (3.14) continues to hold step-by-step as one proceeds through the sequence of pure source transferences that converts $\mathbf{N}$ into $\mathbf{N}_c$.

We thus have that (3.14) holds for any $\mathbf{s} \in \mathcal{K}^0$. Observe that $\mathbf{e}_a^{(\nu)} + \mathbf{e}_o$ is a continuous linear functional on $\mathcal{K}$ by virtue of Condition 3.6-5 and the proof of Lemma 3.3-4. So, since $\mathcal{K}^0$ is dense in $\mathcal{K}$, (3.14) holds for all $\mathbf{s} \in \mathcal{K}$ as well. Thus, the right-hand side remains unchanged for each $\mathbf{s} \in \mathcal{K}$ as ν changes, that is, it remains the same whatever be the way pure voltage sources are transferred into resistive branches.

Now, consider the transferring of the pure current sources into resistive branches for the νth way. The sum of the current sources added to any given branch b_j is denoted by $h_{aj}^{(\nu)}$. The resulting branch parameters for branch b_j in $\mathbf{N}_c^{(\nu)}$ are shown in Figure 3.11(a), and Figure 3.11(b) shows the Thevenin equivalent branch, whose total branch voltage source is now

$$e_j^{(\nu)} = e_{aj}^{(\nu)} - h_{aj}^{(\nu)} r_j + e_{oj} - h_{oj} r_j. \qquad (3.15)$$

In Figure 3.11(a), $i_{oj}^{(\nu)}$ is the branch current for the jth resistive branch in $\mathbf{N}$ corresponding to the branch current $i_j^{(\nu)}$ determined by applying Theorem 3.3-5 to $\mathbf{N}_c^{(\nu)}$. We wish to show that $i_{oj}^{(\nu)}$ does not depend on ν.

As was noted before, the resistance operator R for $\mathbf{N}_c^{(\nu)}$ does not depend on ν. Also, for any given $\mathbf{s} \in \mathcal{K}$ for $\mathbf{N}_c^{(\nu)}$, we have from (3.14) that $\langle \mathbf{e}_a^{(\nu)} + \mathbf{e}_o, \mathbf{s} \rangle$ is independent of ν. By Theorem 3.3-5, there is a unique $\mathbf{i}_r^{(\nu)} \in \mathcal{K}$ such that, for all $\mathbf{s} \in \mathcal{K}$,

$$\langle \mathbf{e}_a^{(\nu)} + \mathbf{e}_o, \mathbf{s} \rangle = \langle R \mathbf{i}_r^{(\nu)}, \mathbf{s} \rangle = (\mathbf{s}, \mathbf{i}_r^{(\nu)}), \qquad (3.16)$$

and so $\mathbf{i}_r^{(\nu)}$ is independent of ν too because a knowledge of $(\mathbf{s}, \mathbf{t})$ for all $\mathbf{s} \in \mathcal{K}$ uniquely determines $\mathbf{t}$ as a member of $\mathcal{K}$. Set $\mathbf{i}_r = \mathbf{i}_r^{(\nu)}$.

Next, set $\mathbf{h}_a^{(\nu)} = (h_{a1}^{(\nu)}, h_{a2}^{(\nu)}, \ldots)$ and $\mathbf{h}_o = (h_{o1}, h_{o2}, \ldots)$. By (3.15), $\mathbf{e}^{(\nu)} = \mathbf{e}_a^{(\nu)} - R \mathbf{h}_a^{(\nu)} + \mathbf{e}_o - R \mathbf{h}_o$. By Condition 3.6-5, $\mathbf{e}^{(\nu)}$ is also a continuous linear functional on $\mathcal{K}$, and hence, for every $\mathbf{s} \in \mathcal{K}$,

$$\langle \mathbf{e}^{(\nu)}, \mathbf{s} \rangle = \langle \mathbf{e}_a^{(\nu)} + \mathbf{e}_o - R \mathbf{h}_a^{(\nu)} - R \mathbf{h}_o, \mathbf{s} \rangle = \langle R \mathbf{i}^{(\nu)}, \mathbf{s} \rangle \qquad (3.17)$$

where $\mathbf{i}^{(\nu)} = (i_1^{(\nu)}, i_2^{(\nu)}, \ldots)$ is uniquely determined for each ν according to Theorem 3.3-5 again. So,

$$
\begin{aligned}
\langle \mathbf{e}_a^{(\nu)} + \mathbf{e}_o, \mathbf{s} \rangle &= \langle R(\mathbf{i}^{(\nu)} + \mathbf{h}_a^{(\nu)} + \mathbf{h}_o), \mathbf{s} \rangle \\
&= \langle \mathbf{s}, \mathbf{i}^{(\nu)} + \mathbf{h}_a^{(\nu)} + \mathbf{h}_o \rangle.
\end{aligned}
\tag{3.18}
$$

Upon comparing (3.16) and (3.18), we see that

$$
\mathbf{i}_r = \mathbf{i}^{(\nu)} + \mathbf{h}_a^{(\nu)} + \mathbf{h}_o \in \mathcal{K}.
\tag{3.19}
$$

As is indicated in Figure 3.11, $\mathbf{i}_o^{(\nu)} = \mathbf{i}^{(\nu)} + \mathbf{h}_a^{(\nu)}$, and so $\mathbf{i}_o^{(\nu)} = \mathbf{i}_r - \mathbf{h}_o$. Since $\mathbf{i}_r - \mathbf{h}_o$ does not depend upon ν, neither does $\mathbf{i}_o^{(\nu)}$, which is what we wished to show.

Note that $\mathbf{i}_r$ is the vector of currents in the resistors of $\mathbf{N}$, which are in turn the currents in the resistors of $\mathbf{N}_c^{(\nu)}$ when the branches of $\mathbf{N}_c^{(\nu)}$ have the form shown in Figure 3.11(a), that is, before those branches are converted into their Thevenin's equivalents, shown in Figure 3.11(b).

We have established the following:

Theorem 3.7-1: *Let $\mathbf{N}$ be a 0-network or a 1-network, every branch of which has either a resistive form (Figure 1.1(a)) or a pure source form (Figure 1.2(a) or (b)). Assume Conditions 3.6-1, 3.6-2, and 3.6-5 are satisfied. Let $\mathbf{N}_c^{(\nu)}$ denote the converted network obtained by using the νth way of properly transferring the pure sources in $\mathbf{N}$ into resistive branches in accordance with Condition 3.6-5, and let all the branches of $\mathbf{N}_c^{(\nu)}$ be in their Thevenin form as shown in Figure 3.11(b). Let $i_j^{(\nu)}$ be the current for the jth branch of $\mathbf{N}_c^{(\nu)}$ as dictated by Theorem 3.3-5. Then, the current $i_{oj}^{(\nu)}$ for every resistive branch b_j in $\mathbf{N}$ is uniquely determined and does not depend on the way sources were transferred (i.e., is independent of ν). It is given by $i_{oj}^{(\nu)} = i_j^{(\nu)} + h_{aj}^{(\nu)}$.*

Before leaving this section, it may be worth pointing out that an approximation theorem can be stated for 1-networks with pure sources – even when some of those sources are impressed at infinity. In particular, construct a *shorted* subnetwork by shorting together some of the 0-nodes and/or 1-nodes of the given 1-network and then removing all branches that have been converted into self-loops. Then the voltage-current regime of the 1-network is the limit – at least branchwise – of the voltage-current regimes of a sequence of shorted subnetworks as the set of shorted nodes is reduced toward a void set [180, Theorem 3.4].

3.8 Thomson's Least Power Principle

Another approach to the theory of finite or infinite networks is to identify the voltage-current regime as that regime for which the power dissipated is minimized. This is an old concept dating back to the nineteenth century [131, pages 294–295]. To examine it, we return to the case of a 1-network **N** with no pure-source branches. Assume that all the branches are in their Thevenin form. Let $\mathcal{L}_e$ be the subset of $\mathcal{K}$ defined by

$$\mathcal{L}_e = \{\mathbf{x} \in \mathcal{K} : \langle \mathbf{e}, \mathbf{x} \rangle = \langle \mathbf{e}, \mathbf{i} \rangle\},$$

where $\mathbf{i}$ is the current vector dictated by Theorem 3.3-5. (The condition $\langle \mathbf{e}, \mathbf{x} \rangle = \langle \mathbf{e}, \mathbf{i} \rangle$ does not imply that $x_j = i_j$ on every branch b_j having a nonzero e_j.)

Theorem 3.8-1: *Under Conditions 3.3-1, the current vector $\mathbf{i}$ specified by Theorem 3.3-5 is that member $\mathbf{x} \in \mathcal{L}_e$ for which $\sum x_j^2 r_j$ is a minimum. There is only one such member of $\mathcal{L}_e$.*

Proof: Set $\mathbf{x} = \mathbf{i} + \Delta\mathbf{i}$. So, $\Delta\mathbf{i} \in \mathcal{K}$. Moreover,

$$\langle \mathbf{e}, \mathbf{x} \rangle = \langle \mathbf{e}, \mathbf{i} \rangle + \langle \mathbf{e}, \Delta\mathbf{i} \rangle.$$

Since $\mathbf{x} \in \mathcal{L}_e$,

$$\langle \mathbf{e}, \Delta\mathbf{i} \rangle = 0. \tag{3.20}$$

We shall show that

$$\langle R(\mathbf{i} + \Delta\mathbf{i}), \mathbf{i} + \Delta\mathbf{i} \rangle - \langle R\mathbf{i}, \mathbf{i} \rangle \tag{3.21}$$

is a positive quantity if $\Delta\mathbf{i} \neq \mathbf{0}$. Indeed, (3.21) is equal to

$$\langle R\mathbf{i}, \Delta\mathbf{i} \rangle + \langle R\Delta\mathbf{i}, \mathbf{i} \rangle + \langle R\Delta\mathbf{i}, \Delta\mathbf{i} \rangle = 2\langle R\mathbf{i}, \Delta\mathbf{i} \rangle + \langle R\Delta\mathbf{i}, \Delta\mathbf{i} \rangle.$$

By (3.9), $\langle R\mathbf{i}, \Delta\mathbf{i} \rangle = \langle \mathbf{e}, \Delta\mathbf{i} \rangle$, which by (3.20) is zero. Hence, (3.21) equals $\langle R\Delta\mathbf{i}, \Delta\mathbf{i} \rangle = \sum r_j (\Delta i_j)^2$. This is strictly positive whenever $\Delta\mathbf{i} \neq \mathbf{0}$. ♣

Now, let

$$\mathcal{J}_e = \{\mathbf{x} \in \mathcal{K} : x_j = i_j \text{ whenever } e_j \neq 0\}.$$

$\mathcal{J}_e$ is a subset of $\mathcal{L}_e$, and therefore we have

Corollary 3.8-2: *For any $\mathbf{x} \in \mathcal{J}_e$ and for $\mathbf{i}$ given by Theorem 3.3-5, $\sum x_j^2 r_j \geq \sum i_j^2 r_j$. Strict inequality occurs if $\mathbf{x} \neq \mathbf{i}$.*

A way of understanding this corollary is to turn our attention to the reduction of **N** induced by those branches that do not possess voltage sources. (There are no current sources because all branches are in the Thevenin form.) The currents in the branches with nonzero voltage sources can be taken to be a set of given current sources feeding the

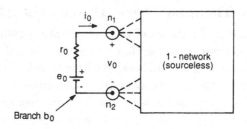

Figure 3.12. The connections for measuring a driving-point resistance $R_0 = e_0/i_0$ or $R_D = v_0/i_0$ for a 1-network. n_1 and n_2 may be either 0-nodes or 1-nodes.

said reduced network at the nodes of those sources. According to the corollary, the currents are distributed throughout the reduced network to minimize the power dissipated therein.

In the dual case the following results can be established in just the same way. Assume now that all the branches are in the Norton form and let

$$\mathcal{M}_h = \{\mathbf{u} \in \mathcal{V} : \langle \mathbf{u}, \mathbf{h} \rangle = \langle \mathbf{v}, \mathbf{h} \rangle\}.$$

Theorem 3.8-3: *Under Conditions 3.5-1, the voltage vector* $\mathbf{v}$ *specified by Theorem 3.5-2 is that member* $\mathbf{u} \in \mathcal{M}_h$ *for which* $\sum u_j^2 g_j$ *is a minimum. There is only one such member of* $\mathcal{M}_h$.

Now, let

$$\mathcal{N}_h = \{\mathbf{u} \in \mathcal{V} : u_j = v_j \text{ whenever } h_j \neq 0\}.$$

Clearly, $\mathcal{N}_h$ is a subset of $\mathcal{M}_h$. Thus, we have

Corollary 3.8-4: *For any* $\mathbf{u} \in \mathcal{N}_h$ *and for* $\mathbf{v}$ *given by Theorem 3.5-2,* $\sum u_j^2 g_j \geq \sum v_j^2 g_j$. *Strict inequality occurs if* $\mathbf{u} \neq \mathbf{v}$.

3.9 The Concavity of Driving-point Resistances

A basic concept for finite electrical networks is that of the "driving-point resistance." It is the ratio of the voltage to the current in a source branch when the rest of the network has no sources. This idea can be extended to 0-networks and 1-networks. For definiteness, consider a 1-network as shown in Figure 3.12. A source branch b_0 is connected between the nodes n_1 and n_2, which may be either 0-nodes or 1-nodes. We have drawn Figure 3.12 as though both n_1 and n_2 are 1-nodes embracing the elementary tips of b_0. Thus, a 1-network can have driving-point resistances even at its infinite extremities. We define two *driving-point resistances*: $R_0 = e_0/i_0$ and $R_D = v_0/i_0 = R_0 - r_0$. R_0 (or R_D) is a function of all the branch resistances r_j including r_0 (respectively, excluding r_0). We shall prove that R_0 and R_D are *concave-downward* functions of those resistances; that is, with r_j and r'_j denoting positive resistances, we have

$$R_0(r_0 + r'_0, r_1 + r'_1, \ldots) \geq R_0(r_0, r_1, \ldots) + R_0(r'_0, r'_1, \ldots), \quad (3.22)$$

$$R_D(r_1 + r'_1, r_2 + r'_2, \ldots) \geq R_D(r_1, r_2, \ldots) + R_D(r'_1, r'_2, \ldots). \quad (3.23)$$

This will extend a result of Shannon and Hagelbarger [93], [121] to 1-networks.

Let $\mathcal{K}$ now be the usual space of allowable branch-current vectors for the network obtained by augmenting the 1-network with the branch b_0, as indicated in Figure 3.12. We will need the condition that $i_0 \neq 0$ whenever $e_0 \neq 0$.

Lemma 3.9-1: *Assume that $e_0 \neq 0$ and that $e_j = 0$ for $j = 1, 2, \ldots$. Then, for the current vector $\mathbf{i} = (i_0, i_1, \ldots)$ dictated by Theorem 3.3-5, $i_0 \neq 0$ if and only if $\mathcal{K}$ contains at least one current vector $\mathbf{f} = (f_0, f_1, \ldots)$ with $f_0 \neq 0$.*

Proof: The "only if" statement being obvious, consider the "if" statement. Suppose $\mathcal{K}$ possesses such an $\mathbf{f}$ and yet $i_0 = 0$. Then, $e_0 i_0 = 0$. Set $\mathbf{s} = \mathbf{i}$ in (3.9). This yields $e_0 i_0 = \sum i_j^2 r_j$. Hence, $i_j = 0$ for every j. Thus, by (3.9) with $\mathbf{s} = \mathbf{f}$, we have $e_0 f_0 = \sum r_j i_j f_j = 0$. Therefore, $e_0 = 0$; this contradicts the hypothesis that $e_0 \neq 0$. ♣

The physical significance of the condition on $\mathcal{K}$ is that the 1-network of Figure 3.12 does not behave like an open-circuit between the nodes n_1 and n_2.

Theorem 3.9-2: *Let $R_0 = e_0/i_0$ be measured in accordance with Figure 3.12, where $e_0 \neq 0$, the 1-network has no sources, and the current regime is dictated by Theorem 3.3-5. Assume that $\mathcal{K}$ has at least one member $\mathbf{f}$ with $f_0 \neq 0$. Then, R_0 is a concave-downward function of the r_j, that is, (3.22) holds for all positive r_j and r'_j.*

Proof: Choose any real nonzero number I_0 and fix it. By Lemma 3.9-1, we can always adjust e_0 to make $i_0 = I_0$; indeed, just multiply $\mathbf{e}$ and $\mathbf{i}$ in (3.9) by I_0/i_0.

Next, choose any collection of positive resistances r_j and r'_j for all $j = 0, 1, 2, \ldots$. Consider three cases where in each case e_0 has been adjusted to make $i_0 = I_0$. In the first case, the branch resistances are

$r_j + r_j'$, the adjusted value of e_0 is denoted by E_0, the branch currents are I_j, and

$$R_0(r_0 + r_0', r_1 + r_1', \ldots) = \frac{E_0}{I_0} = \sum (r_j + r_j')\frac{I_j^2}{I_0^2}.$$

The last equation is obtained by setting $\mathbf{s} = \mathbf{I} = (I_0, I_1, \ldots)$ in (3.9) and then dividing by I_0^2. In the second and third cases, the branch resistances are r_j and respectively r_j', the adjusted values of e_0 are denoted by e_0 and e_0', the branch currents are i_j and i_j', and

$$R_0(r_0, r_1, \ldots) = \frac{e_0}{I_0} = \sum r_j \frac{i_j^2}{I_0^2}$$

respectively

$$R_0(r_0', r_1', \ldots) = \frac{e_0'}{I_0} = \sum r_j' \frac{i_j'^2}{I_0^2}.$$

For the second case, $\mathbf{I}$ is not in general equal to the current regime $\mathbf{i}$ but is a member of $\mathcal{J}_e$ (see Section 3.8). So, by Corollary 3.8-2, $\sum r_j I_j^2 \geq \sum r_j i_j^2$. In the same way, we get $\sum r_j' I_j^2 \geq \sum r_j' i_j'^2$. Hence,

$$R_0(r_0 + r_0', r_1 + r_1', \ldots) = \sum r_j \frac{I_j^2}{I_0^2} + \sum r_j' \frac{I_j^2}{I_0^2}$$

$$\geq \sum r_j \frac{i_j^2}{I_0^2} + \sum r_j' \frac{i_j'^2}{I_0^2} = R_0(r_0, r_1, \ldots) + R_0(r_0', r_1', \ldots),$$

as asserted. ♣

With regard to $R_D = v_0/i_0$, we need merely subtract r_0 from R_0 to obtain

Corollary 3.9-3: *Under the hypothesis of Theorem 3.9-2, (3.23) holds.*

Since we are dealing with an infinite network, the nodes n_1 and n_2 in Figure 3.12 may be effectively shorted together by the network, as we shall see in Section 3.11. Thus, a pure voltage source of value v_0 applied to n_1 and n_2 may yield an infinite i_0. This is why our result on R_D was derived from R_0 rather than directly.

The concavity property continues to hold even when some of the r_j' are zero. The same proof holds except that in the third case we would be dealing with a network some of whose branches have zero resistance. Instead of this, we will short together the nodes of any such branch to make it into a self-loop; in the resulting network those self-loop branches, none of which have sources, will carry zero currents when e_0 is applied as in Figure 3.12. Thus, we can still invoke Corollary 3.8-2 for the latter network to write $\sum r'_j I^2_j \geq \sum r'_j i'^2_j$, as needed.

Corollary 3.9-4: *Theorem 3.9-2 continues to hold even when some of the r'_j are equal to zero.*

Another corollary is *Rayleigh's monotonicity law*, which was first proposed back in the nineteenth century [111, page 94]. For 1-networks we have

Corollary 3.9-5: *Assign two, possibly different, resistance values r_j and r'_j to each branch of the 1-network of Figure 3.12 such that $r_0 = r'_0$ and $r_j \geq r'_j$ for $j = 1, 2, \ldots$. Then,*

$$R_0(r_0, r_1, r_2, \ldots) \geq R_0(r_0, r'_1, r'_2, \ldots)$$

and

$$R_D(r_1, r_2, \ldots) \geq R_D(r'_1, r'_2, \ldots).$$

All these results extend to the dual approach where now the R's and r's are replaced by G's and g's.

3.10 Nondisconnectable 0-Tips

We saw in Example 3.2-1 (Figure 3.2) and Example 3.2-2 (Figure 3.3) that relatively simple infinite graphs can have a continuum of 0-tips. Consequently, the 1-nodes of an infinite graph may also comprise a continuum, although the number of specified nonsingleton 1-nodes may be just a few. On the other hand, a common procedure in the theory of finite networks is to assign voltages to nodes by choosing some voltage for one of the nodes n_0 and then computing voltages at other nodes connected to n_0 by summing branch voltages along paths. Two questions arise: Can a similar procedure be applied to 0-tips and thereby to 1-nodes? If so, how many different 0-tip voltages can there be? The answer to the first question is yes, but a proper response to the second one is more complicated. The number of different 0-tip voltages may be countable or merely finite even though there is a continuum of 0-tips. What can happen is that the network itself forces many of the 0-tips to have the same voltage even though they are not shorted together as parts of a 1-node. In this section we will discuss one way this may occur, and in the next section a more general effect will be examined.

First of all, we should explicate what we mean by a "0-tip voltage" in a 1-network. Any 1-node that embraces a 0-tip t will by definition have the same voltage as t. As before, a 0-path or 1-path P is said to be *perceptible* if $\sum_{j\in\Pi} r_j < \infty$, where Π is the index set for the branches embraced by P. Given any 0-tip t, let x be a 1-node that embraces t. Also, let n be either a 0-node or a 1-node different from x. If there exists a perceptible 0-path or 1-path P that ends at both x and n, then t (and x as well) is said to be *perceptible from n*. Note that P need not contain a representative of t. Assign a voltage u to n and call it a *node voltage*. Then, t (or x) is said to have a *tip voltage* (or a *node voltage*) *with respect to n* if t (or x) is perceptible from n; that tip or node voltage is defined to be

$$u + \sum_{j\in\Pi} \pm v_j \qquad (3.24)$$

The plus (minus) sign is used if the jth branch's orientation agrees (disagrees) with a tracing of P from x to n. This definition has a meaning because the series converges absolutely, as is shown by the proof of Theorem 3.4-3. Moreover, t's voltage is independent of the choice of the path P that meets both x and n so long as the path is perceptible; this follows from Kirchhoff's voltage law which holds for perceptible loops, according to Theorem 3.4-3 again.

Now assume that two 0-tips t_a and t_b are both perceptible from the same node n with a given voltage u. Then, t_a and t_b have voltages v_a and v_b in accordance with (3.24). We shall argue that v_a and v_b must be equal if t_a and t_b have perceptible representatives P_a and P_b respectively that meet infinitely often in the sense that there exists an infinite sequence $\{n_1, n_2, \ldots\}$ of 0-nodes n_k which are all embraced by P_a and also by P_b. When this is so, we shall say that t_a and t_b are *not disconnectable* or *nondisconnectable*. This means in effect that the two representatives cannot be isolated from each other by removing branches. Note also that, if the stated conditions are satisfied by a particular pair P_a and P_b, then they will be satisfied by all pairs of representatives for t_a and t_b.

Theorem 3.10-1: *Let the node n of a 1-network have an assigned voltage. Let t_a and t_b be two 0-tips which are perceptible from n, have perceptible representatives, and are not disconnectable. Then, they both have voltages v_a and v_b with respect to n, and moreover $v_a = v_b$.*

Proof: The existence of v_a and v_b is ensured by the perceptibility of t_a and t_b from n. Remember that the paths along which t_a and t_b are perceptible from n need not contain representatives of t_a and t_b. Nonetheless, we can choose any such representatives P_a and P_b, and they will be perceptible and will meet infinitely often along a sequence $\{n_1, n_2, \ldots\}$

of 0-nodes n_k. The voltages at the n_k are determined from v_a and v_b by formulas such as (3.24), and the absolute convergence of the series in those formulas ensures that both v_a and v_b are the limit of the voltages at the n_k. Thus, $v_a = v_b$. ♣

Example 3.10-2. A consequence of Theorem 3.10-1 is that the number of different 0-tip voltages a 1-network can have may be radically limited by the structure of its graph. A simple example is provided by Figure 3.3; although there is a continuum of 0-tips in that network, all 0-tip voltages must be the same if they exist at all.

More generally, an m-times chainlike 0-graph can have a continuum of 0-tips. However, one can choose m, but no more than m, pairwise node-disjoint, one-ended 0-paths in such a graph. Hence, if a voltage-current regime exists according to Theorem 3.3-5, there can be m (or less) different 0-tip voltages. Any other one-ended 0-path must be nondisconnectable from at least one of the chosen 0-paths. Therefore, by Theorem 3.10-1, there can be no more than m different 0-tip voltages. In fact, when all m 0-tip voltages exist, the 0-node voltages along any one-ended 0-path must either converge to one of the m 0-tip voltages or oscillate indefinitely. ♣

3.11 Effectively Shorted 0-Tips

We turn our attention now to 0-tips that are *disconnectable* (i.e., not nondisconnectable). Such tips may or may not have voltages in accordance with (3.24), but, when they do, those voltages are in general different from one another. However, under certain circumstances, the graph and the resistance values of the network may force two such tip voltages to be the same. In effect, the network behaves as though the two tips have been "shorted together."

Example 3.11-1. Consider the semi-infinite grid of Figure 3.13, wherein all resistance values vary only in the vertical direction and there are only a finite number of voltage sources e_j, which we take to be in the uppermost branches. Assume that all 1-nodes are singletons (i.e., open circuits exist everywhere at infinity). Assign a zero voltage to the infinite 0-node at the top denoted by the "ground" symbols. If the sum $\sum_{l=0}^{\infty} r_l$ of resistances r_l along any vertical path converges, then every 0-tip with a strictly vertical representative has a voltage, that is, the node voltages along any vertical path converge in the downward direction. If in addition every horizontal resistance ρ_l is 1 Ω, then those vertically downward 0-tips must have the same voltage, as we shall show below.

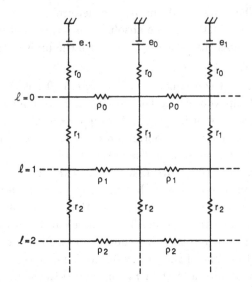

Figure 3.13. A semi-infinite grid. The r_l and ρ_l denote resistances that vary only in the vertical direction. As always, the e_j denote voltage sources.

On the other hand, if the horizontal resistances increase rapidly enough to yield $\sum_{l=0}^{\infty} \rho_l^{-1} < \infty$, then the vertically downward 0-tips can have different voltages [181], as will be shown in Chapter 7. ♣

Let us return to an arbitrary 1-network, where Theorem 3.3-5 is understood to prevail again. Here are some conditions that force two disconnectable 0-tips to have the same voltage.

Theorem 3.11-2: *Let two 0-tips t_a and t_b be perceptible from a 0-node or 1-node n with a given voltage. Let $\{n_1, n_2, \ldots\}$ (and $\{m_1, m_2, \ldots\}$) be an infinite sequence of 0-nodes embraced by a perceptible representative P_a of t_a (respectively, a perceptible representative P_b of t_b). Assume that, for each k, n_k and m_k are connected by a 0-path or 1-path Q_k that terminates at n_k and m_k. Let J_k denote the branch index set for Q_k and assume that $\min J_k = \min \{j : j \in J_k\} \to \infty$ as $k \to \infty$. Assume furthermore that*

$$\sum_{j \in J_k} r_j < A \tag{3.25}$$

where A is a constant independent of k. Then, t_a and t_b have the same voltage with respect to n.

Proof: We proceed as in the proof of Theorem 3.10-1, but now, instead of coincident nodes, we have the paths Q_k. The voltages at the ends of Q_k differ by

$$\sum_{j \in J_k} \pm (e_j - i_j r_j). \qquad (3.26)$$

The plus (minus) sign is chosen if that branch's orientation agrees (disagrees) with a tracing from n_k to m_k along Q_k. By virtue of the proof of Theorem 3.10-1, all we need to show is that (3.26) tends to zero as $k \to \infty$.

Let $\sum_{(k)}$ denote $\sum_{j \in J_k}$. Then,

$$\sum_{(k)} |i_j r_j| \le \sum_{(k)} |i_j| r_j^{1/2} r_j^{1/2} \le \left[\sum_{(k)} i_j^2 r_j \sum_{(k)} r_j \right]^{1/2}.$$

But $\sum_{j=0}^{\infty} i_j^2 r_j < \infty$ because $\mathbf{i} \in \mathcal{K}$ by Theorem 3.3-5. Since $\min J_k \to \infty$ as $k \to \infty$, it follows that $\sum_{(k)} i_j^2 r_j \to 0$ as $k \to \infty$. In view of (3.25), we can conclude that $\sum_{(k)} |i_j r_j| \to 0$.

Similarly, upon setting $g_j = r_j^{-1}$, we may write

$$\sum_{(k)} |e_j| = \sum_{(k)} |e_j| g_j^{1/2} r_j^{1/2} \le \left[\sum_{(k)} e_j^2 g_j \sum_{(k)} r_j \right]^{1/2}.$$

By Conditions 3.3-1, $\sum_{j=0}^{\infty} e_j^2 g_j < \infty$. So, as $k \to \infty$, $\sum_{(k)} e_j^2 g_j \to \infty$ because $\min J_k \to \infty$. Again by (3.25), $\sum_{(k)} |e_j| \to 0$. ♣

Chapter 4

Finite-power Regimes:
The Nonlinear Case

The adjective "nonlinear" will be used inclusively by taking "linear" to be a special case of "nonlinear." As promised, we present in this chapter two different theories for nonlinear infinite networks. The first one is due to Dolezal and is very general in scope – except that it is restricted to 0-networks. It is an infinite-dimensional extension of the fundamental theory for scalar, finite, linear networks [67], [115], [127]. In particular, it examines nonlinear operator networks, whose voltages and currents are members of a Hilbert space $\mathcal{H}$; in fact, infinite networks whose parameters can be nonlinear, multivalued mappings restricted perhaps to subsets of $\mathcal{H}$ are encompassed. As a result, virtually all the different kinds of parameters encountered in circuit theory – resistors, inductors, capacitors, gyrators, transformers, diodes, transistors, and so forth – are allowed. However, there is a price to be paid for such generality: Its existence and uniqueness theorems are more conceptual than applicable, because their hypotheses may not be verifiable for particular infinite networks. (In the absence of coupling between branches, the theory is easy enough to apply; see Corollary 4.1-7 below.) Nonetheless, with regard to the kinds of parameters encompassed, this is the most powerful theory of infinite networks presently available. Dolezal has given a thorough exposition of it in his two books [40], [41]. However, since no book on infinite electrical networks would be complete without some coverage of Dolezal's work, we shall present a simplified version of his theory. Our principal simplification is that the voltages and currents are taken to be real numbers and our parameters are restricted to nonlinear single-valued mappings of the real line R^1 into R^1. Although this radically restricts the generality of Dolezal's approach, our exposition reproduces his key concepts and may serve as a more accessible introduction to a sophisticated theory.

The second theory is of more recent origin and is due to DeMichele and Soardi [37]. Its approach is quite different from Dolezal's in that it mimics Thomson's least-power principle to establish the existence of a unique voltage-current regime. However, instead of minimizing $\sum r_j(i_j)i_j$, where $r_j(\cdot)$ is the nonlinear function corresponding to the jth branch resistance, it minimizes $\sum \int_0^{i_j} r_j(x)\,dx$, an idea suggested by Thomassen [129, Section 2] based on the work of Minty [95] for finite networks. For linear networks the two minimizations amount to the same thing, but they are different in the nonlinear case. This theory is restricted to resistive networks without coupling between branches. Moreover, it uses some results from an unusual part of functional analysis – the theory of modular sequence spaces. Since the literature on that subject is scattered among a few books, journals, and theses, some of which are rather inaccessible, we have chosen to present a self-contained development of just those results that we shall need, before moving on to the DeMichele–Soardi theorem. That theorem was stated for locally finite networks, which implicitly were shorted everywhere at infinity [37]. We will rework the theory to allow both infinite nodes and arbitrary connections at infinity. The DeMichele–Soardi theory does not impose lower bounds on the slopes of all the resistors – as does Dolezal's theory – but it does require that the resistance functions have odd symmetry and imposes a uniform-variation condition in a neighborhood of the origin on all but a finite number of the resistors. In neither theory need upper bounds on the said slopes be imposed. Furthermore, we impose stronger conditions on the branch resistances than do DeMichele and Soardi; in this respect the exposition given herein is not as general as their original work.

4.1 Regular Networks

Dolezal's theory is the subject of this and the next section, throughout which the following Conditions 4.1-1 are assumed. As always, the branches are indexed by $j = 1, 2, \ldots$. It will now be more convenient to denote the range value $\mathbf{v}$ of a nonlinear operator $R : \mathbf{i} \mapsto \mathbf{v}$ by $\mathbf{v} = R\mathbf{i}$ rather than by $\mathbf{v} = R(\mathbf{i})$.

Conditions 4.1-1: *Let* N *be a 0-network having a countably infinite set of branches and a finite or countably infinite set of 0-nodes. Also, let the following conditions be satisfied. Each branch is in its Thevenin form with coupling between branches being allowed. The voltage-source vector* $\mathbf{e} = (e_1, e_2, \ldots)$ *is a member of Hilbert's real coordinate space* l_{2r}, *that is,* $\sum e_j^2 < \infty$. *The branch-voltage vector* $\mathbf{v} = (v_1, v_2, \ldots)$ *is related to* $\mathbf{e}$ *and the branch-current vector* $\mathbf{i} = (i_1, i_2, \ldots)$ *according to* $\mathbf{v} = \mathbf{e} - R\mathbf{i}$, *where the resistance operator* R *is a nonlinear, single-valued mapping of* l_{2r} *into* l_{2r}.

Thus, when $\mathbf{i} \in l_{2r}$, we have $\mathbf{v} = \mathbf{e} - R\mathbf{i} \in l_{2r}$ too. Even though $\mathbf{v}$, $\mathbf{e}$, and $\mathbf{i}$ are one-sided infinite vectors, R need not have an infinite matrix form, but in the special case when it does we can write $R\mathbf{i} = [R_{jm}][i_m]$, where $[i_m]$ denotes the vector $\mathbf{i}$ written in column-vector form. Thus, by the usual rules for matrix multiplication we have that the jth component of the vector $R\mathbf{i}$ is $\sum R_{jm}i_m$, where each $R_{jm} : i_m \mapsto R_{jm}i_m$ is in general a nonlinear function; this means that the jth branch voltage is additive in the branch coupling terms, where R_{jm} is the coupling that relates an additive voltage term in the jth branch to the mth branch current i_m. In the still more specialized case where R is linear, we have $\sum R_{jm}i_m = \sum r_{jm}i_m$, where now r_{jm} is just a real number that multiplies with i_m.

Let us return to the general case where R need not have a matrix form. The problem at hand is to find a branch-current vector $\mathbf{i}$ for which Kirchhoff's two laws are satisfied in some way. For this purpose, we follow the fundamental theory for finite networks more closely than we did in the preceding chapter by introducing now the *incidence matrix* C. This is an infinite matrix with a countable infinity of columns corresponding to the branches b_j and a finite number or countable infinity of rows corresponding to the 0-nodes n_k, $k = 1, 2, \ldots$. The entry C_{kj} in the kth row and jth column of C is $+1$ if b_j is oriented toward n_k, -1 if b_j is oriented away from n_k, and 0 if b_j is either a self-loop or is not incident to n_k. In Dolezal's theory graph-theoretic methods are replaced by operator-theoretic methods by using C, modified to render it applicable to any member of l_{2r}. In particular, let D be a diagonal infinite matrix whose consecutive diagonal entries are $d_1, d_2, \ldots$; set $d_k = 0$ if n_k is an infinite node and choose the remaining d's such that the infinite matrix $a = [a_{kj}] = DC$ has the property that $\sum_k \sum_j a_{kj}^2 = \sum_k d_k^2 \sum_j C_{kj}^2 < \infty$. By virtue of the Schwarz inequality, this ensures that a defines a continuous linear mapping $\hat{a}$ of l_{2r} into R^n or l_{2r} according to the matrix multiplication rule, whereby the kth component of $\hat{a}\mathbf{x}$, $\mathbf{x} \in l_{2r}$, is $\sum_j a_{kj}x_j$. As was stated in Section 1.1, we shall henceforth use the same symbol a to represent both a and $\hat{a}$.

Let $\mathcal{N}_a = \{\mathbf{x} \in l_{2r} : a\mathbf{x} = \mathbf{0}\}$, that is, $\mathcal{N}_a$ is the null space of a. By the continuity of a, $\mathcal{N}_a$ is closed and therefore a Hilbert space in itself. So too is its orthogonal complement $\mathcal{N}_a^\perp$ in l_{2r}. Moreover, $\mathcal{N}_a$ does not depend on the choice of D because $\mathbf{x} \in \mathcal{N}_a$ if and only if its components x_j, when taken as the branch currents, satisfy Kirchhoff's current law at every finite node.

We now let $(\cdot, \cdot)$ denote the inner product for l_{2r}. Under Conditions 4.1-1, we shall say that the network $\mathbf{N}$ is *regular* and that $\mathbf{i}$ *is the current vector corresponding to* $\mathbf{e}$ if, for every $\mathbf{e} \in l_{2r}$, there is a unique current vector $\mathbf{i} \in \mathcal{N}_a$ such that

$$(\mathbf{e} - R\mathbf{i}, \mathbf{s}) = 0 \tag{4.1}$$

for every $\mathbf{s} \in \mathcal{N}_a$. This is equivalent to asserting that $\mathbf{i} \in \mathcal{N}_a$, $\mathbf{e} - R\mathbf{i} \in \mathcal{N}_a^\perp$, and $\mathbf{i}$ is unique. We call the mapping $A : \mathbf{e} \mapsto \mathbf{i}$ the *admittance operator* of the regular network $\mathbf{N}$. Once again, since $\mathcal{N}_a$ does not depend on the choice of D, neither do $\mathbf{i}$ nor A depend upon D. Moreover, by setting $\mathbf{s} = \mathbf{i}$, we see that the total power $(\mathbf{e}, \mathbf{i})$ generated by the voltage sources is finite and is equal to the total dissipated power $(R\mathbf{i}, \mathbf{i})$.

Example 4.1-2. Here's a simple illustration of how infinite nodes are incorporated into the definition of a regular network. Consider the infinite parallel circuit of Figure 1.4(a). Number all the branches by $j = 0, 1, 2, \ldots$ proceeding from left to right. Since both nodes are infinite, a is the $2 \times \aleph_0$ matrix all entries of which are zero. Hence $\mathcal{N}_a = l_{2r}$, which means that (4.1) should hold for every $\mathbf{s} \in l_{2r}$. Consequently, $\mathbf{e} - R\mathbf{i} = \mathbf{0}$ or equivalently $e_0 = r_0 i_0$ and $r_j i_j = 0$ for $j = 1, 2, \ldots$. Since e_0 and all resistors have unit values, $i_0 = 1$ A and $i_j = 0$ A for $j = 1, 2, \ldots$. This is precisely the solution obtained in Example 3.4-2. ♣

Let $\tilde{R}$ denote the restriction of R to $\mathcal{N}_a$, and let P be the orthogonal projection of l_{2r} onto $\mathcal{N}_a$. The composed operator PR, obtained by first applying R and then P, maps l_{2r} into $\mathcal{N}_a$. Its restriction $P\tilde{R}$ to $\mathcal{N}_a$ may or may not be invertible on $\mathcal{N}_a$. It turns out that invertibility is a necessary and sufficient condition for $\mathbf{N}$ to be regular. The invertibility of $P\tilde{R}$ on $\mathcal{N}_a$ means that PR is a bijection from $\mathcal{N}_a$ onto $\mathcal{N}_a$.

Theorem 4.1-3: *Under Conditions 4.1-1, $\mathbf{N}$ is regular if and only if the restricted operator $P\tilde{R} : \mathcal{N}_a \rightsquigarrow \mathcal{N}_a$ possesses an inverse $(P\tilde{R})^{-1} : \mathcal{N}_a \rightsquigarrow \mathcal{N}_a$. In this case, the admittance operator is given by*

$$A = (P\tilde{R})^{-1}P. \tag{4.2}$$

Proof: *If*: Assume that the restricted operator $P\tilde{R}$ has an inverse $(P\tilde{R})^{-1}$ on $\mathcal{N}_a$. With A defined by (4.2), set $\mathbf{i} = A\mathbf{e}$, whatever be $\mathbf{e} \in l_{2r}$. Since A maps l_{2r} into $\mathcal{N}_a$, we have that $\mathbf{i} \in \mathcal{N}_a$. Moreover, for any $\mathbf{s} \in \mathcal{N}_a$ we have $P\mathbf{s} = \mathbf{s}$ and

$$(\mathbf{e} - R\mathbf{i}, \mathbf{s}) = (\mathbf{e} - \tilde{R}(P\tilde{R})^{-1}P\mathbf{e}, P\mathbf{s}).$$

Now, an orthogonal projection is self-adjoint, and therefore the right-hand side can be rewritten as

$$(P\mathbf{e} - P\tilde{R}(P\tilde{R})^{-1}P\mathbf{e}, \mathbf{s}) = 0.$$

Thus, $\mathbf{i} = A\mathbf{e}$ satisfies (4.1).

We have yet to show that $\mathbf{i}$ is unique. Suppose $\mathbf{i}$ and $\mathbf{x}$ are both in $\mathcal{N}_a$ and satisfy (4.1). It follows that $(R\mathbf{i} - R\mathbf{x}, \mathbf{s}) = 0$ for all $\mathbf{s} \in \mathcal{N}_a$. Moreover, since $\mathbf{i}$ and $\mathbf{x}$ are both in $\mathcal{N}_a$, we also have

$$0 = (\tilde{R}\mathbf{i} - \tilde{R}\mathbf{x}, \mathbf{s}) = (\tilde{R}\mathbf{i} - \tilde{R}\mathbf{x}, P\mathbf{s}) = (P\tilde{R}\mathbf{i} - P\tilde{R}\mathbf{x}, \mathbf{s}).$$

But $P\tilde{R}\mathbf{i}$ and $P\tilde{R}\mathbf{x}$ are also in $\mathcal{N}_a$, and therefore we can set $\mathbf{s} = P\tilde{R}\mathbf{i} - P\tilde{R}\mathbf{x}$ to conclude that $P\tilde{R}\mathbf{i} = P\tilde{R}\mathbf{x}$. Upon applying $(P\tilde{R})^{-1}$, we obtain $\mathbf{i} = \mathbf{x}$, as desired. We have shown that **N** is regular and has the admittance operator (4.2).

Only if: Now assume that **N** is regular and therefore has an admittance operator $A : \mathbf{e} \mapsto \mathbf{i}$, where, for any given $\mathbf{e} \in l_{2r}$, $\mathbf{i}$ is in $\mathcal{N}_a$ and is the only member of $\mathcal{N}_a$ for which $\mathbf{e} - R\mathbf{i} \in \mathcal{N}_a^{\perp}$. We are going to prove that $P\tilde{R} : \mathcal{N}_a \rightsquigarrow \mathcal{N}_a$ is invertible. The above "if" argument will then ensure that A is given by (4.2).

Suppose that $\mathbf{i}$ and $\mathbf{x}$ are in $\mathcal{N}_a$ and that $P\tilde{R}\mathbf{i} = P\tilde{R}\mathbf{x} = \mathbf{y}$. Thus, $\mathbf{y} \in \mathcal{N}_a$ too. Then, for each $\mathbf{s} \in \mathcal{N}_a$,

$$(\mathbf{y} - R\mathbf{i}, \mathbf{s}) = (\mathbf{y} - R\mathbf{i}, P\mathbf{s}) = (P\mathbf{y} - P\tilde{R}\mathbf{i}, \mathbf{s}) = (\mathbf{y} - P\tilde{R}\mathbf{i}, \mathbf{s}) = 0.$$

In the same way, $(\mathbf{y} - R\mathbf{x}, \mathbf{s}) = 0$ too. This shows that $\mathbf{i}$ and $\mathbf{x}$ are both current vectors corresponding to $\mathbf{y}$. By the regularity of **N**, $\mathbf{i} = \mathbf{x}$. Hence, $P\tilde{R}$ is injective on $\mathcal{N}_a$.

To show that $P\tilde{R}$ is surjective onto $\mathcal{N}_a$, choose any $\mathbf{y} \in \mathcal{N}_a$ and set $\mathbf{i} = A\mathbf{y}$, where A is the admittance operator for **N**. Again by the regularity of **N**, $\mathbf{i} \in \mathcal{N}_a$ and $(\mathbf{y} - R\mathbf{i}, \mathbf{s}) = 0$ for all $\mathbf{s} \in \mathcal{N}_a$. Consequently,

$$0 = (\mathbf{y} - R\mathbf{i}, P\mathbf{s}) = (P\mathbf{y} - P\tilde{R}\mathbf{i}, \mathbf{s}) = (\mathbf{y} - P\tilde{R}\mathbf{i}, \mathbf{s}).$$

We can set $\mathbf{s} = \mathbf{y} - P\tilde{R}\mathbf{i} \in \mathcal{N}_a$ to conclude that $\mathbf{y} = P\tilde{R}\mathbf{i}$. This completes the argument that $P\tilde{R}$ is surjective, indeed, invertible. ♣

Corollary 4.1-4: *Let* **N** *satisfy Conditions 4.1-1 and be regular. If R is linear, then A is linear; if R is continuous and linear, then A is continuous and linear.*

Proof: The inverse of a linear operator on a Hilbert space is linear [94, page 101]. Our first assertion follows from this fact. Also, boundedness and continuity amount to the same thing for a linear operator on a Hilbert space [94, page 408]. So, assume that R is both linear and bounded. Now, $||P|| = 1$ since P is a projection. Hence, $P\tilde{R}$ is bounded and linear on $\mathcal{N}_a$. But the inverse of a bounded linear operator is also bounded and linear [94, page 416]. Thus, $(P\tilde{R})^{-1}$ and therefore $A = (P\tilde{R})^{-1}P$ are also bounded linear mappings. ♣

We now turn to two different sufficient conditions on PR, each of which ensure the invertibility of $P\tilde{R}$ on $\mathcal{N}_a$ and thereby the regularity of **N**. For this purpose, we introduce the notation $\text{Lip}_\alpha[\mathcal{H}, \mathcal{H}]$, where α is a real positive number and $\mathcal{H}$ is a real Hilbert space. $\text{Lip}_\alpha[\mathcal{H}, \mathcal{H}]$ denotes the set of all nonlinear (possibly linear) operators $F : \mathcal{H} \rightsquigarrow \mathcal{H}$ that satisfy the *Lipschitz condition* $||Fx - Fy|| \le \alpha||x - y||$ for all $x, y \in \mathcal{H}$.

Theorem 4.1-5: *Under Conditions 4.1-1, assume that there exist two real numbers μ and λ with $\mu \ne 0$ and $0 < \lambda < 1$ such that*

$$I - \mu PR \in \text{Lip}_\lambda[\mathcal{N}_a, \mathcal{N}_a] \tag{4.3}$$

*where I is the identity operator on $\mathcal{N}_a$. Then, **N** is regular, and its admittance operator A is in $\text{Lip}_\beta[l_{2r}, l_{2r}]$, where $\beta = |\mu|/(1 - \lambda)$.*

Proof: In accordance with Theorem 4.1-3, we want to show that, for any given $y \in \mathcal{N}_a$, there exists a unique $x \in \mathcal{N}_a$ such that $PRx = y$. To this end, define the operator $Q_y : \mathcal{N}_a \rightsquigarrow \mathcal{N}_a$ by $Q_y s = (I - \mu PR)s + \mu y$. So, for any $s, t \in \mathcal{N}_a$,

$$||Q_y s - Q_y t|| = ||(I - \mu PR)s - (I - \mu PR)t|| \le \lambda||s - t||$$

according to (4.3). Since $0 < \lambda < 1$, Q_y is contractive on $\mathcal{N}_a$ and hence has a unique fixed point $x \in \mathcal{N}_a$ [94, page 314]. Thus, $x = Q_y x = (I - \mu PR)x + \mu y$, from which it follows that $PRx = y$, where x is the only such member of $\mathcal{N}_a$. So truly, $P\tilde{R}$ is invertible on $\mathcal{N}_a$, and **N** is regular.

To obtain the assertion about the admittance operator A, let $x_1, x_2 \in \mathcal{N}_a$. A property of any norm is that $||s|| - ||t|| \le ||s - t||$. So,

$$||x_1 - x_2|| - ||\mu PRx_1 - \mu PRx_2|| \le ||x_1 - x_2 - (\mu PRx_1 - \mu PRx_2)||.$$

By (4.3), the right-hand side is no larger than $\lambda||x_1 - x_2||$. Hence,

$$(1 - \lambda)||x_1 - x_2|| \le |\mu| \, ||PRx_1 - PRx_2||.$$

Since $P\tilde{R}$ is invertible on $\mathcal{N}_a$, we may set $\mathbf{x}_1 = (P\tilde{R})^{-1}\mathbf{y}_1$ and $\mathbf{x}_2 = (P\tilde{R})^{-1}\mathbf{y}_2$, where $\mathbf{y}_1, \mathbf{y}_2 \in \mathcal{N}_a$. This yields

$$\|(P\tilde{R})^{-1}\mathbf{y}_1 - (P\tilde{R})^{-1}\mathbf{y}_2\| \leq \frac{|\mu|}{1-\lambda}\|\mathbf{y}_1 - \mathbf{y}_2\|.$$

Now, for any $\mathbf{w}_1, \mathbf{w}_2 \in l_{2r}$, we may write $\mathbf{y}_1 = P\mathbf{w}_1$, $\mathbf{y}_2 = P\mathbf{w}_2$, and $A = (P\tilde{R})^{-1}P$ to obtain

$$\|A\mathbf{w}_1 - A\mathbf{w}_2\| \leq \frac{|\mu|}{1-\lambda}\|P\mathbf{w}_1 - P\mathbf{w}_2\| \leq \frac{|\mu|}{1-\lambda}\|\mathbf{w}_1 - \mathbf{w}_2\|$$

since $\|P\| \leq 1$. ♣

The condition (4.3) asserts in effect that $P\tilde{R}$ is not too different from a multiple of the identity operator on $\mathcal{N}_a$. An alternative set of sufficient conditions for **N** to be regular places a Lipschitz condition directly on $P\tilde{R}$ in conjunction with a strong monotonicity condition on $\tilde{R}$ itself. In particular, an operator F that maps a real Hilbert space $\mathcal{H}$ into $\mathcal{H}$ is said to be *strongly monotone* if there is a real positive number γ such that, for all $x, y \in \mathcal{H}$,

$$(Fx - Fy, x - y) \geq \gamma\|x - y\|^2.$$

Theorem 4.1-6: *Under Conditions 4.1-1, assume there exist two positive numbers γ and ω such that, for all $\mathbf{x}_1, \mathbf{x}_2 \in \mathcal{N}_a$,*

$$(R\mathbf{x}_1 - R\mathbf{x}_2, \mathbf{x}_1 - \mathbf{x}_2) \geq \gamma\|\mathbf{x}_1 - \mathbf{x}_2\|^2 \tag{4.4}$$

and

$$PR \in \mathrm{Lip}_\omega[\mathcal{N}_a, \mathcal{N}_a]. \tag{4.5}$$

*Then, **N** is regular, and its admittance operator A is in $\mathrm{Lip}_\beta[l_{2r}, l_{2r}]$, where*

$$\beta = \frac{1}{\gamma\omega}\left(\omega + \sqrt{\omega^2 - \gamma^2}\right).$$

Proof: We first show that $I - \mu PR$ is contractive on $\mathcal{N}_a$ for a suitably chosen μ. For all $\mathbf{x}_1, \mathbf{x}_2 \in \mathcal{N}_a$ and by (4.4) and (4.5),

$$0 \leq \gamma\|\mathbf{x}_1 - \mathbf{x}_2\|^2 \leq (R\mathbf{x}_1 - R\mathbf{x}_2, \mathbf{x}_1 - \mathbf{x}_2) = (PR\mathbf{x}_1 - PR\mathbf{x}_2, \mathbf{x}_1 - \mathbf{x}_2)$$

$$\leq \|PR\mathbf{x}_1 - PR\mathbf{x}_2\|\,\|\mathbf{x}_1 - \mathbf{x}_2\| \leq \omega\|\mathbf{x}_1 - \mathbf{x}_2\|^2.$$

Hence, $\omega \geq \gamma$. We are free to choose $\omega > \gamma$. This we do. Moreover,

$$\|(I - \mu PR)\mathbf{x}_1 - (I - \mu PR)\mathbf{x}_2\|^2$$

$$= (\mathbf{x}_1 - \mathbf{x}_2 - \mu(PR\mathbf{x}_1 - PR\mathbf{x}_2), \mathbf{x}_1 - \mathbf{x}_2 - \mu(PR\mathbf{x}_1 - PR\mathbf{x}_2))$$

$$= \|\mathbf{x}_1 - \mathbf{x}_2\|^2 - 2\mu(PR\mathbf{x}_1 - PR\mathbf{x}_2, \mathbf{x}_1 - \mathbf{x}_2) + \mu^2\|PR\mathbf{x}_1 - PR\mathbf{x}_2\|^2$$

$$\leq (1 - 2\mu\gamma + \mu^2\omega^2)\|\mathbf{x}_1 - \mathbf{x}_2\|^2.$$

Let us choose μ to minimize the right-hand side; this occurs when $\mu = \gamma/\omega^2$. So, $I - \mu PR$ satisfies (4.3) with

$$\lambda = (1 - 2\mu\gamma + \mu^2\omega^2)^{1/2} = (1 - \gamma^2/\omega^2)^{1/2}.$$

Since $0 < \gamma < \omega$, we have $0 < \lambda < 1$. We can now invoke Theorem 4.1-5 to conclude that $\mathbf{N}$ is regular again.

That theorem also states that $A \in \mathrm{Lip}_\beta[l_{2r}, l_{2r}]$, where $\beta = |\mu|/(1-\lambda)$. Substituting $\mu = \gamma/\omega^2$ and $\lambda = (1 - \gamma^2/\omega^2)^{1/2}$, we obtain the asserted value for β. ♣

For a network $\mathbf{N}$ satisfying Conditions 4.1-1 but without coupling between branches, the resistance operator R has a diagonal-matrix form; that is, $R = [R_{jm}]$, where $R_{jm} = 0$ if $j \neq m$. Set $R_{jj} = R_j$ for every j. Thus, $R\mathbf{i} = (R_1 i_1, R_2 i_2, \dots)$ for every $\mathbf{i} \in l_{2r}$. Assume that $R_j 0 = 0$. Then, each R_j is described by a curve in the (i, v) plane, which passes through the origin and has a unique $v_j = R_j i_j \in R^1$ for each $i_j \in R^1$. We are going to show that, if the slopes of all these curves are uniformly bounded in a certain way, then $\mathbf{N}$ is regular no matter what its incidence matrix is; that is, its regularity does not now depend on its graph.

Corollary 4.1-7: *Let the 0-network $\mathbf{N}$ satisfy Conditions 4.1-1 and have no coupling between branches. Under the notation just defined, assume that $R_j 0 = 0$ for every j. Assume in addition that, whatever be j, every chord of the curve for R_j has a slope bounded below by γ and above by ω, where γ and ω are independent of j and $0 < \gamma < \omega < \infty$. Then, $\mathbf{N}$ is regular.*

Proof: For any $\mathbf{i}, \mathbf{x} \in l_{2r}$ and any integer $n > 0$,

$$\sum_{j=1}^{n} (R_j i_j - R_j x_j)^2 \leq \sum_{j=1}^{n} \omega^2 (i_j - x_j)^2 \leq \omega^2 \|\mathbf{i} - \mathbf{x}\|^2. \qquad (4.6)$$

Since $R_j 0 = 0$, $\sum_{j=1}^{n} (R_j i_j)^2 \leq \omega^2 \|\mathbf{i}\|^2$. Upon taking $n \to \infty$, we conclude that R maps l_{2r} into l_{2r}. Moreover,

$$\|R\mathbf{i} - R\mathbf{x}\|^2 = \sum (R_j i_j - R_j x_j)^2.$$

So, on taking $n \to \infty$ in (4.6), we see that $R \in \mathrm{Lip}_\omega[l_{2r}, l_{2r}]$. Since $\|P\| \leq 1$, $PR \in \mathrm{Lip}_\omega[\mathcal{N}_a, \mathcal{N}_a]$. Thus, (4.5) is satisfied. We may also write, for any $\mathbf{i}, \mathbf{x} \in l_{2r}$,

$$(R\mathbf{i} - R\mathbf{x}, \mathbf{i} - \mathbf{x}) = \sum (R_j i_j - R_j x_j)(i_j - x_j) \geq \sum \gamma (i_j - x_j)^2 = \gamma \|\mathbf{i} - \mathbf{x}\|^2.$$

Thus, (4.4) is satisfied too. Hence, the conditions of Theorem 4.1-6 are fulfilled, whatever be a. ♣

4.2 The Fundamental Operator

In the theory of finite electrical networks, the loop-impedance operator [34, page 724], [127] plays a fundamental role in that its invertibility ensures the existence of a voltage-current regime. Our restricted operator $P\tilde{R}$ plays the same role, but it does not have the form of a loop-impedance matrix. Our objective in this section is to construct an operator that is more closely analogous to the said matrix and to establish conditions on it that ensure the regularity of $\mathbf{N}$.

We start by choosing arbitrarily but then fixing an orthonormal basis $\{\mathbf{s}_1, \mathbf{s}_2, \ldots\}$ in $\mathcal{N}_a$. Since $\mathcal{N}_a$ is a Hilbert space in itself, such a basis exists. Thus, each $\mathbf{s}_m$ is a vector in l_{2r}; $a\mathbf{s}_m = \mathbf{0}$ for every m; $(\mathbf{s}_m, \mathbf{s}_n) = 0$ if $m \neq n$; $\|\mathbf{s}_m\| = 1$ for all m; and for any given member $\mathbf{x} \in \mathcal{N}_a$ we have the Fourier expansion $\mathbf{x} = \sum_m (\mathbf{x}, \mathbf{s}_m)\mathbf{s}_m$, where the summation either has a finite number of terms or converges under the l_{2r} norm. Moreover, we have Parseval's equation $\|\mathbf{x}\|^2 = \sum_m (\mathbf{x}, \mathbf{s}_m)^2$.

Let c denote the cardinality of $\{\mathbf{s}_1, \mathbf{s}_2, \ldots\}$; thus, c is either a positive integer or $\aleph_0$. We construct an infinite matrix X having a countable infinity of rows and c columns by choosing $\mathbf{s}_1, \mathbf{s}_2, \ldots$ as the consecutive columns. Moreover, we let l^c be either l_{2r} if $c = \aleph_0$ or c-dimensional Euclidean space if c is finite. In either case, we might try to define X as a mapping on l^c through the matrix multiplication of X with any $\mathbf{t} \in l^c$, if indeed the matrix product $X\mathbf{t}$ has a sense. (As always, to simplify notation, we use the same symbol X to represent both the matrix and the corresponding mapping.) Actually, $X\mathbf{t}$ does make sense, as can be seen by expanding $X\mathbf{t}$ through the rule for matrix multiplication, then expanding $\|X\mathbf{t}\|^2$, and finally invoking the orthonormality of the columns of X. This yields $\|X\mathbf{t}\| = \|\mathbf{t}\|_c$, where $\|\cdot\|_c$ denotes the norm in l^c. Consequently, X is a continuous linear mapping of l^c into l_{2r}. This and other results relating to X are listed in the next lemma.

Lemma 4.2-1: *X and its adjoint X^* have the following properties:*

(i) X *is a continuous linear mapping of l^c into l_{2r}, and $\|X\mathbf{t}\| = \|\mathbf{t}\|_c$ for every $\mathbf{t} \in l^c$.*

(ii) $aX = 0$ *on l^c.*

(iii) X *is a bijection from l^c onto $\mathcal{N}_a$.*

(iv) X^* *is a continuous linear mapping of l_{2r} into l^c, and $X^*X = I$, where I is the identity operator on l^c. Moreover, in matrix form X^* is the transpose X^T of X.*

(v)   *The null space $\mathcal{N}_{X^*}$ of X^* is the orthogonal complement $\mathcal{N}_a^\perp$ of $\mathcal{N}_a$, and X^* maps $\mathcal{N}_a$ onto l^c bijectively.*

(vi) $XX^* = P$, where as before P is the orthogonal projection of l_{2r} onto $\mathcal{N}_a$.

Proof: We have already argued (i), so consider (ii). For any $t \in l^c$, $Xt = \sum t_m s_m$ where the series converges in l_{2r} by virtue of the orthonormality of the s_m and the equality $\|Xt\| = \|t\|_c$. Since $s_m \in \mathcal{N}_a$ and $\mathcal{N}_a$ is closed in l_{2r}, $Xt \in \mathcal{N}_a$ too. Hence, $(aX)t = a(Xt) = 0$, which proves (ii) and also shows that X maps l^c into $\mathcal{N}_a$.

As for (iii), the equality $\|Xt\| = \|t\|_c$ for all $t \in l_c$ implies that X is injective from l^c into $\mathcal{N}_a$. On the other hand, for any $x \in \mathcal{N}_a$, the Fourier expansion $x = \sum t_m s_m$, $t_m = (x, s_m)$, in terms of the orthonormal basis $s_1, s_2, \ldots$ of $\mathcal{N}_a$, coupled with Parseval's equation shows that $t = (t_1, t_2, \ldots)$ is in l^c. Moreover, $x = \sum t_m s_m$ is simply the matrix product Xt. Hence, X is surjective from l^c onto $\mathcal{N}_a$.

Next, consider the adjoint operator X^* which is defined on l_{2r} by

$$(X^* x, t)_c = (x, Xt), \quad x \in l_{2r}, \ t \in l^c.$$

Here, $(\cdot, \cdot)_c$ denotes the inner product for l^c. Since X is a continuous linear mapping of l^c into l_{2r}, we have the standard result [94, page 422] that X^* is a continuous linear mapping of l_{2r} into l^c. That X^* is the transpose X^T of X is another standard result obtained by applying X^* and X to vectors having 1 as one component and 0's everywhere else. Thus, $X^* X = X^T X$. By the orthonormality of the columns of X, $X^T X = I$. Thereby we have (iv).

To obtain (v) we invoke another standard result [23, page 133], namely, the orthogonal complement of the null space of X^* is the closure of the range of X on l^c, that is, $\mathcal{N}_{X^*}^\perp = \overline{Xl^c}$, where the overbar denotes closure. By (iii) and the fact that $\mathcal{N}_a$ is closed, $\overline{Xl^c} = \mathcal{N}_a$. Thus, $\mathcal{N}_a^\perp = (\mathcal{N}_{X^*}^\perp)^\perp = \mathcal{N}_{X^*}$. Furthermore, the facts that X maps l^c onto $\mathcal{N}_a$ bijectively and that $X^* X = I$ on l^c imply that X^* must map $\mathcal{N}_a$ onto l^c bijectively.

With regard to (vi), consider XX^*. By (v), X^* maps $\mathcal{N}_a^\perp$ onto $\{0\}$ and bijectively maps $\mathcal{N}_a$ onto l^c. By (i), $X0 = 0$. By (iii), X maps l^c onto $\mathcal{N}_a$ bijectively. Thus, XX^* is the orthogonal projection P of l_{2r} onto $\mathcal{N}_a$. ♣

Conclusion (v) provides still another way of specifying the unique current vector $i \in l_{2r}$ corresponding to a given voltage-source vector $e \in l_{2r}$ in a regular network N: $i \in \mathcal{N}_a$ and $e - Ri \in \mathcal{N}_{X^*}$.

Using X, we define the operator $W = X^* RX$, which maps l^c into l^c according to Lemma 4.2-1(i) and (iv). W will play a role analogous to that of the loop-impedance matrix of a finite network. The critical condition will be that W be invertible on l^c, that is, a bijection from l^c onto l^c.

Theorem 4.2-2: *Under Condition 4.1-1,* **N** *is regular if and only if* W *is invertible on* l^c. *In this case, the admittance operator of* **N** *is* $A = XW^{-1}X^*$; *that is, for any* $\mathbf{e} \in l_{2r}$, $\mathbf{i} = XW^{-1}X\mathbf{e}$ *is the unique member of* l_{2r} *for which* $\mathbf{i} \in \mathcal{N}_a$ *and* $\mathbf{e} - R\mathbf{i} \in \mathcal{N}_a^{\perp}$.

Proof: *If:* We invoke various parts of Lemma 4.2-1. By (iii) we can let $Y : \mathcal{N}_a \rightsquigarrow l^c$ be the inverse of $X : l^c \rightsquigarrow \mathcal{N}_a$. Then, $X^*R = (X^*RX)Y = WY$ is a bijection from $\mathcal{N}_a$ onto l^c. Therefore, by (iii) again, XX^*R is a bijection from $\mathcal{N}_a$ onto $\mathcal{N}_a$. But by (vi), $XX^* = P$. So, by Theorem 4.1-3, **N** is regular.

Only if: This is argued simply by reversing the "if" argument.

Now consider $XW^{-1}X^*$. Choose $\mathbf{e} \in l_{2r}$ arbitrarily and set $\mathbf{i} = XW^{-1}X^*\mathbf{e}$. By (ii), (iv), and the hypothesis, we have $\mathbf{i} \in \mathcal{N}_a$. Moreover, $X^*(\mathbf{e} - R\mathbf{i}) = X^*\mathbf{e} - WW^{-1}X^*\mathbf{e} = 0$. Thus, $\mathbf{e} - R\mathbf{i} \in \mathcal{N}_{X^*}$. Also, $\mathcal{N}_{X^*} = \mathcal{N}_a^{\perp}$ according to (v). So, $\mathbf{i}$ is the current vector corresponding to $\mathbf{e}$. By the regularity of **N**, this $\mathbf{i}$ is unique. Hence, $A = XW^{-1}X^*$. ♣

We now rework Theorems 4.1-5 and 4.1-6 in terms of our fundamental operator W. The results will be two theorems more closely analogous to the fundamental theorem for finite networks. I will now denote the identity operator on whatever Hilbert space is at hand.

Theorem 4.2-3: *Let* **N** *satisfy Conditions 4.1-1. Assume that there exist two real numbers* μ *and* λ *with* $\mu \neq 0$ *and* $0 < \lambda < 1$ *such that*

$$I - \mu W \in \text{Lip}_{\lambda}[l^c, l^c]. \tag{4.7}$$

Then, **N** *is regular, and* $A \in \text{Lip}_{\beta}[l_{2r}, l_{2r}]$, *where* $\beta = |\mu|/(1 - \lambda)$.

Proof: We shall prove that $I - \mu P\tilde{R} \in \text{Lip}_{\lambda}[\mathcal{N}_a, \mathcal{N}_a]$, which in conjunction with Theorem 4.1-5 will prove our assertion. Choose any $\mathbf{x}_1, \mathbf{x}_2 \in \mathcal{N}_a$. By Lemma 4.2-1(iii), there exist $\mathbf{y}_1, \mathbf{y}_2 \in l^c$ such that $\mathbf{x}_1 = X\mathbf{y}_1$ and $\mathbf{x}_2 = X\mathbf{y}_2$. We now invoke the equations $\mathbf{x}_1 = P\mathbf{x}_1$, $\mathbf{x}_2 = P\mathbf{x}_2$, and Conditions (vi), (i), and (iv) of Lemma 4.2-1 to write

$$\|\mathbf{x}_1 - \mathbf{x}_2 - \mu P(R\mathbf{x}_1 - R\mathbf{x}_2)\| = \|XX^*[\mathbf{x}_1 - \mathbf{x}_2 - \mu(R\mathbf{x}_1 - R\mathbf{x}_2)]\| =$$
$$\|X^*[X\mathbf{y}_1 - X\mathbf{y}_2 - \mu(RX\mathbf{y}_1 - RX\mathbf{y}_2)]\|_c = \|\mathbf{y}_1 - \mathbf{y}_2 - \mu(W\mathbf{y}_1 - W\mathbf{y}_2)\|_c.$$

By (4.7), the last expression is bounded by $\lambda\|\mathbf{y}_1 - \mathbf{y}_2\|_c$. Moreover, by Lemma 4.2-1(i), $\|\mathbf{y}_1 - \mathbf{y}_2\|_c = \|X(\mathbf{y}_1 - \mathbf{y}_2)\| = \|\mathbf{x}_1 - \mathbf{x}_2\|$. Combining these results, we get $I - \mu P\tilde{R} \in \text{Lip}_{\lambda}[\mathcal{N}_a, \mathcal{N}_a]$, as desired. ♣

Theorem 4.2-4: *Let* **N** *satisfy Conditions 4.1-1. Assume there exist two positive numbers* γ *and* ω *such that, for all* $\mathbf{y}_1, \mathbf{y}_2 \in l^c$,

$$(W\mathbf{y}_1 - W\mathbf{y}_2, \mathbf{y}_1 - \mathbf{y}_2) \geq \gamma\|\mathbf{y}_1 - \mathbf{y}_2\|_c^2 \tag{4.8}$$

and moreover

$$W \in \mathrm{Lip}_\omega[l^c, l^c]. \tag{4.9}$$

Then, $\mathbf{N}$ *is regular, and* $A \in \mathrm{Lip}_\beta[l_{2r}, l_{2r}]$ *where*

$$\beta = \frac{1}{\gamma\omega}\left(\omega + \sqrt{\omega^2 - \gamma^2}\right).$$

Proof: As before, choose any $\mathbf{x}_1, \mathbf{x}_2 \in \mathcal{N}_a$ and set $\mathbf{x}_1 = X\mathbf{y}_1$ and $\mathbf{x}_2 = X\mathbf{y}_2$, where $\mathbf{y}_1, \mathbf{y}_2 \in l^c$. By (4.8),

$$
\begin{aligned}
(R\mathbf{x}_1 - R\mathbf{x}_2, \mathbf{x}_1 - \mathbf{x}_2) &= (RX\mathbf{y}_1 - RX\mathbf{y}_2, X(\mathbf{y}_1 - \mathbf{y}_2)) \\
&= (X^*RX\mathbf{y}_1 - X^*RX\mathbf{y}_2, \mathbf{y}_1 - \mathbf{y}_2)_c \\
&= (W\mathbf{y}_1 - W\mathbf{y}_2, \mathbf{y}_1 - \mathbf{y}_2) \geq \gamma\|\mathbf{y}_1 - \mathbf{y}_2\|_c^2.
\end{aligned}
$$

As in the last proof, $\|\mathbf{y}_1 - \mathbf{y}_2\|_c = \|\mathbf{x}_1 - \mathbf{x}_2\|$. So, (4.4) is satisfied. Furthermore, by (4.9) and conditions (vi) and (i) of Lemma 4.2-1,

$$\|P(R\mathbf{x}_1 - R\mathbf{x}_2)\| = \|XX^*(RX\mathbf{y}_1 - RX\mathbf{y}_2)\| =$$
$$\|X^*RX\mathbf{y}_1 - X^*RX\mathbf{y}_2\|_c = \|W\mathbf{y}_1 - W\mathbf{y}_2\|_c \leq \omega\|\mathbf{y}_1 - \mathbf{y}_2\|_c = \omega\|\mathbf{x}_1 - \mathbf{x}_2\|.$$

So, (4.5) is satisfied too. Hence, we may invoke Theorem 4.1-6 to complete the proof. ♣

The inequality (4.8) is in effect a lower bound on the slopes of the resistance functions, and (4.9) acts as an upper bound. Actually, (4.9) can be replaced by a weaker continuity condition [41, Theorem 5.8].

Example 4.2-5. In accordance with the comment just after Conditions 4.1-1, let us assume that the nonlinear resistance operator R has an infinite matrix representation $R = [R_{jm}]$. This means that the voltage v_j of the jth branch is additive in the self and coupling voltages $R_{jm}i_m$ induced in that branch by currents i_m in that and other branches; i.e., $v_j = R_{j1}i_1 + R_{j2}i_2 + \cdots$, where the individual R_{jm} may be nonlinear. Even in this case, we cannot in general obtain $W = X^*RX$ as a matrix by invoking the rule for multiplying matrices because this would require that each entry of RX be an additive function of the components of any vector $\mathbf{y} \in l^c$ to which it is applied; that is, for every j and m, $R_{jm}(X_{m1}y_1 + X_{m2}y_2 + \cdots)$ should be expandable into $R_{jm}X_{m1}y_1 + R_{jm}X_{m2}y_2 + \cdots$. Any nonadditivity of R_{jm} will prevent this. On the other hand, if every R_{jm} is linear so that R_{jm} is just multiplication by a constant, we can proceed to obtain $W = X^*RX$ as a linear operator simply by applying the rule for matrix multiplication. ♣

Let us mention in passing that all this theory extends immediately to complex Hilbert spaces. In this case, R is replaced by the branch impedance matrix Z incorporating possibly complex branch-coupling terms. If Z is again a continuous linear mapping of the complex Hilbert coordinate space l_2 into l_2, then we have a foundational theory for the phasor analysis of an infinite RLC network in the AC steady state.

4.3 The Modular Sequence Space l_M

We now prepare ourselves for the DeMichele–Soardi theory by delineating some parts of the theory of modular sequence spaces [88, pages 166–175], [99], [149]. These spaces are generalizations of the Orlicz sequence spaces [75], [85]–[88, pages 137–166], which in turn are generalizations of the the classical l_p spaces.

Later on, the function $r_j : f_j \mapsto v_j = r_j(f_j)$, $j = 1, 2, \ldots$, will denote the jth nonlinear resistance in an infinite network and f_j will be the current flowing in the jth resistor when every branch is in its Norton form. (Previously we used i_r to denote that current; see Figure 1(a).) We now revert to the use of parentheses to indicate nonlinear dependence, with linearity being allowed as a special case. Throughout the rest of this chapter, we assume

Conditions 4.3-1: *For every j, r_j is a continuous, strictly increasing, odd mapping of R^1 into R^1 with $r_j(f) \to \infty$ as $f \to \infty$.*

Thus, $r_j(0) = 0$, $r_j(f) = -r_j(-f)$, and r_j has the form indicated in Figure 4.1. For every $f \in R^1$, set

$$M_j(f) = \int_0^f r_j(x)\, dx.$$

M_j is a continuous function on R^1 with $M_j(0) = 0$ and $M_j(f) > 0$ if $f \neq 0$. Moreover, it is even: $M_j(f) = M_j(-f)$. Finally, it is *strictly convex*, which means that, for $f_1, f_2, \alpha \in R^1$ with $f_1 \neq f_2$ and $0 < \alpha < 1$,

$$M_j(\alpha f_1 + (1 - \alpha)f_2) < \alpha M_j(f_1) + (1 - \alpha)M_j(f_2). \qquad (4.10)$$

We can define the inverse function g_j for r_j by setting $g_j(v) = f$ when $v = r_j(f)$; g_j possesses all the properties listed for r_j in Conditions 4.3-1. Next, set

$$M_j^*(v) = \int_0^v g_j(u)\, du. \qquad (4.11)$$

M_j^* has the same properties as M_j. M_j^* is called the function *complementary to M_j*, and conversely. We will shortly define some Banach spaces by using the M_j. All the definitions and their consequences will hold equally well when the M_j are replaced by the M_j^*.

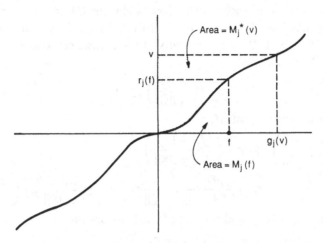

Figure 4.1. A typical curve for $r_j : v \mapsto r_j(f)$ and the complementary area functions M_j and M_j^*. g_j is the inverse function for r_j.

A standard result, which is evident from Figure 4.1, is *Young's inequality*: For any $f, v \in R^1$ not necessarily related by the mapping r_j,

$$|fv| \leq M_j(f) + M_j^*(v). \tag{4.12}$$

When $v = r_j(f)$, we have equality in (4.12), even when the absolute value signs are dropped by virtue of the oddness of the r_j.

Under Conditions 4.3-1, let $\mathbf{M}$ denote the sequence $(M_1, M_2, \ldots)$. The *modular sequence space* l_M is defined as the set of all sequences $\mathbf{f} = (f_1, f_2, \ldots)$ of real numbers such that $\sum M_j(f_j/t) < \infty$ for some $t > 0$. l_M is a linear space. Indeed, if $\mathbf{f} \in l_M$, then clearly $\alpha \mathbf{f} \in l_M$ too for every $\alpha \in R^1$. Now, let $\mathbf{f}, \mathbf{x} \in l_M$ and consider $\mathbf{f} + \mathbf{x}$. By the convexity of the M_j,

$$\sum M_j \left(\frac{f_j + x_j}{2t} \right) \leq \frac{1}{2} \sum M_j \left(\frac{f_j}{t} \right) + \frac{1}{2} \sum M_j \left(\frac{x_j}{t} \right),$$

and the right-hand side is finite for t large enough. Hence, $\mathbf{f} + \mathbf{x} \in l_M$ too.

The following norm is assigned to l_M.

$$\|\mathbf{f}\| = \inf \left\{ t : t > 0, \sum M_j \left(\frac{f_j}{t} \right) \leq 1 \right\}. \tag{4.13}$$

To check the norm axioms, first note that $\|\mathbf{f}\| \geq 0$ and that, for any $f_j \neq 0$, $M_j(f_j/t) \to \infty$ as $t \to 0$. Hence, the infimum in (4.13) must be positive if $\mathbf{f} \neq \mathbf{0}$. Second, that $\|\alpha \mathbf{f}\| = |\alpha| \|\mathbf{f}\|$ is clear. Finally, the triangle inequality $||\mathbf{f}+\mathbf{x}|| \leq ||\mathbf{f}||+||\mathbf{x}||$ is obvious if $\mathbf{f}=\mathbf{0}$ or $\mathbf{x}=\mathbf{0}$. So, assume that neither $\mathbf{f}$ nor $\mathbf{x}$ is $\mathbf{0}$. Note that, for all J and for $t \geq ||\mathbf{f}||$, $\sum_{j=1}^{J} M_j(f_j/t) \leq 1$. By the continuity of the M_j, we can take $t \to ||\mathbf{f}||+$ and then $J \to \infty$ to get

$$\sum M_j \left(\frac{f_j}{||\mathbf{f}||} \right) \leq 1. \qquad (4.14)$$

Now, by the convexity of M_j,

$$M_j \left(\frac{f_j + x_j}{||\mathbf{f}|| + ||\mathbf{x}||} \right) \leq M_j \left(\frac{|f_j| + |x_j|}{||\mathbf{f}|| + ||\mathbf{x}||} \right)$$

$$\leq \frac{||\mathbf{f}||}{||\mathbf{f}|| + ||\mathbf{x}||} M_j \left(\frac{f_j}{||\mathbf{f}||} \right) + \frac{||\mathbf{x}||}{||\mathbf{f}|| + ||\mathbf{x}||} M_j \left(\frac{x_j}{||\mathbf{x}||} \right).$$

Upon summing over j and invoking (4.14), we obtain

$$\sum M_j \left(\frac{f_j + x_j}{||\mathbf{f}|| + ||\mathbf{x}||} \right) \leq 1.$$

So, by (4.13), $||\mathbf{f}+\mathbf{x}|| \leq ||\mathbf{f}||+||\mathbf{x}||$. Truly, (4.13) defines a norm on $l_{\mathbf{M}}$.

Actually, $l_{\mathbf{M}}$ is complete under this norm and is thereby a Banach space. To show this, we will construct another Banach space $l_{\mathbf{M}}^{\#}$ of sequences and then will show that $l_{\mathbf{M}}$ and $l_{\mathbf{M}}^{\#}$ are isomorphic.

4.4 The Space $l_{\mathbf{M}}^{\#}$

With $\mathbf{M}$ as before, $l_{\mathbf{M}}^{\#}$ is defined as the set of all sequences $\mathbf{f}=(f_1, f_2, \dots)$ such that $\sum |f_j v_j| < \infty$ for every sequence $\mathbf{v} = (v_1, v_2, \dots)$ of real numbers satisfying $\sum M_j^*(v_j) \leq 1$. Here, M_j^* is the function complementary to M_j.

$l_{\mathbf{M}}^{\#}$ is clearly a linear space. It is supplied with the norm

$$|||\mathbf{f}||| = \sup \left\{ \sum |f_j v_j| : \sum M_j^*(v_j) \leq 1 \right\}. \qquad (4.15)$$

It is understood here that the supremum is over all $\mathbf{v}$ satisfying the stated inequality. Let us check the norm axioms again. Obviously, $|||\mathbf{f}||| \geq 0$. Assume that $|||\mathbf{f}||| = 0$. For any j_0, we can choose the v_j such that $v_j = 0$ for $j \neq j_0$, $v_{j_0} \neq 0$, and $M_j^*(v_{j_0}) < 1$. Hence, $0 = |||\mathbf{f}||| \geq |f_{j_0} v_{j_0}|$. Whence, $f_{j_0} = 0$. Therefore, $\mathbf{f} = \mathbf{0}$. The second axiom, $|||\alpha\mathbf{f}||| = |\alpha| \, |||\mathbf{f}|||$ is obvious again. The third axiom is also obtained easily:

$$|||\mathbf{f}+\mathbf{x}||| = \sup \left\{ \sum |(f_j + x_j)v_j| : \sum M_j^*(v_j) \leq 1 \right\}$$

$$\leq \sup \left\{ \sum |f_j v_j| + \sum |x_j v_j| : \cdots \right\}$$

$$\leq \sup \left\{ \sum |f_j v_j| : \cdots \right\} + \sup \left\{ \sum |x_j v_j| : \cdots \right\}$$

$$\leq |||\mathbf{f}||| + |||\mathbf{x}|||$$

Thus, $l_M^\#$ is a normed linear space.

Lemma 4.4-1: *Convergence in $l_M^\#$ implies componentwise convergence.*

Proof: This follows directly from (4.15) by choosing a $\mathbf{v}$ such that one of its components is positive and the others are all 0. ♣

Theorem 4.4-2: $l_M^\#$ *is a Banach space.*

Proof: All that needs to be proven is completeness. Let $\{\mathbf{f}_m\}_{m=1}^\infty$ be a Cauchy sequence in $l_M^\#$. Set $\mathbf{f}_m = (f_{m1}, f_{m2}, \dots)$. Thus, given any $\epsilon > 0$, there exists an integer N such that, for all $m, n > N$,

$$\sup \sum_{j=1}^\infty |f_{mj} - f_{nj}|\,|v_j| \leq \epsilon, \qquad (4.16)$$

where again the supremum is taken over all $\mathbf{v}$ for which $\sum M_j^*(v_j) \leq 1$. By Lemma 4.4-1 and the completeness of R^1, there is a sequence $\mathbf{f} = (f_1, f_2, \dots)$ of real numbers such that $\mathbf{f}_m \to \mathbf{f}$ componentwise as $m \to \infty$. We have to show that $\mathbf{f} \in l_M^\#$ and that $|||\mathbf{f}_m - \mathbf{f}||| \to 0$ as $m \to \infty$. Just apply to (4.16) the standard argument used in the proof of Lemma 3.1-3. ♣

Lemma 4.4-3: *Let $\mathbf{f} \in l_M^\#$.*

  (i) If the sequence $\mathbf{v}$ is such that $\sum M_j^*(v_j) \leq 1$, then $\sum |f_j v_j| \leq |||\mathbf{f}|||$.

  (ii) If the sequence $\mathbf{v}$ is such that $\sum M_j^*(v_j) > 1$, then

$$\sum |f_j v_j| \leq |||\mathbf{f}||| \sum M_j^*(v_j).$$

Proof: (i) follows directly from the definition (4.15) of the norm $|||\cdot|||$. As for (ii), first note that, by the convexity of the M_j^*, we have $M_j^*(\alpha v_j) \leq \alpha M_j^*(v_j)$ whenever $0 \leq \alpha \leq 1$. In this case, we may choose $\alpha = [\sum M_j^*(v_j)]^{-1} < 1$. Then, $\sum M_j^*(\alpha v_j) \leq \alpha \sum M_j^*(v_j) = 1$. So, we may apply (i) to $\alpha \mathbf{v}$ to get (ii). ♣

Two normed linear spaces $\mathcal{X}$ and $\mathcal{Y}$ are said to be *isomorphic* if there is a linear bijection between them that preserves Cauchy sequences, that is, a sequence is Cauchy in $\mathcal{X}$ if and only if its image under the bijection is Cauchy in $\mathcal{Y}$. We denote the isomorphism by $\mathcal{X} \cong \mathcal{Y}$. We now set about proving that l_M and $l_M^\#$ are isomorphic by using the identity mapping. Theorem 4.4-2 will then imply that l_M is complete.

Lemma 4.4-4: *If $\mathbf{f} \in l_M^\#$ and $|||\mathbf{f}||| \leq 1$, then $\sum M_j^*(r_j(f_j)) \leq 1$.*

Proof: The conclusion is true if $\mathbf{f} = 0$. So, assume $\mathbf{f} \neq 0$. Since equality holds in (4.12) when $v = r_j(f)$, we may write

$$0 < \sum M_j^*(r_j(f_j)) < \sum [M_j(f_j) + M_j^*(r_j(f_j))] = \sum f_j r_j(f_j). \quad (4.17)$$

Now, suppose that the conclusion is not true. Then, by Lemma 4.4-3(ii), the right-hand side of (4.17) is no larger than $|||\mathbf{f}||| \sum M_j^*(r_j(f_j))$. Hence, $|||\mathbf{f}||| > 1$, in contradiction to the hypothesis. ♣

Since $|||(\mathbf{f}/|||\mathbf{f}|||)||| = 1$ when $\mathbf{f} \neq \mathbf{0}$, it follows immediately that, for every nonzero $\mathbf{f} \in l_\mathbf{M}^\#$,

$$\sum M_j^* \left(r_j \left(\frac{f_j}{|||\mathbf{f}|||} \right) \right) \leq 1. \tag{4.18}$$

Lemma 4.4-5: *If* $\mathbf{f} \in l_\mathbf{M}^\#$, *then* $\sum M_j(f_j/|||\mathbf{f}|||) \leq 1$.

Proof: Set $\mathbf{x} = \mathbf{f}/|||\mathbf{f}||| \neq \mathbf{0}$. By (4.18), $\sum M_j^*(r_j(x_j)) \leq 1$. Again, by (4.12), $M_j(x_j) \leq M_j(x_j) + M_j^*(r_j(x_j)) = x_j r_j(x_j)$. So, by the definition of $||| \cdot |||$, $\sum M_j(x_j) \leq \sum x_j r_j(x_j) \leq |||\mathbf{x}||| = 1$. ♣

Lemma 4.4-6: $\mathbf{f} \in l_\mathbf{M}$ *if and only if* $\mathbf{f} \in l_\mathbf{M}^\#$. *Moreover,* $||\mathbf{f}|| \leq |||\mathbf{f}||| \leq 2||\mathbf{f}||$.

Proof: *If* : Assume $\mathbf{f} \in l_\mathbf{M}^\#$. By Lemma 4.4-5 and the definition (4.13) of $|| \cdot ||$, $\mathbf{f} \in l_\mathbf{M}$ and $||\mathbf{f}|| \leq |||\mathbf{f}|||$.

Only if : Assume $\mathbf{f} \in l_\mathbf{M}$. Choose $t > 0$ such that $\sum M_j(f_j/t) < \infty$. By Young's inequality (4.12), we may write

$$\sum |f_j v_j| = t \sum |f_j/t| \, |v_j| \leq t \left[\sum M_j(f_j/t) + \sum M_j^*(v_j) \right].$$

On the right-hand side, the first summation is finite because of the choice of t, and the second summation will be finite when we choose $\mathbf{v}$ to make $\sum M_j^*(v_j) \leq 1$. It follows that $\mathbf{f} \in l_\mathbf{M}^\#$.

Thus, $l_\mathbf{M}$ and $l_\mathbf{M}^\#$ have the same members. We have $|||\mathbf{f}||| \leq 2||\mathbf{f}||$ left to prove. Invoking Young's inequality (4.12) as well as (4.14), we may write

$$|||\mathbf{f}||| = ||\mathbf{f}|| \sup \left\{ \sum \frac{|f_j v_j|}{||\mathbf{f}||} : \sum M_j^*(v_j) \leq 1 \right\}$$

$$\leq ||\mathbf{f}|| \sup \left\{ \sum M_j \left(\frac{f_j}{||\mathbf{f}||} \right) + \sum M_j^*(v_j) : \cdots \right\}$$

$$\leq 2||\mathbf{f}||. \quad ♣$$

We have proven that $|| \cdot ||$ and $||| \cdot |||$ are equivalent norms on the same linear space. Thus, a sequence is Cauchy under $|| \cdot ||$ if and only if it is Cauchy under $||| \cdot |||$. Since $l_\mathbf{M}^\#$ is complete (Theorem 4.4-2), so too is $l_\mathbf{M}$. In fact, we have established

Theorem 4.4-7: $l_\mathbf{M}$ *and* $l_\mathbf{M}^\#$ *are isomorphic Banach spaces; symbolically,* $l_\mathbf{M} \cong l_\mathbf{M}^\#$.

4.5 The Space c_M

We shall now examine a subspace c_M of l_M, whose dual plays an important role in this theory for nonlinear networks. In the next section, we will impose a certain uniformity condition on the slopes of the r_j, which will make c_M identical to l_M. It will follow from this that the dual of l_M is l_{M^*}. As a result, the voltage vector for the infinite network considered later on will be in l_{M^*} whenever the current vector is in l_M.

c_M is the subset of l_M consisting of all sequences $\mathbf{f}$ for which $\sum M_j(f_j/t) < \infty$ for every $t > 0$. The same argument as the one used for l_M shows that c_M is a linear space.

Lemma 4.5-1: c_M *is closed in* l_M.

Proof: Assume that $\mathbf{f}_m = (f_{m1}, f_{m2}, \dots) \in c_M$ for each $m = 1, 2, \dots$ and that $\mathbf{f}_m \to \mathbf{f}$ in l_M as $m \to \infty$. We wish to show that $\sum M_j(f_j/t) < \infty$ for every $t > 0$. Choose $t > 0$ arbitrarily and fix it. Set $\mathbf{x}_m = \mathbf{f}_m/t$ and $\mathbf{x} = \mathbf{f}/t$. If we can show that $\sum M_j(x_j) < \infty$, we will have $\sum M_j(f_j/t) < \infty$ whatever be $t > 0$. Since $\mathbf{x}_m \to \mathbf{x}$ in l_M, we can choose a fixed m so large that $\|\mathbf{x} - \mathbf{x}_m\| \leq 1/2$. Hence,

$$\sum M_j(2x_j - 2x_{mj}) \leq \sum M_j\left(\frac{x_j - x_{mj}}{\|\mathbf{x} - \mathbf{x}_m\|}\right) \leq 1,$$

where the last inequality is (4.14). So, by the convexity of the M_j,

$$\sum M_j(x_j) = \sum M_j\left(\frac{1}{2}(2x_j - 2x_{mj} + 2x_{mj})\right)$$

$$\leq \frac{1}{2}\sum M_j(2x_j - 2x_{mj}) + \frac{1}{2}\sum M_j(2x_{mj})$$

$$< \infty \quad \clubsuit$$

It is henceforth understood that c_M is assigned the same norm as that of l_M. By virtue of the last lemma, c_M is a Banach space in itself.

For each $j = 1, 2, \dots$, let $\mathbf{u}_j$ be the vector having 1 for the jth component and 0 for every other component. These vectors are called the *unit vectors*. Clearly, any linear combination of the $\mathbf{u}_j$ is in c_M.

Lemma 4.5-2: *For each* $\mathbf{f} \in c_M$, $\sum f_j \mathbf{u}_j$ *converges in* c_M *to* $\mathbf{f}$.

Proof: Given any $\epsilon > 0$, choose the integer K so large that

$$\sum_{j=K}^{\infty} M_j(f_j/\epsilon) \leq 1.$$

For any integer $J \geq K$,

$$\left\|\mathbf{f} - \sum_{j=1}^{J} f_j \mathbf{u}_j\right\| = \inf\left\{t : t > 0, \; \sum_{j=J+1}^{\infty} M_j\left(\frac{f_j}{t}\right) \leq 1\right\}$$

$$\leq \inf\left\{t : t > 0, \; \sum_{j=K}^{\infty} M_j\left(\frac{f_j}{t}\right) \leq 1\right\}$$

$$\leq \epsilon. \quad \clubsuit$$

The dual $\mathcal{X}^*$ of a real Banach space $\mathcal{X}$ is the space of continuous linear functionals on $\mathcal{X}$ (i.e., continuous linear mappings of $\mathcal{X}$ into R^1). $\mathcal{X}^*$ is also a real Banach space when to each $y \in \mathcal{X}^*$ is assigned the norm [94, page 357]

$$\|y\| = \sup\{|y(x)| : x \in \mathcal{X}, \|x\| = 1\}.$$

As was mentioned in Section 4.3, we can define and analyze $l_{\mathbf{M}^*}$ simply by replacing $\mathbf{M}$ by $\mathbf{M}^*$ in the preceding discussion. All the results concerning $l_{\mathbf{M}}$ hold for $l_{\mathbf{M}^*}$ as well. We define a pairing between any $\mathbf{f} \in l_{\mathbf{M}}$ and any $\mathbf{v} \in l_{\mathbf{M}^*}$ by

$$\langle \mathbf{v}, \mathbf{f} \rangle = \sum v_j f_j \tag{4.19}$$

whenever the right-hand side exists. This may define $\mathbf{v}$ as a functional on $l_{\mathbf{M}}$ or perhaps just on $c_{\mathbf{M}}$. We will now show that in this case $l_{\mathbf{M}^*}$ is isomorphic to the dual space $c_{\mathbf{M}}^*$ of $c_{\mathbf{M}}$.

Theorem 4.5-3: *Under the duality defined by the pairing (4.19), $c_{\mathbf{M}}^* \cong l_{\mathbf{M}^*}$.*

Proof: Let $\mathbf{v}$ be any fixed member of $l_{\mathbf{M}^*}$. We can define a mapping $V_{\mathbf{v}}$ from $c_{\mathbf{M}}$ into R^1 by $V_{\mathbf{v}}(\mathbf{f}) = \sum f_j v_j$. Indeed, let us establish the existence of (4.19) for any $\mathbf{f} \in c_{\mathbf{M}}$. As in (4.14), $\sum M_j^*(v_j/\|\mathbf{v}\|) \leq 1$. So, by Lemma 4.4-3(i) and Lemma 4.4-6,

$$\sum |f_j v_j| \leq \|\mathbf{f}\| \|\mathbf{v}\| \leq 2 \|\mathbf{f}\| \|\mathbf{v}\| < \infty. \tag{4.20}$$

Thus, (4.19) is a finite number. The mapping $V_{\mathbf{v}}$ is clearly linear, and (4.20) establishes its boundedness with $\|V_{\mathbf{v}}\| \leq 2\|\mathbf{v}\|$. Thus, $V_{\mathbf{v}} \in c_{\mathbf{M}}^*$.

We now have in turn a mapping $S : \mathbf{v} \mapsto V_{\mathbf{v}}$, which maps $l_{\mathbf{M}^*}$ into $c_{\mathbf{M}}^*$. Again, the linearity of S is clear. Moreover, S is bounded because $\|S\mathbf{v}\| = \|V_{\mathbf{v}}\| \leq 2\|\mathbf{v}\|$; in fact, $\|S\| \leq 2$.

S is injective: Indeed, let $\mathbf{v}, \mathbf{z} \in l_{\mathbf{M}^*}$ and assume that $V_{\mathbf{v}} = V_{\mathbf{z}}$. Thus, $\sum f_j v_j = \sum f_j z_j$. Using the unit vectors for $\mathbf{f}$, we see that $v_j = z_j$ for all j, that is, $\mathbf{v} = \mathbf{z}$.

Next, we show that S is surjective. As before, let $\mathbf{u}_j$ denote the unit vectors in c_M. Let $\mathbf{w}$ be any member of c_M^*. Set $v_j = \mathbf{w}(\mathbf{u}_j)$ and $\mathbf{v} = (v_1, v_2, \ldots)$. We shall first establish that $\mathbf{v} \in l_{M^*}^\#$, which by Lemma 4.4-6 means that $\mathbf{v} \in l_{M^*}$. We shall then show that $\mathbf{w} = S\mathbf{v}$, thereby proving the surjectivity of S.

Let $\mathbf{x}$ be such that $\sum M_j(x_j) \leq 1$. Set $\omega_j = \mathrm{sgn}(x_j v_j)$. Then, for every positive integer m,

$$\sum_{j=1}^m |x_j v_j| = \sum_{j=1}^m \omega_j x_j v_j = \mathbf{w}\left(\sum_{j=1}^m \omega_j x_j \mathbf{u}_j\right)$$

$$\leq \|\mathbf{w}\| \, \|\sum_{j=1}^m \omega_j x_j \mathbf{u}_j\| = \|\mathbf{w}\| \inf\left\{t : \sum_{j=1}^m M_j\left(\frac{x_j}{t}\right) \leq 1\right\}$$

$$\leq \|\mathbf{w}\| \inf\left\{t : \sum_{j=1}^\infty M_j\left(\frac{x_j}{t}\right) \leq 1\right\} = \|\mathbf{w}\| \, \|\mathbf{x}\|.$$

Since m is arbitrary, $\sum |x_j v_j| \leq \|\mathbf{w}\| \, \|\mathbf{x}\|$. Thus, for all $\mathbf{x} \in l_M$ such that $\sum M_j(x_j) \leq 1$, we have $\|\mathbf{x}\| \leq 1$ and $|\sum x_j v_j| \leq \sum |x_j v_j| \leq \|\mathbf{w}\|$. By (4.15) and the complementarity between M_j and M_j^*, we have proven that $\mathbf{v} \in l_{M^*}^\#$. Thus, $\mathbf{v} \in l_{M^*}$.

By Lemma 4.5-2 and the fact that $\mathbf{w} \in c_M^*$, we may write, for any $\mathbf{x} \in c_M$,

$$V_{\mathbf{v}}(\mathbf{x}) = \sum x_j v_j = \sum x_j \mathbf{w}(\mathbf{u}_j) = \mathbf{w}\left(\sum x_j \mathbf{u}_j\right) = \mathbf{w}(\mathbf{x}).$$

Hence, $\mathbf{w} = S\mathbf{v}$, that is, S is surjective.

Finally, we need merely note that a continuous linear bijection of one Banach space onto another has a continuous linear inverse [94, pages 415–416]. Thus, the bijection S establishes the asserted isomorphism between l_M^* and c_{M^*}. ♣

4.6 Some Uniform-variation Conditions

We will now place some restrictions on the variations of r_j and g_j on intervals around the origin – at least for all but a finite number of the indices j. They produce the important results that $c_M = l_M$, $c_{M^*} = l_{M^*}$, and l_M is reflexive. A certain normalized form of them is called the *uniform Δ_2 condition* [149, page 276].

Conditions 4.6-1: *There are a positive integer j_0 and two positive numbers p and q, both greater than one, such that for every $j \geq j_0$*

(i) $fr_j(f) \leq pM_j(f)$ *for* $0 \leq f \leq a_j$ *where a_j is the unique positive number for which $M_j(a_j) = 1$, and*

(ii) $vg_j(v) \leq qM_j^*(v)$ *for* $0 \leq v \leq d_j$ *where d_j is the unique positive number for which $M_j^*(d_j) = 1$.*

Throughout this section, we always assume that Conditions 4.3-1 and 4.6-1 are fulfilled. The restriction (ii) is equivalent to $fr_j(f) \geq sM_j(f)$ for some $s > 1$ and for $0 \leq f \leq g_j(d_j)$. (It should be noted that the original DeMichele–Soardi paper is more general in this regard, for it does not impose Condition 4.6-1(ii).) One result of Conditions 4.6-1 is that, as $j \to \infty$, the r_j do not approach a jump from the abscissa, and similarly for the g_j. But even more is implied. Since r_j is the derivative of M_j, we have, for $j \geq j_0$ and $0 \leq f \leq a_j$,

$$-\ln M_j(f) = \int_f^{a_j} \frac{r_j(x)}{M_j(x)}\, dx$$

$$\leq \int_f^{a_j} \frac{p}{x}\, dx = p\ln a_j - p\ln f.$$

Hence, $M_j(f) \geq (f/a_j)^p$ for $j \geq j_0$ and $0 \leq f \leq a_j$.

Let us normalize the M_j by setting $N_j(f)=M_j(a_j f)$. Hence, $N_j(1)=1$, and Condition 4.6-1(i) becomes

Condition 4.6-2: *For $j \geq j_0$ and $0 \leq f \leq 1$, we have $f\rho_j(f) \leq pN_j(f)$, where $p > 1$, p is independent of j, and $\rho_j = dN_j/dt = a_j r_j(a_j f)$.*

Lemma 4.6-3: *Under Condition 4.6-2, $N_j(f) \geq f^p$ for $j \geq j_0$ and $0 \leq f \leq 1$. If in addition $\sum N_j(f_j) < \infty$, then $f_j \to 0$ as $j \to \infty$.*

Proof: The first assertion has already been established with regard to M_j. By the hypothesis of the second assertion, $N_j(f_j) \to 0$ as $j \to \infty$. Moreover, the N_j are even functions and positive everywhere except for $N_j(0) = 0$. So, whatever be the signs of the f_j, the first assertion implies the second. ♣

A consequence of Condition 4.6-2 is the following restriction on the variations of the N_j in an interval containing the origin.

Lemma 4.6-4: *For every number $\omega > 1$ there exist numbers $\alpha(\omega) > 0$ and $K(\omega) > 1$, as well as a positive integer j_0 that is independent of ω, such that $N_j(\omega f) \leq K(\omega)N_j(f)$ for $0 \leq f \leq \alpha(\omega)$ and $j \geq j_0$.*

Proof: Set $\omega_0 = p/(p-1) > 1$. First consider the case where $1 < \omega \leq \omega_0$. Set $\alpha(\omega) = f_0/\omega$ and $K(\omega) = \omega/(p - \omega p + \omega)$. Note that $K(\omega) > 1$.

For $0 \le f \le \alpha(\omega)$, we have $f < \omega f \le \omega \alpha(\omega) = f_0$. Also, for $j \ge j_0$, we can invoke the mean-value theorem to find a β_j with $f < \beta_j < \omega f$ such that

$$\frac{N_j(\omega f) - N_j(f)}{\omega f - f} \le \rho_j(\beta_j).$$

By Condition 4.6-2, we have $\rho_j(\beta_j) \le p N_j(\beta_j)/\beta_j$. By the facts that $N_j(0) = 0$, N_j is convex, and $\beta_j < \omega f$, we also have that $N_j(\beta_j)/\beta_j \le N_j(\omega f)/\omega f$. Combining these inequalities and rearranging the result, we get $(\omega - \omega p + p) N_j(\omega f) \le \omega N_j(f)$, that is, $N_j(\omega f) \le K(\omega) N_j(f)$ for $0 \le f \le \alpha(\omega)$ and $j \ge j_0$, as asserted.

Now consider the case where $\omega > \omega_0$. There exists a smallest positive integer m such that $\omega_0^m \ge \omega$ because $\omega_0 = p/(p-1) > 1$. Set $K(\omega) = [K(\omega_0)]^m$, where $K(\omega_0) = \omega_0/(p - \omega_0 p + \omega_0)$. Also, set $\alpha(\omega) = \alpha(\omega_0)\omega_0^{-m}$. Then, for $j \ge j_0$ and $0 \le f \le \alpha(\omega)$, we have $\omega_0^{m-1} f \le \omega_0^{m-1}\alpha(\omega) = \omega_0^{m-1}\alpha(\omega_0)\omega_0^{-m} = \alpha(\omega_0)/\omega_0 < \alpha(\omega_0)$ and therefore $\omega_0^n f < \alpha(\omega_0)$ for $n = 0, 1, \ldots, m-1$. Hence, by our prior result,

$$N_j(\omega f) \le N_j(\omega_0^m f) \le K(\omega_0) N_j(\omega_0^{m-1} f)$$

$$\le \cdots \le [K(\omega_0)]^m N_j(f) = K(\omega) N_j(f). \quad \clubsuit$$

Theorem 4.6-5: *Under Conditions 4.3-1 and 4.6-1, $l_N = c_N$.*

Proof: Since c_N is a subspace of l_N, we need merely show that any given member **f** of l_N is also a member of c_N. By definition of l_N, we can fix upon a $t > 0$ such that $\sum N_j(f_j/t)$ converges. The same convergence occurs if t is replaced by any $s \ge t$ because the N_j are positive even functions that are increasing for positive arguments. So, assume $s < t$ and fix s. We have to show that $\sum N_j(f_j/s)$ converges.

Set $\omega = t/s > 1$ in Lemma 4.6-4. Consequently, there exist a j_1, an $\alpha(t/s) > 0$, and a $K(t/s) > 1$ such that $N_j(tx/s) \le K(t/s) N_j(x)$ for $0 \le x \le \alpha(t/s)$ and $j > j_1$. Here, j_1 is independent of t/s. By the second assertion of Lemma 4.6-2, $f_j/t \to 0$ and therefore $f_j \to 0$ as $j \to \infty$. So, there is a j_2 such that $x = |f_j|/t < \alpha(t/s)$ for all $j > j_2$. Let $j_0 = \max\{j_1, j_2\}$. Then,

$$\sum N_j\left(\frac{f_j}{s}\right) = \sum_{j=1}^{j_0} N_j\left(\frac{f_j}{s}\right) + \sum_{j=j_0+1}^{\infty} N_j\left(\frac{t}{s} \cdot \frac{f_j}{t}\right).$$

On the right-hand side, the first sum is finite. Setting $x = |f_j|/t$, we see that the second sum is bounded by $K(t/s) \sum N_j(f_j/t)$, which is also finite. Since s is arbitrary, $\mathbf{f} \in c_N$. $\clubsuit$

Corollary 4.6-6: *Under Conditions 4.3-1 and 4.6-1, $l_M = c_M$.*

Proof: Set $\mathbf{a} = (a_1, a_2, \ldots)$, where the a_j are specified in Condition 4.6-1(i). For any $\mathbf{f} \in l_{\mathbf{M}}$, let $\mathbf{f}/\mathbf{a}$ denote the vector $(f_1/a_1, f_2/a_2, \ldots)$. Thus, for each j, $N_j(f_j/a_j) = M_j(f_j)$. Letting $\| \cdot \|_{\mathbf{M}}$ and $\| \cdot \|_{\mathbf{N}}$ denote the norms for $l_{\mathbf{M}}$ and $l_{\mathbf{N}}$, we see from (4.13) that $\|\mathbf{f}\|_{\mathbf{M}} = \|\mathbf{f}/\mathbf{a}\|_{\mathbf{N}}$. In fact, the mapping: $\mathbf{f} \mapsto \mathbf{f}/\mathbf{a}$ is a linear bijection of $l_{\mathbf{M}}$ onto $l_{\mathbf{N}}$ and preserves Cauchy sequences. Moreover, it carries $c_{\mathbf{M}}$ onto $c_{\mathbf{N}}$. Our conclusion now follows from Theorem 4.6-5. ♣

$\mathcal{X}^*$ denotes the dual of any Banach space $\mathcal{X}$. So far, we have shown that $c_{\mathbf{M}}^* \cong l_{\mathbf{M}^*}$ (Theorem 4.5-3) and $l_{\mathbf{M}} = c_{\mathbf{M}}$ (Corollary 4.6-6). Consequently,

$$l_{\mathbf{M}}^* = c_{\mathbf{M}}^* \cong l_{\mathbf{M}^*}. \tag{4.21}$$

Note that the statements (i) and (ii) of Conditions 4.6-1 are exactly the same except that r_j and M_j are replaced by g_j and M_j^*. Thus, we may normalize M_j^* and proceed exactly as above to arrive at (4.21) with $\mathbf{M}$ replace by $\mathbf{M}^*$, and conversely. In particular, since $(\mathbf{M}^*)^* = \mathbf{M}$, we have $(l_{\mathbf{M}}^*)^* \cong (l_{\mathbf{M}^*})^* \cong l_{(\mathbf{M}^*)^*} = l_{\mathbf{M}}$. Thus, we have

Corollary 4.6-7: *Under Conditions 4.3-1 and 4.6-1, $l_{\mathbf{M}}$ is a reflexive space, that is, the dual of its dual is isomorphic to $l_{\mathbf{M}}$.*

4.7 Current Regimes in $l_{\mathbf{M}}$

We are now ready to consider nonlinear infinite networks whose current regimes are members of $l_{\mathbf{M}}$. Using the definitions of Section 3.2 regarding 1-graphs, we can generalize the theory of DeMichele and Soardi by allowing arbitrary connections at infinity. Their treatment assumed that the network was locally finite and tacitly allowed shorts everywhere at infinity – in other words, there were no infinite 0-nodes and a single 1-node embraced all the 0-tips of the network. We allow a variety of 1-nodes – and perhaps none at all. We also allow infinite 0-nodes.

As always, we index the branches by $j = 1, 2, \ldots$. Moreover, the following is assumed throughout the rest of this chapter.

Conditions 4.7-1: *Let $\mathbf{N}$ be a 0-network or a 1-network, every branch b_j of which is in the Norton form with a nonlinear (possibly linear) resistance r_j in parallel with a pure current source h_j (possibly equal to 0). Assume that all the r_j satisfy Conditions 4.3-1 and 4.6-1 and that there is no coupling between branches. Assume furthermore that $\mathbf{h} = (h_1, h_2, \ldots) \in l_{\mathbf{M}}$.*

Figure 4.2 illustrates a typical branch. The current source may be absent, that is, h_j may be 0. Also, f_j is the current flowing through

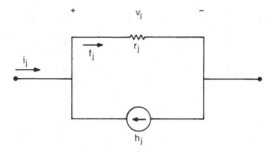

Figure 4.2. A typical branch under Hypothesis 4.7-1. f_j is the current flowing through the nonlinear resistor r_j, and i_j is the branch current.

r_j, i_j is the total branch current, and v_j is the branch voltage. Thus, $v_j = r_j(f_j) = r_j(i_j + h_j)$. The resistance operator R for the network is defined by $R(\mathbf{f}) = (r_1(f_1), r_2(f_2), \ldots)$. Hence, $\mathbf{v} = R(\mathbf{f})$.

Lemma 4.7-2: R *maps* l_M *into* l_{M^*}.

Proof: We use the facts that the r_j are odd functions and the M_j and M_j^* are even functions. Since there is equality in (4.12) when $v = r_j(f)$,

$$f_j r_j(f_j) = M_j(f_j) + M_j^*(r_j(f_j))$$

for all j. By Condition 4.6-1(i), $f_j r_j(f_j) \leq p M_j(f_j)$ for $0 \leq |f_j| \leq a_j$ and $j \geq j_0$. Hence, for f_j and j so restricted,

$$M_j^*(r_j(f_j)) \leq f_j r_j(f_j) \leq p M_j(f_j).$$

Assume $\mathbf{f} \in l_M$. By Corollary 4.6-6, $\sum M_j(f_j) < \infty$. This implies that $|f_j| < a_j$ for all j sufficiently large because $M_j(x) \geq 1$ for $|x| \geq a_j$. Therefore, $\sum M_j^*(r_j(f_j)) \leq p \sum M_j(f_j) < \infty$, which means that $R(\mathbf{f}) \in l_{M^*}$. ♣

We now set up the space $\mathcal{L}$ of allowable current regimes in the given network $\mathbf{N}$ in much the same way as $\mathcal{K}$ was constructed in Section 3.3. The principal difference is that we now use the space l_M in place of $\mathcal{I}$. Note that $\mathbf{f} \in l_M$ if and only if $\mathbf{i} \in l_M$ because $\mathbf{f} = \mathbf{i} + \mathbf{h}$ and $\mathbf{h} \in l_M$ by Conditions 4.7-1.

Also note that, if $\mathbf{i}$ is a 0-loop current, then $\sum M_j(i_j/t)$ has only a finite number of nonzero terms and is therefore finite. Hence, every 0-loop current is a member of l_M. On the other hand, a proper 1-loop current may or may not be a member of l_M. Moreover, an infinite superposition of proper 1-loop currents may be in l_M even when every

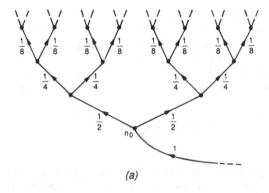

(a)

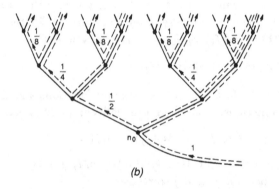

(b)

Figure 4.3. (a) An infinite electrical network all branches of which have 1-Ω resistors and no current sources. The numbers indicate branch currents, which comprise a vector in l_M. (b) The decomposition of that current flow into a countable sum of 1-loop currents. The latter are indicated by the dotted lines along with their values.

one of those 1-loop currents is not in l_M. As an illustration of this, consider

Example 4.7-3. The graph of the network shown in Figure 4.3(a) consists of the infinite binary tree with an appended branch connected between the 0-node n_0 and a 1-node that embraces all the 0-tips. In effect, all of infinity is shorted together and is connected to n_0 through a single branch. Assume that the resistance of every branch – including the appended branch – is 1 Ω. There are no current sources. It is easy to check that this 1-network satisfies all of Conditions 4.7-1.

Now, consider the branch currents indicated by the numbers in Figure

4.3(a). It is also easy to check that the vector of these branch currents is a member of l_M. That vector can be decomposed into a countable sum of proper 1-loop currents in accordance with Figure 4.3(b). ♣

As in Section 3.3, we define a *basic current* to be either a 0-loop current or a branch-current vector of the form $\mathbf{i} = \sum \mathbf{i}_m$ such that the four conditions given in Section 3.3 are satisfied. One consequence is that a basic current fulfills Kirchhoff's current law at every ordinary 0-node. The current flow examined in Example 4.7-3 is a basic current.

With a given network $\mathbf{N}$ satisfying Conditions 4.7-1, we let $\mathcal{L}^0$ denote the span of all basic currents in l_M and let $\mathcal{L}$ be the closure of $\mathcal{L}^0$ in l_M. $\mathcal{L}$ is a linear subspace of l_M and is a Banach space in itself when it has the same norm as l_M. By Lemma 4.4-1 and Theorem 4.4-7, convergence in l_M implies componentwise convergence. Hence, every member of $\mathcal{L}$ satisfies Kirchhoff's current law at every finite ordinary node.

As for Kirchhoff's voltage law, it will be shown later on that the voltage-current regime dictated by Theorem 4.8-4 satisfies

$$\langle R(\mathbf{i} + \mathbf{h}), \mathbf{s} \rangle = 0 \tag{4.22}$$

for all $\mathbf{s} \in \mathcal{L}$, where $\mathbf{h} \in l_M$ is the given current-source vector and $\mathbf{i} \in \mathcal{L}$ is the branch-current vector. Since $\mathbf{v} = R(\mathbf{i} + \mathbf{h})$ is the branch-voltage vector and since every 0-loop current is in $\mathcal{L}$, this implies that Kirchhoff's voltage law is satisfied around every 0-loop. Of course, (4.22) implies more, for $\mathcal{L}$ contains much more than just 0-loop currents.

4.8 A Minimization Principle

The approach used by DeMichele and Soardi is in the spirit of Thomson's least power principle (Section 3.8), but instead of minimizing the total dissipated power, what is minimized is the related quantity

$$W(\mathbf{f}) = \sum M_j(f_j) = \sum \int_0^{f_j} r_j(x)\, dx. \tag{4.23}$$

In accordance with Figure 4.2, we shall restrict $\mathbf{f}$ to the set

$$\{\mathbf{f} \in l_M : \mathbf{f} = \mathbf{i} + \mathbf{h}, \mathbf{i} \in \mathcal{L}\},$$

where $\mathbf{h} \in l_M$ is fixed. Thus, what we will be minimizing is the functional $U(\mathbf{i}) = W(\mathbf{f}) = W(\mathbf{i} + \mathbf{h})$ for all $\mathbf{i} \in \mathcal{L}$. Recall that $\mathcal{L}$ is a subspace of l_M and is a Banach space in itself having the same norm as l_M. Corollary 4.6-6 implies that $W(\mathbf{f})$ is finite for every $\mathbf{f} \in l_M$.

A number of standard definitions and results from nonlinear functional analysis will be needed. For ready reference, we gather them here. In the following, $\mathcal{X}$ is a real Banach space, and $F : \mathcal{X} \rightsquigarrow R^1$ is a (not necessarily linear) functional on $\mathcal{X}$. Also, $\mathcal{X}^*$ denotes the *dual* of $\mathcal{X}$. It is assumed that $\mathcal{X}$ is *reflexive*, that is, the dual of the dual of $\mathcal{X}$ is isomorphic to $\mathcal{X}$.

Strictly convex functionals: As with ordinary functions, a real functional $F : \mathcal{X} \rightsquigarrow R^1$ is called *strictly convex* if, for any $x, y, \in \mathcal{X}$ with $x \neq y$ and for $0 < \lambda < 1$, we have

$$F(\lambda x + (1 - \lambda)y) < \lambda F(x) + (1 - \lambda)F(y).$$

Weak topologies: A sequence $\{x_1, x_2, \ldots\}$ in $\mathcal{X}$ is said to *converge weakly to a limit* $x_0 \in \mathcal{X}$ if $x^*(x_k) \to x^*(x_0)$ as $k \to \infty$ whenever $x^* \in \mathcal{X}^*$.

Lemma 4.8-1 [48, page 68]: *The limit x_0 of a weakly convergent sequence $\{x_1, x_2, \ldots\}$ in $\mathcal{X}$ lies in the closure of the span of the $x_1, x_2, \ldots$.*

F is said to be *weakly lower-sequentially semicontinuous* on the space $\mathcal{X}$ if $F(x_0) \leq \lim \inf F(x_k)$ whenever $\{x_1, x_2, \ldots\}$ converges weakly to x_0 in $\mathcal{X}$. A subset A of $\mathcal{X}$ is called *bounded* if there is a finite constant larger than $\|x\|$ for all $x \in A$. Also, A is called *weakly sequentially compact* if every infinite sequence in A has a subsequence that converges weakly to a point in $\mathcal{X}$.

Lemma 4.8-2 [48, pages 67–68]: *A subset A of the reflexive Banach space $\mathcal{X}$ is weakly sequentially compact if and only if A is bounded.*

Gateaux differentials [94, page 458]: F is called *Gateaux differentiable* at some point $x_0 \in \mathcal{X}$ if there is a mapping $dF(x_0, \cdot) : \mathcal{X} \rightsquigarrow R^1$ such that, for $t \in R^1$ and for each $y \in \mathcal{X}$,

$$\lim_{t \to 0} \frac{F(x_0 + ty) - F(x_0)}{t} = dF(x_0, y).$$

$dF(x_0, y)$ is called the *Gateaux differential of F at x_0 with increment y*. If $F(x) \geq F(x_0)$ for all $x \in \mathcal{X}$, then x_0 is called a *minimum point for F on $\mathcal{X}$*.

Lemma 4.8-3 [94, page 464], [134, page 91]: *Assume that F is Gateaux differentiable at x_0. If x_0 is a minimum point for F on $\mathcal{X}$, then $dF(x_0, y) = 0$ for all $y \in \mathcal{X}$. If F is strictly convex on $\mathcal{X}$ and if $dF(x_0, y) = 0$ for all $y \in \mathcal{X}$, then x_0 is a minimum point of F.*

We are now ready to assert and prove the existence of a unique voltage-current regime for the nonlinear infinite network **N** that minimizes (4.23). The current-source vector **h** is assumed to be an arbitrary but fixed member of $l_{\mathbf{M}}$.

Theorem 4.8-4: *Assume* **N** *satisfies Conditions 4.7-1. Then, there exists a unique* $\mathbf{i} \in \mathcal{L}$ *such that*

$$U(\mathbf{i}) = \min\{U(\mathbf{x}) : \mathbf{x} \in \mathcal{L}\}. \tag{4.24}$$

This **i** *satisfies*

$$\langle R(\mathbf{i} + \mathbf{h}), \mathbf{s} \rangle = 0 \qquad (4.25)$$

for all $\mathbf{s} \in \mathcal{L}$*; moreover, (4.25) is not satisfied for at least one* $\mathbf{s} \in \mathcal{L}$ *when* **i** *is replaced by any other member of* $\mathcal{L}$*.*

Proof: First note again that, by Corollary 4.6-6, $W(\mathbf{f})$ is finite for all $\mathbf{f} \in l_{\mathbf{M}}$. Let $\mathcal{F} = \{\mathbf{h} + \mathbf{i} : \mathbf{i} \in \mathcal{L}\}$. Thus, $\mathcal{F}$ is a nonvoid subset of $l_{\mathbf{M}}$. For each integer n, set $\mathcal{F}_n = \{\mathbf{f} \in \mathcal{F} : |||\mathbf{f}||| \leq n\}$. Recall that we set $U(\mathbf{i}) = W(\mathbf{f})$, where $\mathbf{f} = \mathbf{i} + \mathbf{h}$. The proof now proceeds in five steps.

1. *U is a strictly convex functional on* $\mathcal{L}$*.*

Indeed, as was noted in (4.10), every M_j is a strictly convex function on R^1. Hence, the assertion follows from the definition (4.23).

2. *U is weakly lower-sequentially semicontinuous on* $l_{\mathbf{M}}$*.*

Let $\{\mathbf{f}_1, \mathbf{f}_2, \ldots\}$ converge weakly in $l_{\mathbf{M}}$ to $\mathbf{f} \in l_{\mathbf{M}}$. Since $l_{\mathbf{M}^*}$ is isomorphic to the dual of $l_{\mathbf{M}}$ according to the pairing (4.19) (see also (4.21)), this means that, for every $\mathbf{w} \in l_{\mathbf{M}^*}$, $\langle \mathbf{w}, \mathbf{f}_k \rangle \to \langle \mathbf{w}, \mathbf{f} \rangle$ as $k \to \infty$. We can also choose $\mathbf{w}$ to be any unit vector. Consequently, $\mathbf{f}_k$ converges to $\mathbf{f}$ componentwise. Set $\mathbf{f}_k = (f_{k1}, f_{k2}, \ldots)$ and $\mathbf{f} = (f_1, f_2, \ldots)$. Then, $W(\mathbf{f}_k) = \sum M_j(f_{kj}) < \infty$ for every k, where as always $\sum$ without any limits denotes a summation over all j. Moreover, each $M_j(f_{kj})$ is nonnegative. Hence, Fatou's lemma [48, page 152] asserts that

$$\liminf_{k \to \infty} \sum M_j(f_{kj}) \geq \sum \liminf_{k \to \infty} M_j(f_{kj}).$$

By the componentwise convergence of the $\mathbf{f}_k$ and the continuity of the M_j, the right-hand side is equal to $\sum M_j(f_j) = W(\mathbf{f})$. Thus,

$$\liminf_{k \to \infty} W(\mathbf{f}_k) \geq W(\mathbf{f}).$$

With $\mathbf{i}_k = \mathbf{f}_k - \mathbf{h}$ and $\mathbf{i} = \mathbf{f} - \mathbf{h}$, we have $U(\mathbf{i}_k) = W(\mathbf{f}_k)$ and $U(\mathbf{i}) = W(\mathbf{f})$, whence the assertion.

3. *At each* $\mathbf{i} \in \mathcal{L}$*, U is Gateaux differentiable with the Gateaux differential* $dU(\mathbf{i}, \mathbf{s}) = \sum r_j(i_j + h_j)s_j$ *for all* $\mathbf{s} \in \mathcal{L}$*.*

We shall prove that

$$\lim_{t \to 0} \frac{1}{t}[W(\mathbf{f} + t\mathbf{s}) - W(\mathbf{f})] = \sum r_j(f_j)s_j \qquad (4.26)$$

whatever be $\mathbf{f}, \mathbf{s} \in l_{\mathbf{M}}$. This will imply the assertion.

For all j,

$$\lim_{t \to 0} \frac{1}{t}[M_j(f_j + ts_j) - M_j(f_j)] = s_j \lim_{t \to 0} \frac{1}{ts_j} \int_{f_j}^{f_j + ts_j} r_j(x)\,dx$$

$$= s_j r_j(f_j) \qquad (4.27)$$

because r_j is a continuous function. Since r_j is an odd, strictly increasing function, a similar manipulation shows that, for $|t| \le 1$,

$$\frac{1}{|t|} |M_j(f_j + ts_j) - M_j(f_j)| \le \frac{1}{|t|} \int_{|f_j|}^{|f_j|+|ts_j|} r_j(x)\, dx$$

$$\le |s_j| r_j(|f_j| + |s_j|).$$

Now, let $|\mathbf{x}|$ denote the vector $(|x_1|, |x_2|, \ldots)$. We have that $|\mathbf{f}| + |\mathbf{s}| \in l_{\mathbf{M}}$. Therefore, $R(|\mathbf{f}| + |\mathbf{s}|) \in l_{\mathbf{M}^*}$ according to Lemma 4.7-2. So, by (4.21), the pairing (4.19), and the fact that $|\mathbf{s}| \in l_{\mathbf{M}}$, we have

$$\sum |s_j| r_j(|f_j| + |s_j|) < \infty.$$

This allows us to invoke the Lebesgue theorem of dominated convergence [48, page 151] in order to sum both sides of (4.27) and interchange the summation with the limiting process. The result is (4.26). Finally, $dU(\mathbf{i}, \mathbf{s}) = dW(\mathbf{i} + \mathbf{h}, \mathbf{s})$.

   4. *There is a unique minimum point* $\mathbf{i}_0$ *for* U *on* $\mathcal{L}$.

   For any $\mathbf{f} \in \mathcal{F} \subset l_{\mathbf{M}}$ and any $\mathbf{v} \in l_{\mathbf{M}^*}$, we have from Young's inequality (4.12) that

$$|||\mathbf{f}||| = \sup\{\sum |f_j v_j| : \sum M_j^*(v_j) \le 1\} \le W(\mathbf{f}) + 1.$$

Consequently, there is an integer n such that

$$\inf\{W(\mathbf{f}) : \mathbf{f} \in \mathcal{F}\} = \inf\{W(\mathbf{f}) : \mathbf{f} \in \mathcal{F}_n\}. \qquad (4.28)$$

This infimum exists because $W(\mathbf{f}) \ge 0$. Thus, we need seek the minimum point $\mathbf{f}_0 = \mathbf{i}_0 + \mathbf{h}$ for W only within $\mathcal{F}_n$. This will yield the minimum point $\mathbf{i}_0$ for U on $\mathcal{L}$.

   Since $\mathcal{F}_n$ is a bounded set in $l_{\mathbf{M}}$ and $l_{\mathbf{M}}$ is a reflexive Banach space, $\mathcal{F}_n$ is weakly sequentially compact according to Lemma 4.8-2. Let α denote the infimum (4.28). Either there exists an $\mathbf{f}_0 \in \mathcal{F}_n$ such that $W(\mathbf{f}_0) = \alpha$ or we can choose a sequence $\{\mathbf{f}_1, \mathbf{f}_2, \ldots\}$ in $\mathcal{F}_n$ such that $W(\mathbf{f}_k) \to \alpha$ as $k \to \infty$. Consider the latter case. By the weak-sequential compactness of $\mathcal{F}_n$, there is a subsequence $\{\mathbf{f}_{k_m} : m = 1, 2, \ldots\}$ that converges weakly to a point $\mathbf{f}_0 \in l_{\mathbf{M}}$. By step 2, $W(\mathbf{f}_0) \le \alpha$. Moreover, $\{\mathbf{f}_{k_m} - \mathbf{h}\}$ converges weakly to $\mathbf{f}_0 - \mathbf{h}$. By definition of $\mathcal{F}$, all the $\mathbf{f}_{k_m} - \mathbf{h}$ are in $\mathcal{L}$, which is a closed subspace of $l_{\mathcal{M}}$. Therefore, by Lemma 4.8-1, $\mathbf{f}_0 - \mathbf{h} \in \mathcal{L}$. Hence, $\mathbf{f}_0 \in \mathcal{F}$. Whence, $W(\mathbf{f}_0) \ge \alpha$. Consequently, $W(\mathbf{f}_0) = \alpha$. Thus, we have found a minimum point $\mathbf{f}_0$ for W on $\mathcal{F}$.

   That point is unique. This follows from the strict convexity of W (step 1 above). Indeed, suppose there are two minimum points **x** and **y** for W on $\mathcal{F}$. Then,

$$W\left(\frac{\mathbf{x}+\mathbf{y}}{2}\right) = \sum M_j \left(\frac{x_j+y_j}{2}\right)$$
$$< \frac{1}{2}\sum M_j(x_j) + \frac{1}{2}\sum M_j(y_j) = \alpha,$$

which is a contradiction. Thus, the first conclusion of the theorem is proven.

5. *The second conclusion of the theorem is true.*

We have $\mathbf{i}_0 = \mathbf{f}_0 - \mathbf{h}$ as the unique minimum point of U on $\mathcal{L}$. Moreover, U is strictly convex on $\mathcal{L}$ according to step 1. Also, the left-hand side of (4.25) is $dU(\mathbf{i}, \mathbf{s})$ according to step 3. Consequently, the second conclusion follows from Lemma 4.8-3 and the uniqueness of the minimum point for U. ♣

By the last two paragraphs of Section 4.7, we have

Corollary 4.8-5: *The voltage-current regime dictated by Theorem 4.8-4 satisfies Kirchhoff's current law at every finite ordinary node and Kirchhoff's voltage law around every 0-loop (i.e., finite loop) and also around any 1-loop having a unit flow in $\mathcal{L}$.*

Let us now justify the title of this chapter.

Corollary 4.8-6: *For the voltage-current regime dictated by Theorem 4.8-4, the power dissipated in all the resistors is finite and is equal to the power supplied by all the current sources.*

Proof: With $\mathbf{f} = \mathbf{i} + \mathbf{h}$ as always, $U(\mathbf{i}) = W(\mathbf{f}) = \sum M_j(f_j) < \infty$ because $\mathbf{f} \in l_{\mathbf{M}} = c_{\mathbf{M}}$ (Corollary 4.6-6). This implies that $|f_j| < a_j$ for all j sufficiently large. Therefore, by Condition 4.6-1(i), there is a j_0 such that,

$$\sum_{j \geq j_0} f_j r_j(f_j) \leq p \sum_{j \geq j_0} M_j(f_j) < \infty.$$

On the other hand, $\sum_{j < j_0} f_j r_j(f_j)$ is also finite, and hence so too is $\sum f_j r_j(f_j) = \langle R(\mathbf{f}), \mathbf{f} \rangle$, which is the total power dissipated in all the resistors.

Now, set $\mathbf{s} = \mathbf{i} = \mathbf{f} - \mathbf{h}$ in (4.25). This yields $\langle R(\mathbf{f}), \mathbf{f} \rangle = \langle R(\mathbf{f}), \mathbf{h} \rangle$. The power supplied by all the sources is $\langle R(\mathbf{f}), \mathbf{h} \rangle$. ♣

Finally, the linear framework discussed in Chapter 3 is a special case of this nonlinear theory. Indeed, when **N** satisfies Conditions 3.3-1 and a Norton-to-Thevenin transformation is made on all branches, it is readily seen that l_M becomes $\mathcal{I}$, $\mathcal{L}$ becomes $\mathcal{K}$, (4.22) becomes (3.9), and the following holds.

Corollary 4.8-7: *Assume* **N** *is a linear network and satisfies Conditions 3.3-1. Then,* **N** *– with all its branches converted to the Norton form – satisfies Conditions 4.7-1, and the branch currents dictated by Theorem 4.8-4 are the same as the branch currents dictated by Theorem 3.3-5.*

Chapter 5

Transfinite Electrical Networks

We have already seen that the idea of "connections at infinity" can be encompassed within electrical network theory through the invention of 1-nodes. A consequence is the genesis of transfinitely connected graphs, that is, graphs having pairs of nodes that are connected through transfinite paths but not through finite ones. An example of this is provided by Figure 3.4; there is no finite path connecting a node of branch a to a node of branch α, but there are 1-paths that do so. The maximal, finitely connected subnetworks of that figure are the 0-sections of the 1-graph, and the 1-nodes described in Example 3.2-5 connect those infinitely many 0-sections into a 1-graph.

There is an incipient inductive process arising here. Just as 0-nodes connect branches together to produce a 0-graph, so too do 1-nodes connect 0-sections together to produce 1-graphs. The purpose of the present chapter is to pursue this induction. While doing so, we will discover an infinite hierarchy of transfinite graphs. It turns out that most of the electrical network theory discussed so far can be transferred to such graphs to obtain an infinite hierarchy of transfinite networks. Thus, an electrical parameter in one branch can affect the voltage-current pair of another branch, even when every path connecting the two branches must pass through an "infinity of infinite extremities."

To think about this in another way, let each 0-section of Figure 3.4 be replaced by replicates of the entire 1-graph of that figure. Those 1-graphs now become "1-sections" connected to each other through "2-nodes" to yield a "2-graph." The process might (and will) be repeated indefinitely to obtain an "infinite square grid of p-sections" comprising a "$(p + 1)$-graph," whatever be the natural number p. But that's not all: p can even become a transfinite ordinal. The mathematics carries us onward to electrical networks that boggle the mind. We find ourselves proving results about transfinite networks whose graphs can hardly be imagined.

The main difficulty in making these extensions is to find a consistent set of definitions on which to base the construction of transfinite graphs. Once this is accomplished, our theory for electrical 1-networks can be lifted to electrical k-networks without much alteration.

Does this gambit have any practical significance? No. (Well – perhaps not yet.) But so what? Electrical network theory, a root of our massive electrical industry, is also a part of mathematics. Patience with abstractions far removed from engineering will be helpful to a reading of this chapter.

5.1 p-Graphs

A generic form of a phrase we shall often use is "k-entity," where k is either a natural number (i.e., a nonnegative integer) or a transfinite ordinal [1], [116]. In place of "entity" we will actually have "tip," "node," "path," "section," or "network." We shall refer to k as the *rank* of the entity. In this section p is a fixed natural number greater than 1. As with 1-graphs, we will need a number of new definitions; to make their locations more apparent, we will display the principal ones in separate paragraphs.

We now apply recursion to the definitions given in Sections 1.3 and 3.2. Assume that for each $q = 0, 1, \ldots, p-1$ the q-graphs $(\mathcal{B}, \mathcal{N}^0, \ldots, \mathcal{N}^q)$ have been defined for a given countable branch set $\mathcal{B}$ and specified sets $\mathcal{N}^q$ of q-nodes, and also defined are the q-paths P^q, q-connectedness, and q-sections, along with the terminology pertaining to these ideas. This has explicitly been done for $q = 0$ and $q = 1$, and the constructions of this section will extend those definitions by induction to every natural number p.

Consider the $(p - 1)$-graph $\mathcal{G}^{p-1} = (\mathcal{B}, \mathcal{N}^0, \ldots, \mathcal{N}^{p-1})$.

Definition of a $(p-1)$-tip: Two one-ended $(p-1)$-paths in $\mathcal{G}^{p-1}$ are called *equivalent* if they differ at most by a finite number of $(p-1)$-nodes and $(p-2)$-paths. This equivalence relationship partitions the set of all one-ended $(p-1)$-paths in $\mathcal{G}^{p-1}$ into equivalence classes, called $(p-1)$-*tips*. A *representative* of a $(p-1)$-tip is any one of its members.

This $(p-1)$-graph $\mathcal{G}^{p-1}$ need not contain any one-ended $(p-1)$-paths and therefore any $(p-1)$-tips, but in the event that it does, we may partition the set T^{p-1} of all $(p-1)$-tips into subsets T_τ^{p-1} to get $T^{p-1} = \cup T_\tau^{p-1}$, where again τ denotes the index of a subset. No T_τ^{p-1} is void. Furthermore, for each τ let $\mathcal{N}_\tau^{p-1}$ be either the void set or a singleton whose element is a q-node, where $0 \leq q \leq p-1$.

Definition of a p-node: For each τ, the set

$$x_\tau^p = T_\tau^{p-1} \cup \mathcal{N}_\tau^{p-1} \tag{5.1}$$

is called a *p-node*.

Definition of an exceptional element: The element of $\mathcal{N}_\tau^{p-1}$, if it exists, is called the *exceptional element* of x_τ^p.

We impose one more condition, which the exceptional elements, taken together, are required to satisfy.

Condition 5.1-1: *If a p-node contains a q-node (q < p) as an exceptional element, then that q-node does not appear as the exceptional element of any other m-node, where $q < m \leq p$.*

Definition of "to embrace" for p-nodes: A p-node x^p is said to *embrace* itself, and all its elements, and all the elements of its exceptional element x^q if it has one, and all the elements of the exceptional element of x^q if that exists, and so forth for the exceptional elements of decreasing ranks that arise in this way. However, we take it that a p-node does *not* embrace any other entity; in particular, it does not embrace the representatives of its tips nor the representatives of the tips in the aforementioned exceptional elements. If x_1^q and x_2^n are nodes of ranks q and n respectively, where $n \leq q \leq p$, x_1^q is said to *embrace* x_2^n if x_1^q embraces all the elements embraced by x_2^n including x_2^n itself.

Definition of "totally disjoint" for nodes: The q-node x_1^q and the m-node x_2^m are called *totally disjoint* if their sets of embraced elements have a void intersection.

Lemma 5.1-2: *If x_0^q and y_0^p are respectively a q-node and a p-node with $0 \leq q \leq p$ and if x_0^q and y_0^p embrace a common node, then y_0^p embraces x_0^q. If in addition $p = q$, then $x_0^q = y_0^p$.*

Proof: Let z^n denote an n-node that is embraced by both x_0^q and y_0^p. If $n = q$, then, since by definition x_0^q does not embrace another node of the same rank q but does embrace itself, we must have that $z^n = x_0^q$, and so y_0^p embraces $z^n = x_0^q$.

Now assume that $n < q \leq p$. Let x_{-1} be the unique exceptional element in x_0^q, and let x_{-k} be the unique exceptional element in x_{-k+1}

for $k = 2, 3, \ldots$. Thus, we have a finite sequence of nodes $x_{-1}, x_{-2}, \ldots$ of strictly decreasing ranks, one of which is the n-node z^n. Similarly, let $y_{-1}, y_{-2}, \ldots$ comprise the sequence of unique exceptional elements of strictly decreasing ranks such that $\cdots \in y_{-2} \in y_{-1} \in y_0^p$; z^n is also one of those elements. Suppose that y_0^p does not embrace x_0^q. It follows that there will be a node $w = x_{-i} = y_{-j}$ appearing in both sequences such that its predecessors x_{-i+1} and y_{-j+1} $(i, j \geq 1)$ are not the same. This violates Condition 5.1-1. We can conclude that y_0^p embraces x_0^q.

If $q = p$, we must have that $x_0^q = y_0^p$ because again a p-node cannot embrace another p-node. ♣

Definition of a p-graph: A *p-graph* is a $(p+2)$-tuplet

$$\mathcal{G}^p = (\mathcal{B}, \mathcal{N}^0, \ldots, \mathcal{N}^p) \tag{5.2}$$

where $\mathcal{B}$ is a set of branches and each $\mathcal{N}^q$ for $q = 0, \ldots, p$ is a specified nonvoid set of q-nodes.

For each q the specification of $\mathcal{N}^q$ occurs when the partition $\cup \mathcal{T}_r^{q-1}$ and the $\mathcal{N}^{q-1}$ are chosen. This can be done only after the $\mathcal{N}^m$ for $m = 0, \ldots, q-1$ have been specified. As we shall see, in order for $\mathcal{N}^p$ to be nonvoid, the $\mathcal{N}^q$, where $q = 0, \ldots, p-1$, must be infinite sets.

Given $\mathcal{G}^p$ with $p > 1$, we can examine the corresponding $\mathcal{G}^{p-1} = (\mathcal{B}, \mathcal{N}^0, \ldots, \mathcal{N}^{p-1})$. Once again, let $\mathcal{B}_*$ be a subset of $\mathcal{B}$. Let $\mathcal{G}_*^{p-1} = (\mathcal{B}_*, \mathcal{N}_*^0, \ldots, \mathcal{N}_*^{p-1})$ be the reduction of $\mathcal{G}^{p-1}$ induced by $\mathcal{B}_*$. (This has been explicitly defined for $p - 1 = 1$.) We assume for now that $\mathcal{N}_*^0, \ldots, \mathcal{N}_*^{p-1}$ are all nonvoid. Corresponding to each $(p-1)$-tip t^{p-1} of $\mathcal{G}^{p-1}$ there is a unique $(p-1)$-tip of $\mathcal{G}^{p-1}$, namely, the one containing the representatives of t^{p-1}. However, there may be $(p-1)$-tips in $\mathcal{G}^{p-1}$ that do not exist in $\mathcal{G}_*^{p-1}$ because $\mathcal{G}_*^{p-1}$ may not contain any representatives of those $(p-1)$-tips. Let x^p be any p-node of the given p-graph $\mathcal{G}^p$. Remove every $(p-1)$-tip in x^p that does not exist as a $(p-1)$-tip of $\mathcal{G}_*^{p-1}$. Similarly, if x^p has an exceptional element x_0^q, where $0 \leq q \leq p-1$, all of its $(q-1)$-tips – and therefore it too – may not exist in $\mathcal{G}_*^{p-1}$ for the same reason. If so, remove it from x^p as well. If the resulting set x_*^p has at least one $(p-1)$-tip, it is called a *reduced p-node* (*induced by* $\mathcal{B}_*$). Let $\mathcal{N}_*^p$ be the set of all reduced p-nodes. If $\mathcal{N}_*^p$ is not void, let $\mathcal{G}_*$ be the p-graph $(\mathcal{B}_*, \mathcal{N}_*^0, \ldots, \mathcal{N}_*^p)$; if $\mathcal{N}_*^p$ is void, let $\mathcal{G}_*$ be the $(p-1)$-graph $\mathcal{G}_*^{p-1} = (\mathcal{B}_*, \mathcal{N}_*^0, \ldots, \mathcal{N}_*^{p-1})$.

Another situation may arise if $\mathcal{B}_*$ is chosen arbitrarily. There may be some integer q with $0 \leq q \leq p-1$ for which the $\mathcal{N}_*^{q+1}, \ldots, \mathcal{N}_*^p$ are all void when the construction of the preceding paragraph is applied recursively with p replaced by $m = 1, \ldots, p$. Indeed, if $\mathcal{N}_*^{q+1}$ is void, then so too will be $\mathcal{N}_*^m$ for $m = q + 2, \ldots, p$ because representatives of m-tips will not exist. In this case, we let $\mathcal{G}_*$ be the q-graph $(\mathcal{B}_*, \mathcal{N}_*^0, \ldots, \mathcal{N}_*^q)$ where q is the largest natural number for which $\mathcal{N}_*^q$ is not void.

Definition of a reduced graph: $\mathcal{G}_*$ is called the *reduction of* $\mathcal{G}^p$ *with respect to* $\mathcal{B}_*$ or the *reduced graph induced by* $\mathcal{B}_*$.

Now, consider the $(p-1)$-path

$$P^{p-1} = \{\ldots, x_m^{p-1}, P_m^{p-2}, x_{m+1}^{p-1}, P_{m+1}^{p-2}, \ldots\} \qquad (5.3)$$

which is an alternating sequence of $(p-1)$-nodes x_m^{p-1}, $(p-2)$-paths P_m^{p-2}, and possibly a terminal element – a q-node with $q \leq p-1$ – to the left and/or to the right. (Here too, this has been explicitly defined for $p-1 = 1$. In a moment, we shall complete our definition of a higher-rank path by stating the conditions such a path must fulfill. All that need be known right now is that there are entities, called "paths of higher ranks," that are alternating sequences as stated.) P^{p-1} is called *nontrivial* if it has at least three elements. For larger p, each P_m^{p-2} can be expanded into another alternating sequence of nodes and paths of still lower rank, and so forth repeatedly.

Definition of "to embrace" for a $(p-1)$-path: We say that a $(p-1)$-path *embraces* itself, and all its elements, as well as the paths of lower ranks, and ultimately the branches arising in this repeated expansion of paths. We also say that it *embraces* all the elements embraced by its nodes and by the nodes arising in this repeated expansion of paths. However, it does not embrace any other entity.

Definition of "to meet": If a p-node x^p embraces a 0-node x_0^0 belonging to a branch b, we say that b *meets* x^p *at* x_0^0. Now let the integers n and q be no larger than $p-1$. If an n-path P^n and a q-node x^q embrace a node in common, then P^n and x^q are said to *meet at* that node. Also, with regard to (5.1) and (5.3), if a one-ended or endless $(p-1)$-path P^{p-1} contains as a subsequence a representative of a $(p-1)$-tip t^{p-1} in a p-node x^p, then P^{p-1} is said to *meet* x^p *with* t^{p-1}.

For the last situation, it should be noted that, even though P^{p-1} contains a representative of t^{p-1} as a subsequence, t^{p-1} is not an element of P^{p-1}, and any node containing t^{p-1} is not embraced by P^{p-1}.

Definition of "totally disjoint" and "terminally incident": The q-node x^q and the $(p-1)$-path P^{p-1} are called *totally disjoint* if P^{p-1} does not meet x^q. Also, x^q and P^{p-1} are said to be *terminally incident* if P^{p-1} terminates at a node that embraces or is embraced by x^q or if, for $q = p$, P^{p-1} meets x^p with a $(p-1)$-tip in x^p. Moreover, x^q and P^{p-1} are called

terminally incident but otherwise totally disjoint if they are terminally
incident and P^{p-1} does not meet x^q at any other node embraced by
P^{p-1} or with any other $(p-1)$-tip. Two $(p-1)$-paths P_1^{p-1} and P_2^{p-1}
are called *totally disjoint* if the set of all nodes embraced by P_1^{p-1} has a
void intersection with the set of all nodes embraced by P_2^{p-1}.

Note that, in order for P_1^{p-1} and P_2^{p-1} to be totally disjoint, it is not
in general sufficient to impose this void-intersection property on just
the sets of 0-nodes embraced by P_1^{p-1} and by P_2^{p-1}. For example, in
Figure 3.1 the two 1-paths, whose embraced branches are $a_2, a_3, a_4, \ldots$
and $f, a_0, b_1, b_2, b_3, \ldots$, embrace nonintersecting 0-node sets. However,
they are not totally disjoint because one of them ends with x, the other
ends with n_0, and x embraces n_0.

We now complete our recursive definition of a path of higher rank by
explicating the conditions that such a path must satisfy.

Definition of a p-path: Consider the alternating sequence

$$\{\ldots, x_m^p, P_m^{p-1}, x_{m+1}^p, P_{m+1}^{p-1}, \ldots\} \tag{5.4}$$

where each P_m^{p-1} is a nontrivial $(p-1)$-path and each x_m^p is a p-node
except possibly when (5.4) terminates on the left and/or on the right,
in which case the terminal element is a q-node where $0 \le q \le p$. Such a
sequence is called a *p-path* if in addition it satisfies

Conditions 5.1-3:

(i)   *The indices m are restricted to the integers and number the el-
      ements of (5.4) consecutively as indicated.*
(ii) *Each P_m^{p-1} is terminally incident to the two nodes immediately
      preceding and succeeding it in (5.4) but is otherwise totally dis-
      joint from those nodes.*
(iii) *Every two elements in (5.4) that are not adjacent therein are
      totally disjoint.*

As in the definition of a 1-path (Section 3.2), we insert condition (i)
to emphasize the fact that the indices of any sequence are not allowed
to extend into the transfinite ordinals.

A p-path is called *nontrivial* if it has at least three elements, and it
is called either *finite*, or *one-ended*, or *endless* if its sequence (5.4) is
respectively finite, or one-way infinite, or two-way infinite. Also, all the
terminology used with (5.3) is carried over to (5.4).

Definition of a p-loop: A *p-loop* is a finite p-path except for the following
requirement: One of its terminal elements embraces its other terminal
element.

Lemma 5.1-4: *Assume that the p-path (5.4) contains at least one p-node, say, x_{m+1}^p that is not a terminal element. Then, at least one of the adjacent paths, P_m^{p-1} or P_{m+1}^{p-1}, meets x_{m+1}^p with a $(p-1)$-tip.*

Proof: The only way the conclusion can be negated is if both P_m^{p-1} and P_{m+1}^{p-1} terminate at the single exceptional element x_0 in x_{m+1}^p in such a way that P_m^{p-1} has a terminal element y_m and P_{m+1}^{p-1} has a terminal element y_{m+1}, each of which embraces or is embraced by x_0. Three cases arise:

1. y_m and y_{m+1} both embrace x_0: By Lemma 5.1-2, either y_m embraces y_{m+1} or y_{m+1} embraces y_m.

2. y_m embraces x_0 and x_0 embraces y_{m+1} (or conversely): By definition, y_m embraces all the elements embraced by x_0. Hence, y_m embraces y_{m+1}. (Conversely, y_{m+1} embraces y_m.)

3. x_0 embraces both y_m and y_{m+1}: We now invoke the fact that x_0 contains as an element of itself no more than one exceptional element w, and the rank of w is lower than the rank of x_0. Moreover, w contains no more than one exceptional element u, and u is of still lower rank. Continuing in this way, we find that x_0 and all its embraced exceptional elements form a sequence $\{x_0, w, u, \ldots\}$ whose elements have strictly decreasing ranks. So, y_m and y_{m+1} must appear in this sequence. This implies that y_m embraces y_{m+1}, or conversely.

In all three cases, we obtain a contradiction to the fact that P_m^{p-1} and P_{m+1}^{p-1} are totally disjoint. ♣

If (5.4) terminates on the left (or right) at x_a, then the $(p-1)$-path P_m^{p-1} of lowest (or highest) index m will be called the *leftmost* (or *rightmost*) *subpath of rank* $p-1$ *embraced by* (5.4). Similarly, the $(p-2)$-path in that leftmost (rightmost) subpath of lowest (or highest) index, if it exists, will be called the *leftmost* (or *rightmost*) *subpath of rank* $p-2$ *embraced by* (5.4). This terminology is extended to subpaths of still lower rank.

Lemma 5.1-5: *If a p-path P^p terminates on the left (right) at a node x_a of rank q where $q < p$, then P^p embraces leftmost (rightmost) subpaths of every rank n, where $n = q - 1, \ldots, p - 1$.*

Proof: Since P^p terminates on, say, the left, it contains a leftmost subpath P^{p-1} of rank $p-1$. If P^{p-1} does not contain a leftmost subpath of rank $p-2$, then it can meet x_a only with a $(p-1)$-tip. Hence, x_a must be of rank p at least. Thus, if x_a's rank is less than p, P^{p-1} must contain a leftmost subpath of rank $p-2$. This argument can be continued inductively to obtain the proposition. ♣

Since P^p may terminate at x_a^q with an i-node y^i such that $i < q$ and x_a^q embraces y^i, the conclusion of the last lemma may also hold for some values of n smaller than $q - 1$.

Definition of "q-connected": Let the integers m, n, and q be no larger than p. Let x_a^n be an n-node and x_b^m be an m-node. x_a^n and x_b^m are said to be *q-connected* if there exists a finite q-path that meets x_a^n and x_b^m. Two branches are called *q-connected* if their nodes are q-connected. A p-graph is called *q-connected* if all its nodes are q-connected.

Lemma 5.1-6: *If two nodes in a p-graph are q-connected, where $q < p$, then they are n-connected for each $n = q + 1, \ldots , p$.*

Proof: Let

$$P^q = \{x_a^i, P_0^{q-1}, x_1^q, P_1^{q-1}, \ldots , P_i^{q-1}, x_b^m\}$$

be a finite q-path connecting the two nodes x_a^i and x_b^m. Thus, both i and m are no larger than q. Then, $P^{q+1} = \{x_a^i, P^q, x_b^m\}$ is a finite $(q + 1)$-path, $P^{q+2} = \{x_a^i, P^{q+1}, x_b^m\}$ is a finite $(q + 2)$-path, and so forth. ♣

Definition of a q-section: For any q with $0 \leq q \leq p$, a *q-section* of a p-graph $\mathcal{G}$ is a reduction of $\mathcal{G}$ induced by a maximal set of branches that are pairwise q-connected.

Lemma 5.1-7: *An n-path can pass from one q-section to another q-section only if $n > q$.*

Proof: Suppose this is not so. Then, there will be an n-path P^n with $n \leq q$ which terminates at both ends at 0-nodes having incident branches b_a and b_b lying in different q-sections. Thus, P_n is a finite n-path. Consequently, b_a and b_b are n-connected and, by Lemma 5.1-6, q-connected. By the maximality condition of q-sections, b_a and b_b lie in the same q-section, a contradiction. ♣

The last proposition implies that, if $n \leq q$, any n-path or n-loop is confined to a single q-section. On the other hand, the condition $n > q$ is not in general sufficient for the existence of an n-path passing through two given q-sections because nodes of rank larger than q may not be suitably located in $\mathcal{G}$.

5.2 ω-Graphs

In order to comprehend the discussion of this and the next two sections, a knowledge of just the most elementary facts about the countable transfinite ordinals is needed. This information appears in a number of introductory books; see, for example, [1], [64], [116]. Rather than attempting one more exposition of that well-documented material, we simply refer the reader to the readily available sources. Alternatively, the reader may skip this and the next two sections and instead think of k as being restricted to the natural numbers during our discussion of k-networks in Section 5.5. As is customary, ω will denote the least transfinite ordinal.

We start by extending the idea of a p-graph to an entity that has p-nodes for every natural number p. This can be achieved by continuing indefinitely the process of constructing a $(p + 1)$-graph by connecting p-graphs together through $(p + 1)$-nodes.

Definition of an $\vec{\omega}$-graph: An $\vec{\omega}$-graph is an entity obtained by repeating without end the recursion through which a p-graph is constructed. It is specified by the infinite set

$$\mathcal{G}^{\vec{\omega}} = (\mathcal{B}, \mathcal{N}^0, \mathcal{N}^1, \ldots)$$

where now the listing of the sets $\mathcal{N}^p$ of p-nodes continues onward through all the natural numbers p. Each $\mathcal{N}^p$ is a nonvoid set (and is in fact an infinite set, for otherwise $\mathcal{N}^{p+1}$ would be void).

Definition of an $\vec{\omega}$-path: An $\vec{\omega}$-path is a one-way infinite sequence of the form

$$P^{\vec{\omega}} = \{x_0^{q_0}, P_0^{p_0-1}, x_1^{p_1}, P_1^{p_1-1}, x_2^{p_2}, P_2^{p_2-1}, \ldots\} \qquad (5.5)$$

where the natural numbers suffice to index the elements consecutively as shown, $x_0^{q_0}$ is a q_0-node, $x_m^{p_m}$ is a p_m-node, $P_m^{p_m-1}$ is a nontrivial $(p_m - 1)$-path, $q_0 \leq p_0$, the p_m are strictly increasing natural numbers (i.e., $p_0 < p_1 < p_2 < \cdots$), and the members of (5.5) are pairwise totally disjoint except for adjacent members, which are terminally incident but otherwise totally disjoint.

These conditions imply that $P_m^{p_m-1}$ meets $x_{m+1}^{p_{m+1}}$ at a node of rank p_m or less and therefore $P_{m+1}^{p_{m+1}-1}$ meets $x_{m+1}^{p_{m+1}}$ with a $(p_{m+1} - 1)$-tip (see Lemma 5.1-4, whose proof applies just as well to the present situation).

Example 5.2-1. A simple example of an $\vec{\omega}$-graph can be obtained by connecting an infinity of branches in series through 0-nodes to obtain an endless 0-path, then connecting an infinity of endless 0-paths in series through 1-nodes to obtain an endless 1-path, and so forth without end. This $\vec{\omega}$-graph contains an $\vec{\omega}$-path.

On the other hand, consider the infinite star $\vec{\omega}$-graph having a single infinite node n_0 at which terminate an infinity of one-ended paths $P^0, P^1, P^2, \ldots$. We take it that all these paths are terminally connected at n_0 but otherwise totally disjoint from each other. Each P^p is a p-path but not a $(p+1)$-path, p being a natural number. This too is an $\vec{\omega}$-graph, for $\mathcal{N}^p$ is nonvoid whatever be p, but it does not contain a one-ended $\vec{\omega}$-path. ♣

Definition of an $\vec{\omega}$-tip: An equivalence class of all one-ended $\vec{\omega}$-paths that pairwise differ by no more than a finite number of elements is called an *$\vec{\omega}$-tip* and denoted by $t^{\vec{\omega}}$. A *representative* of $t^{\vec{\omega}}$ is any member of the equivalence class.

Assume now that the $\vec{\omega}$-graph $\mathcal{G}^{\vec{\omega}}$ has a nonvoid set $T^{\vec{\omega}}$ of $\vec{\omega}$-tips and choose a partition $T^{\vec{\omega}} = \cup T_{\tau}^{\vec{\omega}}$. Also, for each index τ of the partition, let $\mathcal{N}_{\tau}^{\vec{\omega}}$ be either the void set or a singleton whose element is a q-node for some natural number q.

Definition of an ω-node: For each index τ, the set

$$x^{\omega} = T_{\tau}^{\vec{\omega}} \cup \mathcal{N}_{\tau}^{\vec{\omega}}$$

is called an *ω-node*, and, if $\mathcal{N}_{\tau}^{\vec{\omega}}$ is not void, its unique element is called the *exceptional element* of x^{ω}. In addition, every ω-node is required to satisfy Condition 5.2-2, which is a restriction on its exceptional element reading exactly as does Condition 5.1-1 except that p is replaced by ω.

Condition 5.2-2:   *If an ω-node contains a q-node $(q < \omega)$ as an exceptional element, then that q-node does not appear as the exceptional element of any other m-node, where $q < m \leq \omega$.*

Definition of an ω-graph: An *ω-graph* is an infinite totally ordered set

$$\mathcal{G}^{\omega} = (\mathcal{B}, \mathcal{N}^0, \mathcal{N}^1, \ldots, \mathcal{N}^{\omega})$$

having $\omega + 2$ entries, where $\mathcal{B}$ is a set of branches and each $\mathcal{N}^q$ for $q = 0, \ldots, \omega$ is a nonvoid specified set of q-nodes.

As with any p-graph, when q is a natural number, each $\mathcal{N}^q$ can be specified only after the $\mathcal{N}^m$, for $m = 0, \ldots, q-1$, have been specified, and similarly all the $\mathcal{N}^q$ have to be specified before $\mathcal{N}^{\omega}$ can be specified.

A *reduced graph $\mathcal{G}_*$ of an ω-graph induced by* a subset $\mathcal{B}_*$ of $\mathcal{B}$ is defined exactly as is a reduced graph of a p-graph.

Another way of representing a one-ended $\vec{\omega}$-path is obtained by replacing every index m by $-m$ in (5.5) and in the conditions imposed upon (5.5). Furthermore upon appending the result to the left of (5.5) (and striking out the extra $x_0^{q_0}$), we obtain an *endless $\vec{\omega}$-path $P^{\vec{\omega}}$*. All the terminology for p-paths extends to one-ended and endless $\vec{\omega}$-paths.

Our definition of an ω-*path*

$$\{\ldots, x_m^{\omega}, P_m^{\vec{\omega}}, x_{m+1}^{\omega}, P_{m+1}^{\vec{\omega}}, \ldots\} \tag{5.6}$$

is the same as that of a p-path, given in Section 5.1, except that p is replaced by ω and $p - 1$ by $\vec{\omega}$. The conditions that every ω-path is required to satisfy are

Conditions 5.2-3:

(i) *The indices m are restricted to the integers and number the elements of (5.6) consecutively as indicated.*

(ii) *Each $P_m^{\vec{\omega}}$ is terminally incident to the two nodes immediately preceding and succeeding it in (5.6) but is otherwise totally disjoint from those nodes.*

(iii) *Every two elements in (5.6) that are not adjacent therein are totally disjoint.*

[Note that now each $P_m^{\vec{\omega}}$ in (5.6) must be one-ended or endless, not finite.]

An ω-*loop* is a finite ω-path, except that one of its two terminal nodes embraces the other one.

With these alterations, Lemmas 5.1-2, 5.1-4, and 5.1-5 hold as before except for some obvious modifications. For example, in Lemma 5.1-5 the values for n are now $q - 1, q, q + 1, \ldots$, but not $p - 1$ because with $p = \omega$ there is no ordinal $p - 1$. The proof of Lemma 5.1-5 must be altered accordingly; this can be achieved by first noting that a leftmost (rightmost) subpath of rank $q - 1$ exists and then arguing inductively that leftmost (rightmost) subpaths of ranks $q, q + 1, \ldots$ also exist.

As for connectedness, let x_a^n and x_b^q be nodes of ranks n and q respectively, where $0 \le n \le \omega$ and $0 \le q \le \omega$. These nodes are said to be ω-*connected* if there is a finite p-path, where $p \le \omega$, that meets x_a^n and x_b^q. Two branches are ω-*connected* if their nodes are ω-connected. Finally, an ω-*section* is a reduction of the ω-graph induced by a maximal set of branches that are pairwise ω-connected. Such a section will have more significance in k-graphs of ranks k higher than ω.

5.3 Graphs of Still Higher Ranks

With ω-graphs in hand, we can proceed as in Section 5.1 to obtain $(\omega + p)$-graphs for any natural number $p > 0$ by using $(\omega + p - 1)$-tips to define $(\omega + p)$-nodes. Then, the method of Section 5.2 provides $(\omega + \vec{\omega})$-tips, from which $(\omega \cdot 2)$-nodes and $(\omega \cdot 2)$-graphs can be obtained. (Here, the "dot" denotes multiplication of ordinals [116, page 84].) This process can be continued to generate k-graphs where k is any countable ordinal that can be reached through these recursive constructions. The

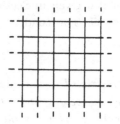

Figure 5.1. An infinite square 0-grid.

procedure of Section 5.1 (Section 5.2) is used when k is a successor ordinal (respectively, limit ordinal).

For example, an ω^2-graph can be constructed as follows: Start with one-ended paths of the form (5.5), where now $p_m = \omega \cdot n_m$ and the n_m are natural numbers indexed by the natural numbers m with $n_0 < n_1 < n_2 < \cdots$. Then define $(\omega \cdot \vec{\omega})$-*tips* $t^{\omega \cdot \vec{\omega}}$ as equivalence classes of such paths, pairwise differing on no more than a finite number of elements. This leads to ω^2-*nodes* of the form

$$x^{\omega^2} = \mathcal{T}_\tau^{\omega \cdot \vec{\omega}} \cup \mathcal{N}_\tau^{\omega \cdot \vec{\omega}}$$

where the single element of $\mathcal{N}^{\omega \cdot \vec{\omega}}$, if it exists, is a node of rank $q < \omega^2$. Then, ω^2-graphs of the form $(\mathcal{B}, \mathcal{N}^0, \ldots, \mathcal{N}^{\omega^2})$ can be defined.

Thus, we can have transfinite graphs of rank k for quite a range of finite or countably transfinite ordinals k.

The idea of connectedness is generalized to k-graphs as follows: Two nodes of arbitrary ranks in the k-graph $\mathcal{G}$ are said to be q-connected, where $q \leq k$, if there is a finite m-path with $m \leq q$ that meets the two nodes. Then, a q-section is any reduction of $\mathcal{G}$ induced by a maximal set of branches whose nodes are q-connected.

Example 5.3-1. We start with the 0-graph of Figure 5.1, which is an infinite square grid and which we shall now refer to as an "infinite square 0-grid." In Example 3.2-5 a 1-graph was constructed out of an infinity of replications of the said 0-grid by arranging the replications into an infinite square pattern as shown in Figure 3.4 and then connecting the 0-grids together at their adjacent horizontal and vertical 0-tips by using two-element 1-nodes. The result may be called an "infinite square 1-grid." Replicate the 1-grid infinitely often, arrange the replications into a square pattern, and connect together their adjacent horizontal and vertical 1-tips by using two-element 2-nodes. The result is a 2-graph; call it an "infinite square 2-grid." Keep repeating this process to get,

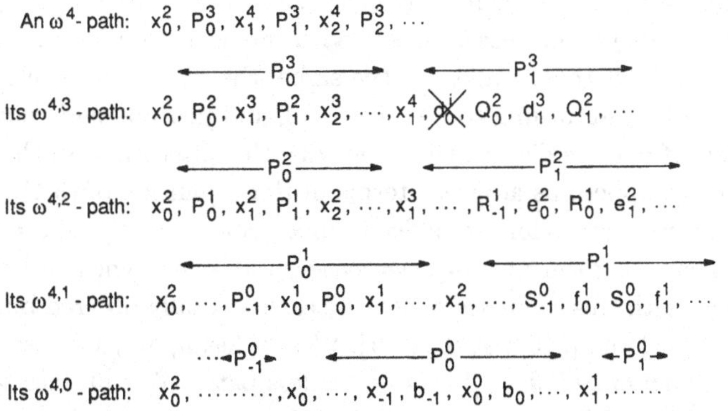

An ω^4-path: $x_0^2, P_0^3, x_1^4, P_1^3, x_2^4, P_2^3, \cdots$

$\longleftarrow\!\!\!-\!\!\!- P_0^3 \longrightarrow$ $\longleftarrow\!\!\!-\!\!\!- P_1^3 \longrightarrow$

Its $\omega^{4,3}$-path: $x_0^2, P_0^2, x_1^3, P_1^2, x_2^3, \cdots, x_1^4, \cancel{d_0^3}, Q_0^2, d_1^3, Q_1^2, \cdots$

$\longleftarrow\!\!\!-\!\!\!- P_0^2 \longrightarrow$ $\longleftarrow\!\!\!-\!\!\!- P_1^2 \longrightarrow$

Its $\omega^{4,2}$-path: $x_0^2, P_0^1, x_1^2, P_1^1, x_2^2, \cdots, x_1^3, \cdots, R_{-1}^1, e_0^2, R_0^1, e_1^2, \cdots$

$\longleftarrow\!\!\!-\!\!\!- P_0^1 \longrightarrow$ $\longleftarrow\!\!\!-\!\!\!- P_1^1 \longrightarrow$

Its $\omega^{4,1}$-path: $x_0^2, \cdots, P_{-1}^0, x_0^1, P_0^0, x_1^1, \cdots, x_1^2, \cdots, S_{-1}^0, f_0^1, S_0^0, f_1^1, \cdots$

$\cdots\!\!\blacktriangleleft P_{-1}^0\blacktriangleright$ $\longleftarrow\!\!\!-\!\!\!- P_0^0 \longrightarrow$ $\blacktriangleleft P_1^0\blacktriangleright$

Its $\omega^{4,0}$-path: $x_0^2, \cdots\cdots\cdots, x_0^1, \cdots, x_{-1}^0, b_{-1}, x_0^0, b_0, \cdots, x_1^1, \cdots\cdots$

Figure 5.2. Illustration of the possible terminal behaviors of a 4-path and its corresponding $(4, q)$-paths, where $q = 0, \ldots, 3$. The 4-path is assumed to terminate on the left at a 2-extremity x_0^2. The d's, e's, f's, and x's denote nodes, and the P's, Q's, R's, and S's denote paths of the indicated ranks. The b's are branches. The terminal element d_0^i ($i \le 3$) on the left-hand side of P_1^3 is deleted in the expansion of P_1^3 in the second line because it is embraced by x_1^4.

for every natural number p, an "infinite square p-grid." Continuing this recursion indefinitely, we get an $\vec{\omega}$-graph having, for every p, p-sections that are infinite square p-grids. Replications of the last $\vec{\omega}$-graph can be arranged into an infinite square pattern and connected together at their adjacent $\vec{\omega}$-tips by two-element ω-nodes to get an "infinite square ω-grid." Continuing recursively onward, we obtain "infinite square k-grids" for a range of countable transfinite ordinals k. ♣

5.4 (k, q)-Paths and Terminal Behavior

By way of more illustration we now examine in greater detail the structure of p-paths, where again p is a natural number larger than 0. Given the p-path (5.4), we can think of each P_m^{p-1} being explicitly written out as a $(p - 1)$-path. This will yield an expanded display of (5.4) involving the p-nodes x_m^p and the possible terminal nodes of (5.4), as well as the $(p - 1)$-nodes, possibly other terminal nodes, and $(p - 2)$-paths arising from the expansions of all the P_m^{p-1} in (5.4). (For an example wherein $p = 4$, see the second line of Figure 5.2.) If a P_m^{p-1} terminates at an i-node d^i ($i \le p - 1$) that is embraced by a p-node, the notation d^i is deleted from the expanded version of (5.4). No such deletion is needed if P_m^{p-1} meets the p-node with a $(p - 1)$-tip. By virtue of Lemma 5.1-4, no more than one such deletion need be made at each p-node. On the

other hand, if P_m^{p-1} is a leftmost (rightmost) subpath, its terminal node
on the left (right) is compared in rank with the terminal node on the left
(right) in (5.4). If those ranks are the same, the two terminal elements
will be identical, according to Lemma 5.1-2, and just one node notation
is retained. If not, we discard the node with the lower rank. In this way,
no two nodes appear as adjacent terms in the expansion of (5.4).

The integers may no longer suffice to index consecutively all the terms
of this expanded form of (5.4). Moreover, its terms, when ordered in
accordance with this sequence of sequences, are totally ordered but may
not be well-ordered [116, pages 20-21]. Well-ordering may be absent, for
example, when one of the P_m^{p-1} is an endless path. We will refer to this
expanded form of (5.4) as a $(p, p-1)$-*path* or as a *transfinite path* and
will denote it by $P^{p,p-1}$.

This process can be repeated, as is indicated in Figure 5.2. An expansion of all the $(p-2)$-paths in the transfinite $(p, p-1)$-path yields a
transfinite $(p, p-2)$-*path* $P^{p,p-2}$. Continuing in this way, we obtain for
$q < p$ the transfinite (p, q)-*path* $P^{p,q}$ and finally a transfinite $(p, 0)$-*path*
$P^{p,0}$, which is totally ordered but not necessarily well-ordered. The elements of $P^{p,0}$ will be branches interspersed with nodes of various ranks.
Two adjacent branches will be separated by the 0-node to which they
are both incident. The higher-order nodes will separate various, finite or
infinite, totally ordered sets.

The $(p, 0)$-*loops* are defined from the p-loops in just the same way and
are called *transfinite loops* if they have more than a finite number of
branches. Note that, for each p-path or p-loop, there is a corresponding,
uniquely defined $(p, 0)$-path or $(p, 0)$-loop.

Assume P^p is a p-path that terminates on the left at a q-node x_0^q where
$q < p$. Then, by Lemma 5.1-5, P^p embraces leftmost subpaths P_0^n of every
rank n, where n varies from $p-1$ down to $q-1$ and perhaps lower. Let m be
the smallest n for which P^p embraces a leftmost node of rank $n-1$. (This
is illustrated in Figure 5.2 for $p = 4$, $q = 2$, and $m = 2$.) In short, there
may be a critical value of n, namely, m such that P^p embraces a leftmost
subpath P_0^{n-1} of rank $n-1$ for every $n = m, \ldots, p$, but not for $n < m$.
This means that P_0^m, $P_0^{m+1}, \ldots, P_0^p$ all terminate at x_0^q with a node, that
P_0^{m-1} meets x_0^q with an $(m-1)$-tip, and that for $n = 0, \ldots, m-2$ there
is no embraced n-path that terminates at or meets x_0^q.

The same kind of pattern will exist at all the p-nodes of P^p at which
$(p-1)$-subpaths terminate.

Similarly, any ω-path can be expanded into nested sequences of paths
and nodes of lower ranks because each $P_m^{\vec{\omega}}$ in (5.6) has the one-ended

form (5.5) or the endless version of (5.5). This yields the (ω, q)-*paths* and ultimately a uniquely defined, totally ordered (but not in general well-ordered) set of branches interspersed with nodes whose ranks vary from 0 to ω; that result will be called an $(\omega, 0)$-*path*.

These ideas extend directly to k-paths, where k is any countable ordinal obtained as indicated in Section 5.3. We obtain thereby (k, q)-paths.

Example 5.4-1. Example 5.3-1 might lead one to believe that the construction of k-graphs of higher ranks require the appending of more branches. This is not so. Some 0-graphs can be converted into k-graphs, whatever be k, simply by shorting the infinite extremities of the 0-graphs with q-nodes of all ranks from 1 to k. Let us indicate how this may be done to the infinite square 0-grid of Figure 5.1.

Consider an endless $(k, 1)$-path $P^{k,1}$ all of whose elements are endless 0-paths and q-nodes, where $q = 1, \ldots, k$. Those endless 0-paths must comprise a countable set because they are pairwise totally disjoint and their branches taken altogether comprise a countable set. Therefore, we can index all those 0-paths by the integers $i = \ldots, -1, 0, 1, \ldots$ [but perhaps not consecutively in the $(k, 1)$-path $P^{k,1}$]. Moreover, we may also number all the horizontal endless paths in Figure 5.1 by i and thereby obtain a one-to-one mapping between the 0-paths in $P^{k,1}$ and the said horizontal paths in Figure 5.1. Now, we need merely make the same q-node connections as those in $P^{k,1}$ to the horizontal extremities of the 0-grid in order to convert it into a k-graph. That k-graph is 0-connected. ♣

5.5 k-Networks

The aim of the preceding definitions about k-graphs is to create the kinds of graphs upon which a foundational theory for transfinite electrical networks might be based. We have accomplished that end. The discussions of Chapter 3 and of the latter part of Chapter 4 can now be transferred directly to k-networks with just a few modifications. Because of this, we will merely sketch through the electrical theory by just stating the principal assumptions and theorems. A more detailed exposition of these extensions is given in [182]. Now, k is finite or transfinite.

A *linear electrical network of rank k* or simply a *linear k-network* is a k-graph every branch of which possesses the analytical representation shown in Figure 1.1(a) with $r > 0$. To derive a foundational theorem by using basic currents as in Section 3.3, we assume that every branch is in its Thevenin form and that the total isolated power $\sum e_j^2 g_j$ is finite, where as always $g_j = 1/r_j$.

The Hilbert space $\mathcal{I}$ is defined exactly as before. To obtain $\mathcal{K}$, however, we must expand the idea of a loop current. Let L denote any oriented q-loop, where $0 \le q \le k$. A *q-loop current* or simply *loop current* is a branch-current vector $\mathbf{i}$ for which $i_j = 0$ if the branch b_j is not embraced by L and $i_j = \pm i$ if b_j is embraced by L; here, $i \in R^1$ is a constant, and the plus (minus) sign is used if b_j's orientation agrees (disagrees) with the q-loop's orientation. If L is a 0-loop, then $\mathbf{i} \in \mathcal{I}$ because there are only finitely many nonzero i_j. On the other hand, if L is a transfinite loop, then $\mathbf{i} \in \mathcal{I}$ if and only if $\sum_{(L)} r_j < \infty$, where $\sum_{(L)}$ denotes a summation over the branch indices for L. L is called *perceptible* whenever $\sum_{(L)} r_j < \infty$.

Our definition of a basic current in a k-network also needs to be expanded. To do this, we need some more preliminary definitions. Let q be any (successor or limit) ordinal such that $0 < q \le k$. As before, a node is called *ordinary* if it is a 0-node that is not embraced by any q-node. A q-loop is called *proper* if it is not an l-loop for any $l < q$. A set of branches is said to have a finite q-diameter if there is a natural number p (possibly $p = 0$) such that every two branches of the set are connected by a q-path having no more than p q-nodes.

A 0-*basic current* will mean a 0-loop current. For $0 < q \le k$ again, a *q-basic current* will be a branch-current vector of the form $\mathbf{i} = \sum \mathbf{i}_m$, where

   (i)   the set of summands $\mathbf{i}_m$ is countable,
   (ii) each $\mathbf{i}_m$ is a q-loop current flowing through a proper q-loop,
   (iii) the support of $\mathbf{i}$ has a finite q-diameter, and
   (iv) given any ordinary 0-node n, the q-loops of only finitely many of the $\mathbf{i}_m$ meet n.

Finally, a *basic current* is any q-basic current with $0 \le q \le k$. Note that this definition encompasses that given in Section 3.3. Moreover, were we to let $q = 0$ in the four conditions, we would obtain a sum $\sum \mathbf{i}_m$ of finitely many 0-loop currents $\mathbf{i}_m$. Another consequence of the definition is that every basic current satisfies Kirchhoff's current law at every ordinary node.

With the k-network $\mathbf{N}$ given, $\mathcal{K}^0$ is the span of all basic currents in $\mathcal{I}$. Thus, $\mathcal{K}^0 \subset \mathcal{I}$. $\mathcal{K}$ is the closure of $\mathcal{K}^0$ in $\mathcal{I}$, and so $\mathcal{K} \subset \mathcal{I}$ too. With the same inner product as that for $\mathcal{I}$, $\mathcal{K}$ is a Hilbert space by itself. Moreover, convergence in $\mathcal{K}$ implies branchwise convergence by virtue of Lemma 3.1-2. Also, we define the *resistance operator R* on $\mathcal{K}$ by $R\mathbf{i} = (r_1 i_1, r_2 i_2, \ldots)$.

As before, the voltage-source vector $\mathbf{e}$ determines a functional on $\mathcal{K}$ through the pairing

$$\langle \mathbf{e}, \mathbf{i} \rangle = \sum e_j i_j, \quad \mathbf{i} \in \mathcal{K}.$$

This functional is a continuous linear mapping of $\mathcal{K}$ into R^1 by virtue of the proof of Lemma 3.3-4, which applies just as well to the present case.

We now state a foundational theorem for k-networks. Its proof is a direct extension of the proofs of Theorem 3.3-5 and Corollary 3.3-6.

Theorem 5.5-1: *Let $\mathbf{N}$ be a linear k-network, every branch of which is in the Thevenin form. Assume that $\sum e_j^2 g_j < \infty$. Then, there exists a unique $\mathbf{i} \in \mathcal{K}$ such that*

$$\langle \mathbf{e} - R\mathbf{i}, \mathbf{s} \rangle = 0 \qquad (5.7)$$

for every $\mathbf{s} \in \mathcal{K}$. Furthermore,

$$\sum i_j^2 r_j = \sum e_j i_j \leq \sum e_j^2 g_j. \qquad (5.8)$$

The relationship (5.8) states that the power dissipated in all the resistors is equal to the power supplied by all the sources, which in turn is no larger than the total isolated power. Another ramification is the reciprocity principle for k-networks, which follows directly from (5.7) as in the proof of Corollary 3.3-7.

Kirchhoff's laws may or may not hold in k-networks, much as they do or do not in 1-networks (see Section 3.4). Once again, a 0-node n is called *restraining* if $\sum_{(n)} g_j < \infty$ where $\sum_{(n)}$ is a summation over the indices of the branches incident to n.

Theorem 5.5-2: *Under the hypothesis of Theorem 5.5-1, Kirchhoff's current law, $\sum_{(n)} \pm i_j = 0$, holds at every ordinary restraining 0-node n, the plus (minus) sign being chosen if the jth branch is incident toward (away from) n. Moreover, $\sum_{(n)} |i_j| < \infty$ even when n is an infinite node.*

Theorem 5.5-3: *Under the hypothesis of Theorem 5.5-1, Kirchhoff's voltage law, $\sum_{(L)} \pm v_j = 0$, holds around every perceptible q-loop L, whatever be $q \leq k$, where the plus (minus) sign is chosen if the jth branch's orientation agrees (disagrees) with L's orientation. Moreover, $\sum_{(L)} |v_j| < \infty$ even when L is transfinite.*

A dual analysis of the given k-network $\mathbf{N}$ can be devised after all the branches are converted into the Norton form. Then, $\mathcal{V}$ is again defined as the space of all branch-voltage vectors $\mathbf{v}$ for which $\sum v_j^2 g_j < \infty$ and also $\langle \mathbf{v}, \mathbf{s} \rangle = \sum v_j s_j = 0$ for every $\mathbf{s} \in \mathcal{K}^0$. This is the same definition as that in Section 3.5 except for the more general form of $\mathcal{K}^0$ that we are now using. With all the other definitions of Sections 3.5 transferred word-for-word to this dual case, we have

Theorem 5.5-4: *Let* **N** *be a linear k-network, every branch of which is in the Norton form. Assume that* $\sum h_j^2 g_j = 0$. *Then, there exists a unique* $\mathbf{v} \in \mathcal{V}$ *such that*

$$\langle \mathbf{w}, \mathbf{h} - G\mathbf{v} \rangle = 0$$

for every $\mathbf{w} \in \mathcal{V}$. *Moreover,*

$$\sum v_j^2 g_j = \sum v_j h_j \leq \sum h_j^2 r_j.$$

Theorem 5.5-5: *Theorems 5.5-1 and 5.5-4 dictate exactly the same voltage-current regime when corresponding branches are related by the Thevenin–Norton conversion.*

Pure sources can also be allowed in k-networks. The expositions of Sections 3.6 and 3.7 are equally valid in this case; the principal alteration is that entities, whose ranks were restricted to 0 or 1, now have various ranks $q \leq k$. Thomson's least power principle (Section 3.8), the concavity of driving-point resistances (Section 3.9), and Rayleigh's monotonicity law (Section 3.9) generalize directly to k-networks as well. Our discussion of the voltages at the extremities of a network given in Sections 3.10 and 3.11 can also be extended to k-networks, but now the definitions of "nondisconnectable p-tips" and of a "sequence of nodes that approach a p-tip" are more complicated [182, Sections 14 and 15].

Finally, let us consider nonlinear k-networks. There does not appear to be any way of extending Dolezal's theory to k-networks, for it employs the matrix of incidences between branches and 0-nodes in an essential way. Our "connections at infinite" given by the nodes of ranks greater than 0 cannot be specified in matrix form it seems.

On the other hand, the DeMichele–Soardi theory can be extended. All that is required is a generalization of the space $\mathcal{L}$ in much the same way as the space $\mathcal{K}$ was generalized. In particular, let $l_\mathbf{M}$ be defined exactly as it was in Section 4.3. The discussions of Sections 4.3 through 4.6 require no changes. We now impose Conditions 5.5-6, which read exactly as do Conditions 4.7-1 except that "0-network or 1-network" is replaced by "k-network."

Conditions 5.5-6: *Let* **N** *be a k-network, every branch b_j of which is in the Norton form with a nonlinear (possibly linear) resistance r_j in parallel with a pure current source h_j (possibly equal to 0). Assume that* all the r_j satisfy Conditions 4.3-1 and 4.6-1 and that there is no coupling between branches. Assume furthermore that $\mathbf{h} = (h_1, h_2, \ldots) \in l_{\mathbf{M}}$.

$\mathcal{L}^0$ is taken to be the span of all the basic currents in $l_{\mathbf{M}}$, and $\mathcal{L}$ is the closure of $\mathcal{L}^0$ in $l_{\mathbf{M}}$. $\mathcal{L}$ is a Banach space by itself when equipped with the norm of $l_{\mathbf{M}}$. The nonlinear resistance operator R is defined as before: $R(\mathbf{f}) = (r_1(f_1), r_2(f_2), \ldots)$. Using the same notation as that employed in Section 4.8, we again define the functional U on any $\mathbf{i} \in l_{\mathbf{M}}$ by

$$U(\mathbf{i}) = \sum \int_0^{f_j} r_j(x)\, dx$$

where $f_j = i_j + h_j$.

Theorem 5.5-7: *Assume that* $\mathbf{N}$ *satisfies the Conditions 5.5-6. Then, there exists a unique* $\mathbf{i} \in \mathcal{L}$ *such that*

$$U(\mathbf{i}) = \min\{U(\mathbf{x}) : \mathbf{x} \in \mathcal{L}\}.$$

This $\mathbf{i}$ *is also uniquely determined as a member of* $\mathcal{L}$ *by the equation*

$$\langle R(\mathbf{i} + \mathbf{h}),\, \mathbf{s} \rangle = 0$$

which is required to hold for all $\mathbf{s} \in \mathcal{L}$. *Furthermore,* $\mathbf{i}$ *satisfies Kirchhoff's current law at every finite ordinary node, and* $\mathbf{v} = R(\mathbf{i} + \mathbf{h})$ *satisfies Kirchhoff's voltage law around every 0-loop and also around every q-loop having a unit flow in* $\mathcal{L}$. *Finally, the power* $\langle R(\mathbf{i} + \mathbf{h}),\, \mathbf{i} + \mathbf{h} \rangle$ *dissipated in all the resistors is finite and is equal to the power* $\langle R(\mathbf{i} + \mathbf{h}),\, \mathbf{h} \rangle$ *supplied by all the current sources.*

We end by remarking that Corollary 4.8-7 also extends to k-networks: Theorem 5.5-1 regarding linear networks is subsumed as a special case of the last theorem – except for the right-hand bound in (5.8).

Chapter 6

Cascades

The preceding chapters were focused on the mathematical foundations of infinite electrical networks, existence and uniqueness theorems being their principal result. Generality was a concomitant aim of those discussions. For the remaining chapters, we shift our attention to particular kinds of networks (namely, the infinite cascades and grids) that are more closely related to physical phenomena. Two examples of this significance were given in Section 1.7, and more will be discussed in Chapter 8. Our proofs will now be constructive, and consequently methods for finding voltage-current regimes will be encompassed. Moreover, various properties of voltage-current regimes will examined.

We must now be specific about any network we hope to analyze. In particular, its graph and element values need to be stipulated everywhere. An easy way of doing this is to impose some regularity upon the network. Most of this chapter (Sections 6.1 to 6.8) is devoted to the simplest of such regularities, the periodic two-times chainlike structures. They are 0-networks appearing in two forms. One form will be called a *one-ended grounded cascade* and is illustrated in Figure 6.1; it has an infinite node as one of its spines, and a one-ended 0-path as the other spine. The other form will be called a *one-ended ungrounded cascade* and is shown in Figure 6.2; in this case, both spines are one-ended 0-paths. The third possibility of both spines being infinite nodes is a degenerate case and will not be discussed. The finite subnetworks N_p comprising the chainlike structures are called *two-ports* in the electrical engineering literature, and for the first form are also called *three-terminal*

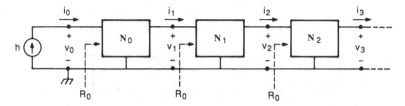

Figure 6.1. A one-ended grounded cascade. The subnetworks $\mathbf{N}_p$ are purely resistive with no sources. There is no branch coupling between different $\mathbf{N}_p$. The entire cascade is driven at the input terminals by a single source, which has been drawn as a pure current source h. The single ground node is drawn as a sequence of nodes all shorted together. R_0 denotes the characteristic resistance when all the $\mathbf{N}_p$ are identical.

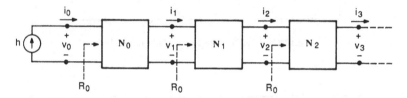

Figure 6.2. A one-ended ungrounded cascade. The $\mathbf{N}_p$ are restricted as in Figure 6.1. R_0 denotes the characteristic resistance again when all the $\mathbf{N}_p$ are identical.

networks. Furthermore, the $\mathbf{N}_p$ are called *stages* of the cascade. We will assume that there is no branch coupling between different $\mathbf{N}_p$. A prototype for the grounded cascade is the *grounded ladder* shown in Figure 6.3, and a prototype for the ungrounded cascade is the *lattice cascade* of Figure 6.4.

When the $\mathbf{N}_p$ have identical graphs and element values, the cascade is called *uniform* and is the subject of the first half of this chapter (Sections 6.1 to 6.6). This uniformity makes infinity imperceptible and frees us from specifying the connections at the network's infinite extremities. All the nodes are taken to be ordinary. Cascades that are not uniform are discussed in the latter half of this chapter (Sections 6.7 to 6.11).

In the next section we assume that the network is linear and derive the classical characteristic-resistance method for determining the voltage-current regime. Nonlinear cascades are examined in the rest of the chapter. A nonlinear generalization of the characteristic resistance is established in Sections 6.2 to 6.6. Our analysis naturally leads us to examine *endless* cascades, that is, cascades wherein the $\mathbf{N}_p$ also

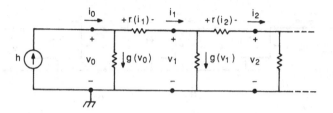

Figure 6.3. A one-ended uniform grounded ladder.

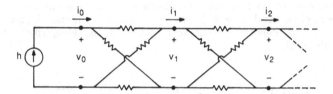

Figure 6.4. A one-ended lattice cascade.

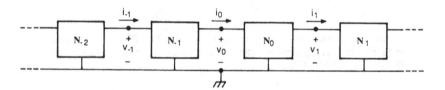

Figure 6.5. An endless grounded cascade.

continue infinitely to the left with $p = -1, -2, \ldots$ and with no source being present; this is illustrated in Figure 6.5 for the grounded case.

As we have already seen in Chapters 3 to 5, infinity may be percep-tible – in the sense that connections at the infinite extremities of the cascade (infinite sources excluded) may affect its voltage-current regime. Such can occur only if the cascade is nonuniform. Moreover, an output pair of terminals, which are either two 1-nodes or a 1-node coupled with a ground node, may be defined for the cascade, and a forward mapping from the input voltage-current pair to the output voltage-current pair may then result. Furthermore, those output terminals may be the input terminals of another one-ended cascade. This process of cascading one-ended cascades can be continued to obtain cascades that are in fact k-networks in the sense of Chapter 5. The point is that input-output systems consisting of transfinite cascades of electrical two-ports exist. This is the subject of Sections 6.9 to 6.11.

6.1 Linear Uniform Cascades

If a cascade is uniform and contains only linear resistors, one may use the ideas of a "characteristic resistance" and a "propagation constant" to calculate the voltages and currents in the cascade. The method is an ancient one in electrical network theory. It was used in the earliest analyses of lumped electrical transmission lines [31], [73], [139], and it also arose in the early theories of mechanical vibrations in repetitive structures [142, page 103]. Yet, until quite recently there had been a lacuna in its reasoning, for it was tacitly assumed that there is only one solution to Kirchhoff's and Ohm's laws for the infinite cascade. In this section we shall show that a justification for the method can be based upon Corollary 3.5-7.

Our argument will employ a convergence theorem for contraction mappings [94, page 314]. Recall that a function F defined on a compact interval I in R^1 is called a *contraction mapping on* I if F maps I into I and if there exists a real number γ with $0 < \gamma < 1$ such that $|F(x) - F(y)| < \gamma|x - y|$ for every $x, y \in I$.

Lemma 6.1-1: *If F is a contraction mapping on the compact interval $I \subset R^1$, then there exists a unique $z \in I$ such that $F(z) = z$. Moreover, $F^k(x) \to z$ as $k \to \infty$ whatever be the choice of $x \in I$.*

We shall also need some standard results [142], [143] from the theory of finite electrical networks with regard to the two-port networks of Figures 6.6 and 6.7. Before stating those results, let us specify the assumptions we impose upon the cascade and its two-ports. By the *terminals* of the cascade or of any stage in the cascade, we mean the external nodes between stages as indicated by the dots of Figures 6.1 and 6.2 and denoted by the n_k in Figures 6.6 and 6.7. In Figure 6.1 the single ground node is drawn split into a sequence of lower terminals all shorted together. This provides an *input* pair n_1, n_3 and an *output* pair n_2, n_4 of terminals for each stage – illustrated by $\mathbf{M}_3$ of Figure 6.6. Similarly, for an ungrounded cascade the terminals for each stage – illustrated by $\mathbf{M}_4$ of Figure 6.7 – are also so paired. These terminal pairs are also called *ports*.

Conditions 6.1-2: $\mathbf{N}$ *is a uniform one-ended cascade of either grounded three-terminal stages* $\mathbf{M}_3$*, illustrated in Figure 6.6, or ungrounded two-port stages* $\mathbf{M}_4$*, illustrated in Figure 6.7. Each* $\mathbf{M}_3$ *or* $\mathbf{M}_4$ *is a finite connected network containing only linear resistors, but no sources. All nodes are ordinary. There is no coupling between branches, neither within any stage nor between different stages. Also, no external terminal of any stage is shorted internally to any other external terminal of the same stage.*

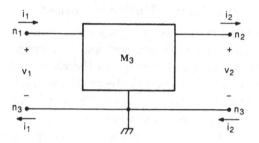

Figure 6.6. A three-terminal network acting as a two-port.

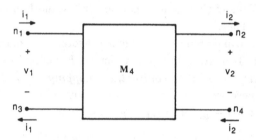

Figure 6.7. A two-port.

Since all nodes are ordinary, all loops are either 0-loops or, in the event there is a short at infinity, 1-loops that pass to infinity along the upper terminals and return along the lower terminals. This implies that, for any member of $\mathcal{K}$, the current passing through an upper terminal of the cascade is equal but opposite in direction to the current passing through the corresponding lower terminal. Consequently, we are truly justified in referring to those terminal pairs as the inputs or outputs of the various stages. Furthermore, we have an *input pair* (v_1, i_1) and an *output pair* (v_2, i_2) of voltages and currents for each stage, and the polarities shown in Figures 6.6 and 6.7 identify them, for power flows into a stage on its input side and out of the stage on its output side if the currents and voltages are all positive.

These voltage-current pairs are related by the *open-circuit resistance parameters* R_{11}, R_{12}, R_{21}, and R_{22} [142, page 53] according to

$$R_{11}i_1 - R_{12}i_2 = v_1, \qquad (6.1)$$

$$R_{21}i_1 - R_{22}i_2 = v_2. \qquad (6.2)$$

Conditions 6.1-2 insure that $R_{11} > 0$ and $R_{22} > 0$. Furthermore, $R_{12} = R_{21}$ by virtue of the reciprocity principle, and, in the case of the three-terminal stage of Figure 6.6, we have $R_{21} > 0$. For the two-port stage of Figure 6.7, R_{21} may be equal to zero – as, for example, when the lattices of Figure 6.4 are balanced; this will yield a degenerate cascade, for which $(v_k, i_k) = (0, 0)$ for all $k = 1, 2, \ldots$ in Figure 6.2, whatever be the driving current h. Henceforth, we will assume that $R_{12} \neq 0$; all our conclusions will obviously hold when $R_{12} = 0$.

We shall also need the *open-circuit voltage-transfer ratio A* [142, page 54], which is defined by $A = v_1/v_2$ when $i_2 = 0$; hence, $A = R_{11}/R_{21}$. Also needed is the *short-circuit current-transfer ratio D* [142, page 54], given by $D = i_1/i_2$ when $v_2 = 0$; thus, $D = R_{22}/R_{21}$, A and D have the same sign, namely, that of R_{21}. In the case of $\mathbf{M}_3$, that sign is positive. It is a standard result that $|A| \geq 1$ and $|D| \geq 1$ [143, page 370]. Actually, we will need a somewhat sharper result; it is proven in [164] and [169, Lemma 3.3].

Lemma 6.1-3: *Under Conditions 6.1-2, at least one of the two true inequalities $|A| \geq 1$ and $|D| \geq 1$ is strict.*

With these facts about finite networks in hand, we turn to the characteristic resistances of one-ended linear uniform cascades. These are denoted by R_0 in Figures 6.1 and 6.2 and represent the ratios v_k/i_k for all k. The classical argument is that these ratios are all the same because the deletion of any initial part of the cascade leaves an identical one-ended cascade. Hence, for any given cascade driven by a current source h at its input, we may calculate v_0 by $v_0 = R_0 h$. Upon replacing all the stages after the first by the single resistor R_0, we may analyze the resulting finite network to obtain v_1 and i_1. Repeating this calculation, we get in turn $(v_2, i_2), (v_3, i_3), \ldots$. Then, the branch voltages and currents can be determined by analyzing each stage separately.

This classical argument overlooks the fact that there are many voltage-current regimes satisfying Kirchhoff's and Ohm's laws. The characteristic resistance method picks out one of them. Which one? Perhaps it is the finite-power regime. But is that regime unique? Our purpose in this section is to answer these questions. The existence and uniqueness theorems of the preceding chapters do not by themselves assure us that the ratios v_k/i_k are independent of k. Some more analysis is needed.

We attack our problem by examining the finite network of Figure 6.8, which is obtained by truncating the one-ended cascade and loading the last stage with a resistance R. We take R to be any member of the extended interval $[0, \infty] = \{R : 0 \leq R \leq \infty\}$; thus, R may also represent a short circuit or an open circuit.

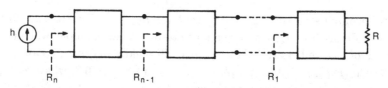

Figure 6.8. A finite cascade terminated in a resistance R or alternatively in an open circuit or a short circuit.

If we can show that the driving-point resistance R_n at the input to this truncated cascade tends to a limit R_0 as $n \to \infty$ and that R_0 is independent of R, we can take R_0 to be by definition the *characteristic resistance* of the infinite cascade; R_0 will then be independent of any load (i.e., either an open circuit, a short circuit, or a positive resistance) that may be connected at infinity. If in addition we can show that the voltage-current regime – obtained by using the characteristic resistance as indicated above – is a finite-power regime, we can invoke Theorem 3.5-6 to conclude that the regime is the same as that dictated by the fundamental Theorem 3.5-2.

Consider the last stage in Figure 6.8. On combining $R_1 = v_1/i_1$ and $R = v_2/i_2$ with (6.1) and (6.2) and using $R_{12} = R_{21} \neq 0$, we obtain

$$R_1 = R_{11} - \frac{R_{21}^2}{R_{22} + R}. \tag{6.3}$$

(When $R = \infty$, we use the convention: $y/R = 0$ for any $y \in R^1$. Also, in the degenerate case where $R_{12} = R_{21} = 0$, we have $R_n = R_{11}$ obviously.) Repeating this calculation, we get $R_n = f^n(R)$, where $f(R)$ denotes the right-hand side of (6.3) and f^n denotes n applications of f.

Set

$$x = R + \frac{R_{22} - R_{11}}{2} \tag{6.4}$$

and

$$F(x) = f(R) + \frac{R_{22} - R_{11}}{2} = r - \frac{R_{21}^2}{r + x}$$

where $r = (R_{11} + R_{22})/2 > 0$. The function f is the same as the three-step composite function: $R \mapsto x$, $x \mapsto F(x)$, $F(x) \mapsto f(R)$. Similarly, f^n is transformed into F^n by an addition and subtraction of $(R_{22} - R_{11})/2$. Moreover, the extended interval $0 \leq R \leq \infty$ is mapped by (6.4) into the extended interval

$$X = \left\{ x : \frac{R_{22} - R_{11}}{2} \leq x \leq \infty \right\}.$$

Furthermore, F maps X into the compact interval $F(X) = [x_0, r]$, where $x_0 > (R_{22} - R_{11})/2$ because $AD > 1$ according to Lemma 6.1-3. Similarly, a straightforward manipulation yields $F^2(X) \subset F(X)$. If we can show that F is a contraction mapping on $F(X)$, we can conclude from Lemma 6.1-1 that, as $n \to \infty$, $F^n(x)$ converges for any $x \in X$. It will then follow that $f^n(R)$ converges to a limit R_0, which is independent of the choice of $R \in [0, \infty]$ according to Lemma 6.1-1 again.

For any $x, y \in F(X)$, we may write

$$F(x) - F(y) = R_{21}^2 \frac{x - y}{(r + x)(r + y)}.$$

Moreover, $(r + x)(r + y) \geq (r + x_0)^2 > R_{22}^2$. Set $\gamma = R_{21}^2/(r + x_0)^2$. Then, $0 < \gamma < R_{21}^2/R_{22}^2 \leq 1$. Consequently, $|F(x) - F(y)| \leq \gamma |x - y|$. We have shown that F truly is a contraction mapping of $F(X)$ into $F^2(X)$. Altogether, we can conclude with

Theorem 6.1-4: *Under Conditions 6.1-2, the one-ended cascade of Figure 6.1 or Figure 6.2 possesses a characteristic resistance R_0. R_0 is independent of any load $R \in [0, \infty]$ that may be connected between the upper and lower 0-tips of the cascade.*

Upon replacing R_1 by R_n and R by R_{n-1} in (6.3) and then letting $n \to \infty$, we obtain

$$R_0 = R_{11} - \frac{R_{21}^2}{R_{22} + R_0} \qquad (6.5)$$

Hence, $R_0 \leq R_{11}$. Moreover, $R_0 \geq 0$ because $R_0 = \lim R_n$ and

$$R_n \geq R_{11} - \frac{R_{21}^2}{R_{22}} > 0.$$

Equation (6.5) is a quadratic equation in R_0 and possesses a negative root and a positive root. The latter is R_0:

$$R_0 = \frac{1}{2}\{R_{11} - R_{22} + [(R_{11} + R_{22})^2 - 4R_{21}^2]^{1/2}\}. \qquad (6.6)$$

This is a standard formula [142, page 107].

Another standard formula [142, pages 106 and 109] is the expression for the parameter

$$T = \frac{v_{k+1}}{v_k} = \frac{i_{k+1}}{i_k}, \quad k = 0, 1, 2, \ldots \qquad (6.7)$$

given by

$$T = K\left[1 - (1 - K^{-2})^{1/2}\right], \qquad K = \frac{A + D}{2} = \frac{R_{11} + R_{22}}{2R_{21}}.$$

It can be obtained by terminating $\mathbf{M}_3$ or $\mathbf{M}_4$ in R_0 and solving for the ratios in (6.7) using the open-circuit resistance parameters. Lemma 6.1-3 states that $|K| > 1$. These definitions yield $v_0 = R_0 h$ and $v_k = T^k v_0$ as the terminal voltages of the cascade.

Lemma 6.1-5: *Under Conditions 6.1-2, $|T| < 1$.*

Proof: As always, assume that $R_{21} \neq 0$. (When $R_{21} = 0$, we have immediately that $|K| = \infty$ and $T = 0$.) By Lemma 6.1-3, $1 < |K| < \infty$. Therefore,

$$0 < 1 - (1 - K^{-2})^{1/2} < 1.$$

So, T has the same sign as K. Since

$$T^{-1} = K \left[1 + (1 - K^{-2})^{1/2} \right],$$

we can also conclude that, if $T > 0$, then $1 < K < \infty$ and $T < T^{-1}$, and, if $T < 0$, then $-\infty < K < -1$ and $T > T^{-1}$. So, in both cases we have $|T| < |T|^{-1}$. ♣

Lemma 6.1-6: *Assume Conditions 6.1-2 and let the branch voltages in the cascade of Figure 6.1 or 6.2 be determined by the characteristic-resistance method for any given $h \in R^1$. Then, the vector of all branch voltages is a member of l_{2r}, Hilbert's real coordinate space.*

Proof: Consider the kth stage. Since that stage is terminated in R_0 and has the input voltage v_k, we can invoke the standard theorems on voltage-transfer ratios [143, page 351] to conclude that any branch voltage v_j within that stage satisfies $|v_j| < |v_k|$. Assume that the stage contains m branches. With $\sum_{(k)}$ denoting the summation over those m branches, we may write $\sum_{(k)} v_j^2 \leq m v_k^2$. We now sum over all the branches of the cascade and invoke (6.7) to write

$$\sum v_j^2 = \sum_{k=0}^{\infty} \sum_{(k)} v_j^2 \leq m \sum_{k=0}^{\infty} v_k^2 = m v_0^2 \sum_{k=0}^{\infty} T^{2k}.$$

By Lemma 6.1-5, the right-hand side is finite. ♣

The characteristic-resistance method yields branch voltages that satisfy Kirchhoff's voltage law around 0-loops and branch currents that satisfy Kirchhoff's current law at all nonground nodes. In view of the last lemma and the uniformity of the cascade, we can invoke Theorem 3.5-6 and Corollary 3.5-7 to obtain the following principal theorem of this section.

Theorem 6.1-7: *Assume that Conditions 6.1-2 hold. Then, the voltage-current regime determined by the characteristic-resistance method is the regime dictated by Theorem 3.5-6 or equivalently by Theorem 3.5-2.*

6.2 Nonlinear Uniform Cascades

We turn now to uniform cascades whose stages are in general nonlinear networks. The reasoning that the deletion of a finite number of initial stages will yield an equivalent cascade suggests that a characteristic resistance may exist even in the nonlinear case. That this is truly so has only recently been established [172], [174] despite the long existence of the characteristic-resistance concept for linear cascades. In this and the next three sections we shall show that uniform (grounded or ungrounded) cascades of nonlinear two-ports satisfying some quite general conditions have characteristic "immittances." By an *immittance* we mean a relation between the voltage and current at a terminal-pair of the cascade specified by a subset of the voltage-current plane. That subset will turn out to be a nonlinear (possibly linear) curve, which may vary quite radically and, in fact, may not be representable by a single-valued resistance or by a single-valued conductance.

As always, we employ the notation specified in Section 1.1, but now (v, i) will denote a point in R^2 with $v, i \in R^1$. If a function f maps $x = (v, i) \in R^2$ into $f(x) \in R^2$, we will also write $f(v, i)$ for $f(x)$ instead of $f((v, i))$. By an *increasing* function $w : R^1 \mapsto R^1$ we will mean a strictly increasing function [i.e., $u_1 < u_2$ implies $w(u_1) < w(u_2)$], and similarly for a *decreasing* function.

Consider the three-terminal network $\mathbf{M}_3$ shown in Figure 6.6 or the two-port $\mathbf{M}_4$ shown in Figure 6.7. The input voltage-current pair (v_1, i_1) and the output voltage-current pair (v_2, i_2) are measured with the polarities shown. Set $x_k = (v_k, i_k) \in R^2$, where $k = 1, 2$. Then, the network's *forward mapping* f is defined by $f(x_1) = x_2$ and its *backward mapping* $b = f^{-1}$ by $b(x_2) = x_1$. We shall impose the restrictions given below in Conditions 6.2-1 upon these mappings. Although they may appear to be rather formidable, they express some natural properties, which are not too restrictive. Example 6.2-2 below shows that quite general ladder networks satisfy them. More generally, a broad class of series-parallel three-terminal networks also satisfy them [172, Sections XI and XII]. So too do certain lattice cascades, as we shall show in Section 6.6.

Set $V = \{(v,0) \in R^2 : v > 0\}$ and $I = \{(0,i) \in R^2 : i > 0\}$. Thus, V and I denote sets in R^2, whereas the voltage v and the current i are one-dimensional variables. For $j = 1, 2, 3,$ or 4, Q_j denotes the open jth quadrant in R^2. Let $a \in R^1, a > 0$, and set $S_a = \{(v,i) \in R^2 : |v| \le a, |i| \le a\}$.

Conditions 6.2-1: *The cascade is endless and uniform (*and may be either grounded or ungrounded*). There is no branch coupling between different stages. The forward mapping f and the backward mapping b satisfy the following restrictions:*

(1) *f is a homeomorphism of R^2 onto R^2 with $f(0) = 0$ and $f(x) \neq x$ for $x \neq 0$. (That is, f is a bijection of R^2 onto R^2 and has exactly one fixed point, the origin; also, f and b are both continuous.)*

(2) *The mapping f has the following directions throughout R^2.*

   a. *If $f(x) \in Q_1$, then $x - f(x) \in Q_1$; if $f(x) \in V$, then $x - f(x) \in Q_1 \cup I$; if $f(x) \in I$, then $x - f(x) \in Q_1 \cup V$.*

   b. *If $x \in Q_2$, then $f(x) - x \in Q_2$; if $x \in -V$, then $f(x) - x \in Q_2 \cup I$; if $x \in I$, then $f(x) - x \in Q_2 \cup (-V)$.*

   c. *If $f(x) \in Q_3$, then $x - f(x) \in Q_3$; if $f(x) \in -V$, then $x - f(x) \in Q_3 \cup (-I)$; if $f(x) \in -I$, then $x - f(x) \in Q_3 \cup (-V)$.*

   d. *If $x \in Q_4$, then $f(x) - x \in Q_4$; if $x \in V$, then $f(x) - x \in Q_4 \cup (-I)$; if $x \in -I$, then $f(x) - x \in Q_4 \cup V$.*

(3) *In some sufficiently small square neighborhood S_a of the origin, the following two conditions hold.*

   a. *If $x, y, f(x), f(y) \in S_a$, if $x \neq y$, and if $y - x \in \bar{Q}_4$, then, $f(y) - f(x) \in Q_4$ and $\|y - x\| < \|f(y) - f(x)\|$.*

   b. *If $x, y, b(x), b(y) \in S_a$, if $x \neq y$, and if $x - y \in \bar{Q}_1$, then, $b(x) - b(y) \in Q_1$ and $\|x - y\| < \|b(x) - b(y)\|$.*

Restrictions 2 and 3a are illustrated in Figure 6.9(a) and restrictions 2 and 3b in Figure 6.9(b). They emulate the behavior of trajectories in a neighborhood of a saddle point of the first kind for a second-order difference dynamical system [66], [106]. Throughout Sections 6.2 to 6.5 we will assume that Conditions 6.2-1 are satisfied without explicitly saying so in every lemma.

Example 6.2-2. Consider an endless uniform grounded ladder. Figure 6.3 can serve as an illustration if we delete the current source h and extend the ladder infinitely to the left, as indicated in Figure 6.5. We assume that the shunt conductances $g : v \mapsto g(v)$ and the series

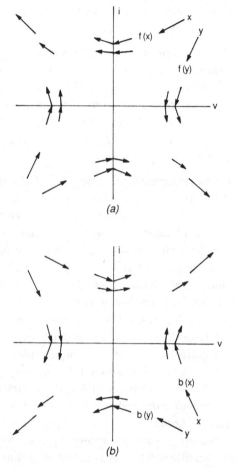

Figure 6.9. Illustration of some of the restrictions in Conditions 6.2-1. (The vectors starting or ending on the v-axis or i-axis may be perfectly vertical or horizontal, respectively.) (a) Restrictions 2 and 3a for the forward mapping f. (b) Restrictions 2 and 3b for the backward mapping b.

resistances $r : i \mapsto r(i)$ are (single-valued) functions on all of R^1. The forward mapping f for this cascade is given by

$$v_{k+1} = v_k - r(i_{k+1}) = v_k - r(i_k - g(v_k)) \tag{6.8}$$

$$i_{k+1} = i_k - g(v_k) \tag{6.9}$$

and the backward mapping $b = f^{-1}$ is given by

$$v_k = v_{k+1} + r(i_{k+1}) \tag{6.10}$$

$$i_k = i_{k+1} + g(v_k) = i_{k+1} + g(v_{k+1} + r(i_{k+1})). \tag{6.11}$$

In this special case of a ladder, Conditions 6.2-1 are entirely equivalent to the following restrictions on g and r.

Conditions 6.2-3:   *The cascade is an endless uniform grounded ladder whose shunt conductances g and series resistances r are continuous mappings of R^1 into R^1. Moreover, they are positive for positive arguments, negative for negative arguments, and zero at the origin (i.e., $vg(v) > 0$ for $v \neq 0$, $ir(i) > 0$ for $i \neq 0$, and $g(0) = r(0) = 0$). Finally, around the origin in some sufficiently small closed interval $[-a, a]$, where $a > 0$, g and r are (strictly) increasing functions.*

It is completely straightforward to show that, if the ladder satisfies Conditions 6.2-3, then it also satisfies Conditions 6.2-1. Conversely, assume that our ladder satisfies Conditions 6.2-1. In view of (6.9) (or (6.10)), g (respectively, r) must be single-valued and continuous if f (respectively, b) is to have these properties. If $g(v) = 0$ for some $v > 0$, then $b(v, 0) = (v, 0)$, that is, b has a fixed point other than the origin, in contradiction to restriction 1 of Conditions 6.2-1. Variations of this argument show that neither g nor r can be zero anywhere on R^1 except at the origin, where they are both zero. By restriction 2 of Conditions 6.2-1, both g and r are positive (negative) for positive (respectively, negative) values of their arguments. Finally, choose $x = (v_x, 0)$ and $y = (v_y, 0)$ on the v-axis with $v_x > v_y$ and such that $x, y, f(x), f(y) \in S_a$. By the continuity of b and the fact that $b(0) = 0$, such an S_a exists. It follows from restriction 3b of Conditions 6.2-1 that $g(v_x) > g(v_y)$. A similar argument with $x = (0, i_x)$, $y = (0, i_y)$, and restriction 3a shows that $r(i_x) > r(i_y)$ whenever $i_x > i_y$. Altogether then, our ladder satisfies Conditions 6.2-3. It is in this way that the nonlinear uniform ladder serves as a prototype for nonlinear uniform grounded cascades.   ♣

Let us now return to the general case of Conditions 6.2-1. Set $f = h + e$ and $b = d + e$, where e is the identity mapping on R^2. Thus, h and d yield the increments between the initial and final values of the forward and backward mappings. (h and e do not now represent sources.) Since f, b, and e are continuous, so too are h and d. Also, $h(0) = d(0) = 0$.

Lemma 6.2-4: *Let D be a compact set in R^2 not containing the origin. Then, there exists a real positive number α (depending on the choice of D) such that $\|h(x)\| > \alpha$ and $\|d(x)\| > \alpha$ for every $x \in D$.*

Proof: The proof for d being no different, consider only h. Since 0 is the only fixed point of f, $h(x) \neq 0$ for every $x \in D$. Now, suppose the lemma is not true. Then, there exists a sequence $\{x_n\} \subset D$ such that $h(x_n) \to 0$. By the compactness of D, there exists a subsequence $\{x_{n_k}\}$ such that $x_{n_k} \to x_\infty \in D$. By the continuity of h, $h(x_\infty) = 0$. We have a contradiction. ♣

6.3 Backward Mappings of the Axes

Consider the following heuristic argument. It is reasonable to expect that a very long but finite uniform cascade will have a driving-point immittance at its input that is close to the characteristic immittance of the corresponding one-ended uniform cascade. Now, if the finite cascade is terminated in a short circuit, the voltage-current pair at the short circuit is a point on the i-axis. So, if we apply the backward mapping many times to the i-axis, we will get a curve whose points are the input voltage-current pairs corresponding to various current values in the output short circuit. That curve ought to be close to the characteristic immittance of the one-ended cascade – and slightly above (below) that characteristic immittance in the first quadrant (respectively, third quadrant) because a short-circuit truncation should increase the absolute value of the input current for any given input voltage according to Rayleigh's monotonicity law (Corollary 3.9-5). By a similar argument with an open-circuit truncation, many backward mappings of the v-axis should yield a curve in the voltage-current plane that is close to the characteristic immittance and slightly below (above) it in the first (third) quadrant. In fact, by letting the number of backward mappings tend to infinity, we should be able to find the characteristic immittance through a limiting process. This is what we shall do.

First, however, let us gather a number of definitions concerning sets in R^2 and then state an important result: Jordan's theorem [38, Chapter VI]. Let $t \in R^1$ be restricted to a compact interval $I = [t_1, t_2]$. Let ν and η be continuous (single-valued) mappings of I into R^1. If the two equations $\nu(t) = \nu(t')$ and $\eta(t) = \eta(t')$ are never satisfied simultaneously for any two points t and t' in I, then the set

$$A = \{(v, i): v = \nu(t), i = \eta(t), t \in I\} \tag{6.12}$$

is called a *Jordan arc*. In more descriptive terms, this is a continuous curve in the (v, i)-plane that does not close on itself. It follows readily that a continuous injective mapping of R^2 into R^2 transforms any Jordan arc into a Jordan arc. $(\nu(t_1), \eta(t_1))$ and $(\nu(t_2), \eta(t_2))$ are called the *end points* of A. Furthermore, a *closed Jordan curve* is defined exactly as is a Jordan arc except that $\nu(t_1) = \nu(t_2)$ and $\eta(t_1) = \eta(t_2)$ are required at the end points t_1 and t_2 of I. An open set S in R^2 is called *connected* if for every two points $x, y \in S$ there is a Jordan arc containing x and y and lying entirely within S.

Lemma 6.3-1 (Jordan's theorem): *A closed Jordan curve C separates R^2 into two open connected sets D and E, one bounded and the other unbounded, such that $\{C, D, E\}$ is a partition of R^2. Moreover, any Jordan arc A having one point in D and another point in E meets C (i.e., $A \cap C \neq \emptyset$).*

The bounded set is called the *interior* of the closed Jordan curve. A connected set D is called *simply* connected if the interior of every closed Jordan curve lying in D is also entirely contained in D.

A variation of Jordan's theorem concerns the separation of the open first quadrant Q_1 by a "one-ended Jordan curve." Now, let J be a semi-infinite interval $[t_1, \infty)$ containing its end point t_1. Let ν and η be continuous mappings of J that define a Jordan arc for each compact subinterval of J. Define B just as A is defined in (6.12) except that I is replaced by J. We shall say that B *approaches infinity* if for every real number α there is a $t_\alpha \in J$ such that $\nu(t)^2 + \eta(t)^2 > \alpha$ for all $t > t_\alpha$. Under these conditions, B is said to be a *one-ended Jordan curve that approaches infinity.* The point $(\nu(t_1), \eta(t_1))$ is the one and only *end point* of B. As before, $Q_1 \cup \{0\}$ denotes the open first quadrant Q_1 in R^2 with the origin appended.

Lemma 6.3-2: *If the one-ended Jordan curve B lies entirely within $Q_1 \cup \{0\}$, approaches infinity, and has the origin as its end point, then B separates Q_1 into two open, simply connected sets D and E such that $\{B, D, E\}$ is a partition of $Q_1 \cup \{0\}$. Moreover, any Jordan arc A in Q_1 having one point in D and another point in E meets B.*

For any real positive number c, let $V_c = \{(v, 0) : 0 < v \leq c\}$ and $I_c = \{(0, i) : 0 < i \leq c\}$. $\bar{V}_c = V_c \cup \{0\}$ is the closure of V_c, and similarly $\bar{V} = V \cup \{0\}$, $\bar{I}_c = I_c \cup \{0\}$, and $\bar{I} = I \cup \{0\}$.

Lemma 6.3-3: *For each $n = 1, 2, 3, \ldots$ and for every $c > 0$, $b^n(\bar{V}_c)$ and $b^n(\bar{I}_c)$ are Jordan arcs with the origin as one of their end points. Also, $b^n(\bar{V})$ and $b^n(\bar{I})$ are contained in $Q_1 \cup \{0\}$. For every $v_0 \in V$ (or $i_0 \in I$), there is at least one point $(v_0, i) \in b^n(\bar{V})$ (respectively, $(v, i_0) \in b^n(\bar{I})$).*

Proof: Consider $b^n(\bar{V}_c)$. We can take $0 \leq v \leq c$ as the compact interval in R^1 that is mapped continuously through $v \mapsto (v, 0) \mapsto b^n(v, 0)$ into a set which resides in $Q_1 \cup \{0\}$ according to restrictions 1 and 2a of Conditions 6.2-1. Since b is injective, that set is a Jordan arc. Since $b^n(0) = 0$, the origin is an end point of the arc.

Similarly, $b^n(\bar{V})$, being the continuous injective image of $\bar{V}$, is a one-ended Jordan arc. By restriction 2a of Conditions 6.2-1 again, for every $(v_0, 0) \in V$, $b^n(v_0, 0) - (v_0, 0) \in Q_1 \cup I$. It follows from this fact and Lemma 6.3-2 that $b^n(\bar{V})$ lies in $Q_1 \cup \{0\}$ and approaches infinity, and, for every $(v_0, 0) \in V$, there is at least one point in $b^n(\bar{V})$ [different from $b^n(v_0, 0)$ in general] having v_0 as its first component.

Similar proofs hold for $b^n(\bar{I}_c)$ and $b^n(\bar{I})$. ♣

$b^n(V)$ is simply $b^n(\bar{V})$ with the end point at the origin deleted, and similarly for $b^n(I)$. In the sequel, we shall often work with $b^n(V)$ and $b^n(I)$.

Let E be a set in R^2 and let D be its projection onto the first-coordinate axis; thus, $D \subset R^1$. If for each $v_0 \in D$ there is exactly one $i_0 \in R^1$ such that $(v_0, i_0) \in E$, then E is the graph of a function that maps each $v_0 \in D$ into that corresponding i_0. To save words, we shall simply say that "E *is* a function of v." Similarly, if D' is the projection of E onto the second-coordinate axis $(D' \subset R^1)$ and if for each $i_0 \in D'$ there is exactly one $v_0 \in R^1$ such that $(v_0, i_0) \in E$, we will again say that "E *is* a function of i." This will simplify our phraseology.

Much of our attention will be confined to the open square $S = (S_a \cap Q_1)^o$, which also depends upon the choice of a. S does not contain any point of the v-axis or i-axis. In the following lemma $b^n(V)$ plays the role of E in our just-defined terminology; $b^n(V)$ is a set in R^2, which defines (we say "is") a function of v, as we shall prove.

Lemma 6.3-4: *Within S and for each $n = 1, 2, \ldots$, $b^n(V)$ is a continuous increasing function of v.*

Proof: Suppose that, within S, $b^n(V)$ is either multivalued as a mapping of v or is a nonincreasing function of v. In either case, there are two distinct points (v, i) and (u, j) on $b^n(V)$ within S such that $(v, i) - (u, j) \in \bar{Q}_4$. By the definition of $b^n(V)$, we have

$$f^n(v, i) \in V \quad \text{and} \quad f^n(u, j) \in V. \tag{6.13}$$

Moreover, by restriction 2a of Condition 6.2-1 and the injectivity of f, $f^n(v, i)$ and $f^n(u, j)$ are distinct points in V_a and all the intermediate points $f^m(v, i)$ and $f^m(u, j)$, where $m = 1, 2, \ldots, n - 1$, are distinct points in S. So, by restriction 3a of Condition 6.2-1, we have $f(v, i) - f(u, j) \in Q_4$, $f^2(v, i) - f^2(u, j) \in Q_4, \ldots, f^n(v, i) - f^n(u, j) \in Q_4$. The last is a contradiction of (6.13). So truly, within S, $b^n(V)$ is an increasing function of v.

Finally, the only type of discontinuity an increasing function can have is a positive jump. But a jump would destroy the connectedness of $b^n(V)$. Hence, $b^n(V)$ is also a continuous function of v. ♣

Virtually the same arguments establish

Lemma 6.3-5: *Within S and for each $n = 1, 2, \ldots, b^n(I)$ is a continuous increasing function of i.*

We shall say that a Jordan arc A *lies above* (or *below*) another Jordan arc B if there is at least one vertical line that meets both A and B and, for every such vertical line L, all the points in $L \cap A$ lie above (respectively, below) all the points in $L \cap B$, that is, if $(v, i) \in L \cap A$ and $(v, j) \in L \cap B$, then $i > j$ (respectively, $i < j$). We shall also say that A *lies to the left* (or *right*) *of B* if there is at least one horizontal line that meets both A and B and, for every such horizontal line H, all the points in $H \cap A$ lie to the left (respectively, right) of all the points in $H \cap B$, that is, if $(v, i) \in H \cap A$ and $(u, i) \in H \cap B$, then $v < u$ (respectively, $v > u$).

Lemma 6.3-6: *Within S and for each $n = 1, 2, 3, \ldots, b^n(I)$ lies above and to the left of $b^{n+1}(I)$, and $b^n(V)$ lies below and to the right of $b^{n+1}(V)$.*

(See Figure 6.10.)

Proof: Since $b^n(I)$ and $b^{n+1}(I)$ both approach the origin through S, we can find a vertical line in S that meets both curves. Moreover, $b^n(I)$ and $b^{n+1}(I)$ cannot meet within S because b is injective and has only the origin as its fixed point.

Suppose $b^n(I)$ lies below $b^{n+1}(I)$. Then, we can choose an $x \in b^{n+1}(I)$ and a $y \in b^n(I)$ such that $y - x \in -I$. By restriction 3a, $f(y) - f(x) \in Q_4, f^2(y) - f^2(x) \in Q_4, \ldots, f^n(y) - f^n(x) \in Q_4$. Moreover, $f^n(y) \in I$. Hence, $f^n(x) \in Q_2$, which implies that $f^{n+1}(x) \in Q_2$, according to restriction 2b. But this cannot be since $x \in b^{n+1}(I)$ so that $f^{n+1}(x) \in I$. Thus, $b^n(I)$ must lie above $b^{n+1}(I)$.

Very similar arguments prove the other three assertions of Lemma 6.3-6. ♣

Lemma 6.3-7: *Within S and for every $n = 1, 2, 3, \ldots, b^n(I)$ lies above and to the left of $b^n(V)$.*

(See Figure 6.10 again.)

Proof: First note again that $b^n(I)$ and $b^n(V)$ cannot meet within S because of the injectivity of b. As before, for each n we can find within S both a horizontal line and a vertical line close enough to the origin to ensure that they each meet both $b^n(I)$ and $b^n(V)$.

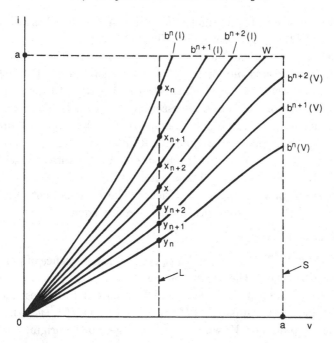

Figure 6.10. The characteristic immittance W as the limit within S of the images of the axes.

Assume that the conclusion is true for some $n \geq 1$. By Lemma 6.3-4, within S, $b^n(V)$ is a continuous increasing function of v. Thus, within S, the vector from any point on $b^n(V)$ to any point on $b^n(I)$ cannot be in Q_4.

Suppose $b^{n+1}(V)$ lies above or to the left of $b^{n+1}(I)$ within S. We can choose two points $x \in b^{n+1}(V)$ and $y \in b^{n+1}(I)$ such that $y - x \in -I$ or $y - x \in V$. Then, by restriction 3a of Condition 6.2-1, $f(y) - f(x) \in Q_4$. But $f(x) \in b^n(V)$ and $f(y) \in b^n(I)$. Thus, we have achieved a contradiction.

So, this lemma holds for $n + 1$ if it holds for n. But the very same argument coupled with the fact that the vector from any point of V to any point of I is not in Q_4 shows that the lemma holds for $n = 1$. ♣

6.4 Trajectories near the Origin

So far, we have that within S the $b^n(I)$ and the $b^n(V)$, where $n = 1, 2, 3, \ldots$, are continuous increasing functions of v, and of i as well. Moreover, each $b^n(I)$ lies above and to the left of $b^{n+1}(I)$, and each

$b^n(V)$ lies below and to the right of $b^{n+1}(V)$. Also, $b^n(I)$ lies above and to the left of $b^n(V)$ for every n. See Figure 6.10 again.

Now, choose a vertical line L in S close enough to the origin to ensure that it intersects one of the $b^n(I)$ and therefore all the $b^{n+1}(I)$, $b^{n+2}(I),\ldots$ as well. As is also illustrated in Figure 6.10, let $x_n, x_{n+1}, x_{n+2},\ldots$ be the corresponding intersection points on L. Their i coordinates are decreasing and bounded below. Hence, these points converge. Let x be the limit point. x is different from every $x_n, x_{n+1}, x_{n+2}, \ldots$. Similarly, let $y_n, y_{n+1}, y_{n+2},\ldots$ be the intersection points of L with the $b^n(V), b^{n+1}(V), b^{n+2}(V), \ldots$. These points converge to a limit y, different from every $y_n, y_{n+1}, y_{n+2}, \ldots$. Clearly, the i component of x is no less than that of y. We shall prove that they are equal.

Lemma 6.4-1: $x = y$.

Proof: $\{f^n(x_n), f^{n+1}(x_{n+1}), f^{n+2}(x_{n+2}), \ldots\}$ is a sequence of points on I. It must converge to the origin, for otherwise $x_n, x_{n+1}, x_{n+2}, \ldots$ could not all be contained in S by virtue of Lemma 6.2-4 and restriction 2 of Conditions 6.2-1. Similarly, $\{f^n(y_n), f^{n+1}(y_{n+1}), f^{n+2}(y_{n+2}), \ldots\}$ is a sequence of points on V, which also converges to the origin.

Suppose, $x \neq y$. Let $\delta = \|x - y\|$. Choose the integer m so large that $\|f^m(x_m) - f^m(y_m)\| < \delta$. We have $\|x_m - y_m\| > \delta$. Moreover, by virtue of restriction 3a of Conditions 6.2-1,

$$\delta < \|x_m - y_m\| < \|f(x_m) - f(y_m)\| < \|f^2(x_m) - f^2(y_m)\|$$
$$< \cdots < \|f^m(x_m) - f^m(y_m)\|.$$

This is a contradiction. ♣

As we vary the location of the vertical line, we obtain different limit points x. We let W denote the set of those limit points generated by all the vertical lines in S that meet one of the $b^n(I)$ in S. Thus, $W \subset S$. W is the graph of a function of v (as before, we shall say that W "is" a function of v) for all v in an interval of the form $(0, v_0)$, where $v_0 \leq a$. Since all the $b^n(I)$ and $b^n(V)$ are increasing functions of v, W is a nondecreasing function of v. Also, the origin is a limit point of W.

It is also true that W is a nondecreasing function of i over some interval $(0, i_0)$, where $i_0 \leq a$. Indeed, for every horizontal line H in S that meets one of the $b^n(V)$, the intersection points $H \cap b^n(I)$ converge, as $n \to \infty$, monotonically toward the right to a limit $z \in S$ and the intersection points $H \cap b^n(V)$ converge monotonically toward the left to the same limit z. This can be seen by repeating the argument of Lemma 6.4-1.

Now choose a vertical line L through z and let

$$x = \lim_{n \to \infty} L \cap b^n(I) = \lim_{n \to \infty} L \cap b^n(V)$$

as before. We must have that $x = z$. Indeed, since the $b^n(I)$ and $b^n(V)$ are continuous increasing functions of v, with the $b^n(I)$ passing to the left and above both x and z and the $b^n(V)$ passing to the right and below both x and z, we would not get at least one of the stated convergences were $x \neq z$. Thus, we have obtained W again – this time as limits of horizontally varying intersection points.

Lemma 6.4-2: *Within S, W is a continuous increasing function of v and of i.*

Proof: We have already seen that W is a nondecreasing function of both v and i. If W as a function of v remains constant over some interval in V_a, then W cannot be a single-valued function of i. Hence, W is an increasing function of v. Reversing the roles of v and i, we see that W is also an increasing function of i. Consequently, it is continuous with respect to both v and i as well. ♣

Sequences of the form $\{x, f(x), f^2(x), \ldots\}$ or $\{\ldots, b(x), x, f(x), \ldots\}$ will be called *forward trajectories*. Similarly, $\{x, b(x), b^2(x), \ldots\}$ and $\{\ldots, f(x), x, b(x), \ldots\}$ will be called *backward trajectories*. We now examine some properties of forward trajectories.

Lemma 6.4-3: *If $x \in W$, then $\{x, f(x), f^2(x), \ldots\}$ converges to the origin within S.*

Proof: As has already been shown, x lies strictly between x_n and y_n, the intersections of the vertical line through x with $b^n(I)$ and $b^n(V)$ respectively. Also, $f^n(x_n) \in I$ and $f^n(y_n) \in V$. Moreover, by restriction 3a, $f^n(x) - f^n(x_n) \in Q_4$ and $f^n(y_n) - f^n(x) \in Q_4$. Therefore, $f^n(x)$ lies in the rectangle $\{(v, i) : 0 < v < f^n(y_n), 0 < i < f^n(x_n)\}$. But, as $n \to \infty$, this rectangle contracts to the origin (as was noted in the proof of Lemma 6.4-1). Hence, $f^n(x)$ tends to the origin too. ♣

By our construction of W, it follows that W separates S. That is, every point of S not on W either lies above and/or to the left of W or it lies below and/or to the right of W. Also, we have already seen that the origin is a limit point of W.

Lemma 6.4-4: *If x lies in S and also below and/or to the right of W, then $\{x, f(x), f^2(x), \ldots\}$ eventually enters Q_4 and diverges to ∞ through Q_4.*

Proof: Let $z \in W$ lie directly above or directly to the left of x. Then, by restriction 3a, $f^n(x) - f^n(z) \in Q_4$ for every n, and $\|f^n(x) - f^n(z)\|$ increases (strictly) as $n \to \infty$. On the other hand, by Lemma 6.4-3, the $f^n(z)$ tend to the origin. It follows that the $f^n(x)$ eventually enter Q_4 and remain therein.

Set $(v_k, i_k) = f^k(x)$. By restriction 2d, after $f^n(x) \in Q_4$ we subsequently have that the $v_n, v_{n+1}, v_{n+2}, \ldots$ increase and the $i_n, i_{n+1},$ $i_{n+2}, \ldots$ decrease monotonically. Suppose both sequences are bounded and therefore converge. Then, by Lemma 6.2-4, the increments $v_{k+1} - v_k$ do not approach zero, and neither do the increments $i_{k+1} - i_k$. But this contradicts the boundedness of both sequences. Hence, at least one of the sequences is unbounded so that $x_n, x_{n+1}, x_{n+2}, \ldots$ must diverge to ∞. ♣

By the same argument, we have

Lemma 6.4-5: *If x lies in S and also above and/or to the left of W, then $\{x, f(x), f^2(x), \ldots\}$ eventually enters Q_2 and diverges to ∞ through Q_2.*

Lemma 6.4-6: *If $x \in W$, then $f(x) \in W$ and $\|f(x)\| < \|x\|$.*

Proof: Within S, $b^m(V)$ is a continuous increasing function of v. Hence, a point $z \in S$ lies above and/or to the left of $b^m(V)$ if and only if there exists a nonzero vector in $\bar{Q}_4$ from z to some point on $b^m(V)$.

Since $x \in W$, x lies directly above or directly to the left of a point on $b^m(V)$ for every $m = n, n+1, n+2, \ldots$, where n is chosen sufficiently large. Let y be such a point on $b^{m+1}(V)$; thus, $y - x \in \bar{Q}_4$. Then, by restriction 3a, $f(y) - f(x) \in Q_4$. Also, $f(y) \in b^m(V)$. So, $f(x)$ lies above and/or to the left of $b^m(V)$, this being true for all $m > n$.

A similar argument shows that, for a possibly different choice of n, $f(x)$ lies below and/or to the right of $b^m(I)$ for all $m > n$. Since the $b^m(V)$ and $b^m(I)$ converge to W in the manner stated above, this proves that $f(x) \in W$.

It now follows from restriction 2a that $\|f(x)\| < \|x\|$. ♣

Lemma 6.4-7: *If $x \in W$ and if $b(x) \in S$, then $b(x) \in W$ and $\|x\| < \|b(x)\|$.*

Proof: Suppose $b(x) \notin W$. Then, we can find a point $z \in W$ such that $b(x) - z \in \bar{Q}_4$ or $z - b(x) \in \bar{Q}_4$. So, by restriction 3a, $x - f(z) \in Q_4$ or $f(z) - x \in Q_4$. By Lemma 6.4-6, $f(z) \in W$. By the hypothesis, $x \in W$ too. This contradicts the fact that, within S, W is a continuous increasing function of i and of v. So, $b(x) \in W$. It now follows from Lemma 6.4-6 that $\|x\| < \|b(x)\|$. ♣

6.5 Characteristic Immittances for Nonlinear Uniform Cascades

Our first objective in this section is to extend W into $Q_1 \backslash S$, the part of the first quadrant outside of S, in such a fashion that the extension continues to represent a characteristic immittance; that is, the extension preserves the properties W has within S – other than its monotonicity. W and its extension will together act as a "separatrix" between the forward trajectories that eventually enter Q_2 and those that eventually enter Q_4. Forward trajectories on the extended W will converge again to the origin. Then, as a second objective, we will establish the existence of analogous characteristic immittances in the other three quadrants.

As always, $\bar{W}$ denotes the closure of W. Thus, $\bar{W}$ contains the origin, and $\bar{W} \subset \bar{S}$. By Lemma 6.4-2, the i-component of $\bar{W}$ is a continuous function of v in some interval $[0, v_0]$, where $0 < v_0 \leq a$. Obviously, so too is the v-component of $\bar{W}$. Hence, $\bar{W}$ is a Jordan arc.

Moreover, for each $n = 1, 2, \dots$, we have that $b^n(\bar{W})$ is a Jordan arc because it is the image of a continuous bijective mapping of the Jordan arc $\bar{W}$. Furthermore, by restriction 2a, $b(\bar{W})$ extends W outside of S, and by Lemma 6.4-7 it coincides with W inside of S; in fact, the upper end point of $b(\bar{W})$ is further away from the origin than is the upper end point of $\bar{W}$. Continuing in this way and invoking the bijectivity of f, we see that $b^{n+1}(\bar{W})$ extends $b^n(\bar{W})$ and coincides with it along all of $b^n(\bar{W})$. Set

$$W_1 = \bigcup_{n=0}^{\infty} b^n(\bar{W}), \tag{6.14}$$

where b^0 denotes the identity mapping. W_1 is the *characteristic immittance* in the first quadrant of a uniform cascade; it takes the role of R_0 in Figure 6.1 or 6.2 when all the stages are identical and possibly nonlinear.

A direct consequence of (6.14) and our prior results is

Lemma 6.5-1: *If* $x \in W_1$, *then* $f(x) \in W_1$ *and* $b(x) \in W_1$.

By applying to $b^k(x)$, where $x \in W_1$, the argument given in the second paragraph of Lemma 6.4-4 and using Lemma 6.4-3 as well, we obtain

Lemma 6.5-2: W_1 *is a one-ended Jordan curve that approaches infinity. Moreover, if* $x \in W_1$, *then* $\lim_{k \to \infty} \|b^k(x)\| = \infty$ *and* $\lim_{k \to \infty} f^k(x) = 0$.

Another consequence of these results and Lemma 6.3-2 is that W_1 separates Q_1 into two open connected sets having W_1 as their common boundary.

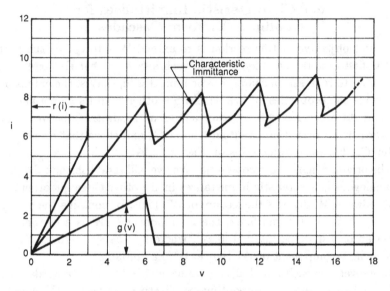

Figure 6.11. The nonlinear parameters r and g and the nonlinear characteristic immittance for the uniform ladder of Example 6.5-3.

Example 6.5-3. For the uniform ladder of Figure 6.3 let us choose

$$g(v) = \begin{cases} 0.5v & \text{for } 0 \leq v \leq 6 \\ 33 - 5v & \text{for } 6 \leq v \leq 6.5 \\ 0.5 & \text{for } 6.5 \leq v \end{cases}$$

and

$$r(i) = \begin{cases} 0.5i & \text{for } 0 \leq i \leq 6 \\ 3 & \text{for } 6 \leq i. \end{cases}$$

These curves are plotted in Figure 6.11. Since they are linear for small v and i, we can apply the formula (6.6) to find the linear characteristic resistance $R_0 = 0.7808$ that represents the slope of the characteristic immittance (i.e., $v = R_0 i$) near the origin. We can then extend this plot toward larger values of v and i by repeatedly applying the backward mapping (6.10) and (6.11). The result is the piecewise-linear curve shown in Figure 6.11. For example, the uppermost point of the initial linear part is (v, i), where $v = 6$ and $i = 7.685$. One application of the backward mapping gives, by (6.10), $v = 6 + 3 = 9$ and, by (6.11), $i = 7.685 + 0.5 = 8.185$. By applying f or b to other points, one can check that the characteristic immittance is mapped onto itself; on the other hand, under f, points effectively below (to the left of) it are mapped effectively below (respectively, to the left of) it and – for repeated mappings – eventually enter the fourth (respectively, the second) quadrant. Note how radically the characteristic immittance keeps varying even though $g(v)$ and $r(i)$ remain constant for large v and i. In fact, the characteristic immittance is neither a single-valued function of v nor a single-valued function of i for many v and i. ♣

Example 6.5-4. Some nonlinear ladders can have perfectly linear characteristic immittances. For instance, consider again the uniform ladder of Figure 6.3, but this time let

$$g(v) = \begin{cases} v & \text{for } 0 \leq v \leq 4 \\ 4 & \text{for } 4 \leq v \leq 6 \\ 1.5v - 5 & \text{for } 6 \leq v \leq 10 \\ 10 & \text{for } 10 \leq v \end{cases}$$

and

$$r(i) = \begin{cases} .5i & \text{for } 0 \leq i \leq 4 \\ 2 & \text{for } 4 \leq i \leq 8 \\ 1.5i - 10 & \text{for } 8 \leq i \leq 10 \\ 5 & \text{for } 10 \leq i \end{cases}$$

These nonlinear relations are illustrated in Figure 6.12. Upon constructing the characteristic immittance as in the preceding example, we find it to be the perfectly linear plot given by $i = 2v$; it is illustrated by the line-with-dots in Figure 6.12. For any given pair (v_0, i_0) on the characteristic immittance, the v_k and i_k do not vary exponentially with $k = 0, 1, 2, \ldots$, as they would for linear ladders, but the characteristic immittance is nonetheless a straight line. ♣

So far, we have restricted v and i to nonnegative values and have obtained the curve W_1, shown in Figure 6.13. Note, however, that Conditions 6.2-1 are effectively the same for all four quadrants, except that the forward and backward mappings reverse their roles in the second and fourth quadrants. Consequently, if we restrict v and i to nonpositive values, we can conclude that there is a continuous curve W_3 in $Q_3 \cup \{0\}$ which possesses properties exactly analogous to those that W_1 possesses in $Q_1 \cup \{0\}$.

On the other hand, if we restrict v to nonpositive values and i to nonnegative values or alternatively if we restrict v to nonnegative values and i to nonpositive values, we transfer our analysis from $\bar{Q}_1$ to $\bar{Q}_2$ or $\bar{Q}_4$, respectively. In this case, the forward mapping f is replaced by the backward mapping b, and conversely. We then obtain a curve W_2 in $Q_2 \cup \{0\}$ and a curve W_4 in $Q_4 \cup \{0\}$ possessing properties analogous to those that W_1 has in $Q_1 \cup \{0\}$.

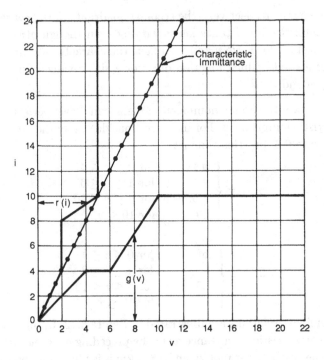

Figure 6.12. The nonlinear parameters r and g and the linear characteristic immittance for the uniform ladder of Example 6.5-4.

These curves are illustrated in Figure 6.13. They separate the (v, i)-plane into four open, simply connected sets (not including the W_j), which we denote by R_N, R_E, R_S, and R_W in the natural way. The W_j act as boundaries or *separatrices* for those sets. The arrows in Figure 6.13 indicate the directions of the increments $f(x) - x$.

It is worth mentioning that, if restrictions 3 hold on all of R^2, then, according to Lemma 6.4-2, $W_1 \cup W_3$ is continuous and increasing everywhere and $W_2 \cup W_4$ is continuous and decreasing everywhere.

Our arguments up to this point have established the following theorem. (See especially Lemmas 6.5-1 and 6.5-2.)

Theorem 6.5-5: *Assume Conditions 6.2-1. Then, the cascade possesses four characteristic immittances W_j, where $j = 1, 2, 3,$ or 4; they are contained in the four quadrants $Q_j \cup \{0\}$ respectively. W_1 is a one-ended Jordan curve that approaches infinity, has its end point at the origin, and separates Q_1 into two open, simply connected sets; in a neighborhood of the origin it is a continuous increasing function of both v and i. The*

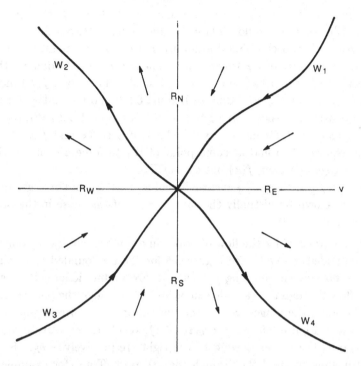

Figure 6.13. The characteristic immittances W_1, W_2, W_3, and W_4 and the four open connected sets R_N, R_E, R_S, and R_W separated by the W_k. The arrows represent forward mappings.

other W_j possess analogous properties. Under the forward mapping f, any trajectory that starts on one of the W_j remains thereon and tends to 0 if $j = 1$ or 3 and to ∞ if $j = 2$ or 4. The same is true under the backward mapping b except that now the trend is to ∞ if $j = 1$ or 3 and to 0 if $j = 2$ or 4.

Also, we can sharpen and extend Lemmas 6.4-4 and 6.4-5 as follows.

Theorem 6.5-6: *Assume Conditions 6.2-1. Let $A = N$, E, S, or W. Under either f or b, any trajectory that starts in R_A remains therein and eventually diverges to ∞. Under f, that divergence occurs within $R_N \cap Q_2$ if the trajectory starts in R_N, within $R_E \cap Q_4$ if it starts in R_E, within $R_S \cap Q_4$ if it starts in R_S, and within $R_W \cap Q_2$ if it starts in R_W. Under b, the divergence occurs within $R_N \cap Q_1$ if the trajectory starts in R_N, within $R_E \cap Q_1$ if it starts in R_E, within $R_S \cap Q_3$ if it starts in R_S, and within $R_W \cap Q_3$ if it starts in R_W.*

Proof: We argue the case where $A = E$ and the mapping is f; the arguments for the other cases are the same. Let $x \in R_E$. $f(x)$ cannot lie on any W_j, for that would violate the injectivity of f. Suppose $f(x) \in R_N \cup R_W$. We can choose a Jordan arc P from x to a point $y \neq 0$ on W_4 with P entirely in R_E except for y. Then, $f(P)$ is a Jordan arc that connects $f(x)$ to another point $f(y) \neq 0$ on W_4. Therefore, $f(P)$ meets either W_1, W_2, or W_3 according to Lemma 6.3-2 and its analogs for the other quadrants. Hence, some point in R_E is mapped onto W_1, W_2, or W_3, a violation of f's injectivity. Similarly, if $x \in R_E$ and $f(x) \in R_S$, we can choose P to end at some point of W_1 to force a contradiction again. Altogether then, $f(x)$ must be in R_E.

That $\{x, f(x), f^2(x), \ldots\}$ eventually enters $R_E \cap Q_4$ and diverges to ∞ therein is shown by virtually the same arguments as those in the proof of Lemma 6.4-4. ♣

Let us now consider the flow of power in a uniform endless (grounded or ungrounded) cascade; see Figure 6.5 for the ungrounded case. Since there is no branch coupling between separate (but identical) stages, power flow through the cascade can occur only through the port connections between the stages. So, under the voltage and current polarities indicated in Figure 6.5, when $(v_k, i_k) \in Q_1$ or Q_3 (alternatively, Q_2 or Q_4), $v_k i_k$ is the power flowing to the right (respectively, $-v_k i_k$ is the power flowing to the left) through the kth port. Thus, for example, if $(v_{k+1}, i_{k+1}) \in \bar{Q}_1 \backslash \{0\}$, then, by restriction 2a, $(v_k, i_k) \in Q_1$; moreover, $v_k i_k - v_{k+1} i_{k+1}$ is the power absorbed in the kth stage.

In particular, if $(v_0, i_0) \in (W_1 \cup W_3) \backslash \{0\}$, then $v_k i_k > 0$ for every k, $v_k i_k \to 0$ as $k \to \infty$, and $v_0 i_0$ is the finite amount of power absorbed in that part of the cascade to the right of the 0th port. However, $v_k i_k \to \infty$ as $k \to -\infty$, and an infinite amount of power is absorbed in any part of the cascade that extends infinitely to the left. On the other hand, if $(v_0, i_0) \in (W_2 \cup W_4) \backslash \{0\}$, then $v_k i_k < 0$ for every k, $v_k i_k \to 0$ as $k \to -\infty$, and $|v_0 i_0|$ is the finite amount of power absorbed in all of the cascade to the left of the 0th port; in this case, $v_k i_k \to -\infty$ as $k \to \infty$, and an infinite amount of power is absorbed in any part of the cascade that extends infinitely to the right.

Furthermore, if $(v_0, i_0) \in R_E \cap Q_1$, then the power injected at the 0th port into the cascade toward the right is positive and remains positive at the kth port ($k > 0$) so long as the trajectory remains in $R_E \cap Q_1$. But, at some $k_0 > 0$, the trajectory will enter $R_E \cap Q_4$, which means that $v_{k_0} i_{k_0} < 0$. Thus, for $k \geq k_0$, the power $|v_k i_k|$ flows to the left at the kth port and increases indefinitely as $k \to \infty$. The physical interpretation of this is that an infinite amount of power is being injected into the

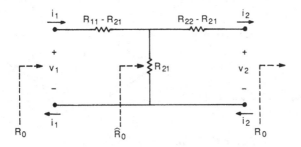

Figure 6.14. The positions of observation for the two characteristic
resistances R_0 and $\hat{R}_0$ discussed in Example 6.5-7.

cascade from infinity at the right and is being dissipated in that part of
the cascade to the right of – and perhaps partly in the stage just before
– the k_0th port.

It is only the trajectories that are confined to one of the W_j and converge to the origin for which power injection at infinity from the right
for $W_1 \cup W_3$ or from the left for $W_2 \cup W_4$ is zero. Because of this physical reason, we refer to $W^f = W_1 \cup W_3$ as the *forward characteristic
immittance of the cascade* and to $W^b = W_2 \cup W_4$ as the *backward characteristic immittance of the cascade*. We have hereby achieved our goal
of constructing the characteristic immittances of a uniform cascade of
nonlinear stages.

Example 6.5-7. Finally, let us relate the nonlinear analysis of this and
the last three sections to that given in Section 6.1 for linear cascades.
Our nonlinear analysis subsumes that linear analysis when R_{21} is positive. To see this, first note that, since $A = R_{11}/R_{21}$ and $D = R_{22}/R_{21}$
have the same sign as R_{21}, it follows that $R_{11} \geq R_{21}$ and $R_{22} \geq R_{21}$,
with one of these inequalities being strict according to Lemma 6.1-3.
Now, the equations (6.1) and (6.2) can be realized by the three-terminal
network of Figure 6.14, where the shunt resistor R_{21} is positive and at
least one of the two series resistors $R_{11} - R_{21}$ and $R_{22} - R_{21}$ is positive too. Furthermore, let us construct a one-ended cascade from this
three-terminal network. We obtain a ladder as shown in Figure 6.3 except that the series resistor $R_{11} - R_{21} \geq 0$ now appears between the
source and the first shunt conductance. The other series resistances
are $r : i \mapsto (R_{11} + R_{22} - 2R_{21})i$, and the shunt conductances are
$g : v \mapsto v/R_{21}$. This obviously satisfies Conditions 6.2-3 (except that
we are now dealing with a one-ended cascade), and therefore the ladder
has the forward characteristic immittance $W^f = W_1 \cup W_3$ specified by

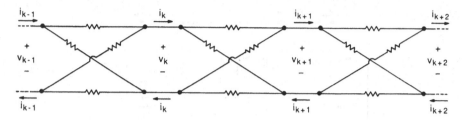

Figure 6.15. An endless lattice cascade.

Theorem 6.5-5. In fact, $W^f = \{(v,i): v = \hat{R}_0 i\}$, where $\hat{R}_0 = R_0 - R_{11} + R_{21}$ and R_0 is given by (6.6). ♣

6.6 Nonlinear Uniform Lattice Cascades

We have examined realizations of nonlinear grounded cascades in the form of ladders (see Examples 6.2-2, 6.5-3, 6.5-4, and 6.5-7) but have not as yet done so for nonlinear ungrounded cascades. This we now do by analyzing a fairly general class of nonlinear uniform endless lattice cascades, illustrated in Figure 6.15. The lattice structure is perhaps the simplest of the intrinsically ungrounded two-ports, but it possesses greater generality because various nonlinear two-ports have lattice representations [33, pages 408–412, 565–572].

We shall achieve our objective of establishing the characteristic immittances for a lattice cascade by showing that Conditions 6.2-1 are satisfied when the lattice of every stage is suitably restricted. Actually, we shall show that the following simpler and stronger Conditions 6.6-1 are satisfied; restriction 1 is the same as before, restriction 2 is obviously stronger than before, and restriction 3 is the same as before except that it is now imposed on all of the (v, i)-plane – instead of in just the vicinity of the origin. (Note also that we have changed our notation somewhat in restriction 3; this will prove to be more convenient.)

Conditions 6.6-1: *The cascade is endless and uniform. There is no branch coupling between different stages. The forward mapping f satisfies the following restrictions.*

(1) *f is a homeomorphism of R^2 onto R^2 with $f(0) = 0$ and $f(x) \neq x$ for $x \neq 0$.*

(2) *f has the following directions throughout R^2.*
   a. *If $f(x) \in \bar{Q}_1 \backslash \{0\}$, then $x - f(x) \in Q_1$.*
   b. *If $x \in \bar{Q}_2 \backslash \{0\}$, then $f(x) - x \in Q_2$.*

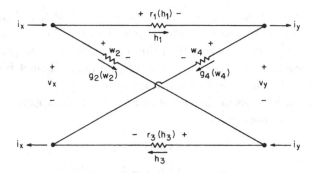

Figure 6.16. The notations and assumed polarities for a typical lattice.

   c.   *If* $f(x) \in \bar{Q}_3 \backslash \{0\}$, *then* $x - f(x) \in Q_3$.

   d.   *If* $x \in \bar{Q}_4 \backslash \{0\}$, *then* $f(x) - x \in Q_4$.

(3) *Let* $f(x) = y$ *and* $f(\xi) = \eta$ *for any* $x, \xi \in R^2$.

   a.   *If* $x - \xi \in \bar{Q}_4 \backslash \{0\}$, *then* $y - \eta \in Q_4$ *and* $\|x - \xi\| < \|y - \eta\|$.

   b.   *If* $y - \eta \in \bar{Q}_1 \backslash \{0\}$, *then* $x - \xi \in Q_1$ *and* $\|x - \xi\| > \|y - \eta\|$.

Consider now any one of the lattices in the uniform cascade. Its voltages, currents, and parameters will be denoted as shown in Figure 6.16; the r's are resistance functions, the g's are conductance functions, the h's are currents, and the w's are voltages. Later on, we will set $x = (v_x, i_x) \in R^2$ and $y = (v_y, i_y) \in R^2$ (now the subscripts x and y do not denote integers). As is indicated in Figures 6.14 and 6.15, we always assume that the current passing through one terminal of any port is equal but opposite in direction to the current through the other terminal of that port. This result can be achieved by assuming that there is an open circuit everywhere at infinity, that is, all 1-nodes are singletons. Our basic objective in this section is to show that the uniform lattice cascade of Figure 6.15 satisfies Conditions 6.6-1 whenever the lattice parameters r_1, g_2, r_3, and g_4 satisfy

Conditions 6.6-2:

(1) *r_1, g_2, r_3, and g_4 are continuous increasing functions that map R^1 into R^1 and assume the value zero at the origin.*

(2) *Let s represent in turn each of r_1, g_2, r_3, g_4. Then, there is a positive constant K_s, depending upon what s represents, such that*

$$|s(t) - s(t')| \leq K_s |t - t'| \qquad (6.15)$$

for every $t, t' \in R^1$. *Moreover,*

$$K_{r_1} K_{g_2} K_{r_3} K_{g_4} < 1. \qquad (6.16)$$

Restriction 2 of Conditions 6.6-2 asserts in effect that the wheatstone bridge inherent in the lattice structure never "balances," that is, we have effectively $r_1'/r_2' < r_4'/r_3'$, where the primes now denote the slopes of the respective resistance functions. As a result, no point distinct from the origin in R^2 is forward-mapped into the origin.

We start our analysis of the lattice of Figure 6.16 by applying Kirchhoff's and Ohm's laws to write

$$v_x = w_2 + r_3(h_3) = w_4 + r_1(h_1) \tag{6.17}$$

$$i_x = h_1 + g_2(w_2) = h_3 + g_4(w_4) \tag{6.18}$$

$$v_y = w_4 - r_3(h_3) = w_2 - r_1(h_1) \tag{6.19}$$

$$i_y = h_3 - g_2(w_2) = h_1 - g_4(w_4). \tag{6.20}$$

Therefore,

$$v_x - v_y = r_1(h_1) + r_3(h_3) \tag{6.21}$$

$$i_x - i_y = g_2(w_2) + g_4(w_4) \tag{6.22}$$

$$v_x + v_y = w_2 + w_4 \tag{6.23}$$

$$i_x + i_y = h_1 + h_3. \tag{6.24}$$

Our first task is to show that any given $x = (v_x, i_x)$ determines uniquely h_1, w_2, h_3, and w_4 and thereby $y = (v_y, i_y)$. Upon combining (6.17) and (6.18) appropriately, we obtain

$$h_1 = i_x - g_2(v_x - r_3(i_x - g_4(v_x - r_1(h_1)))). \tag{6.25}$$

Let us denote the mapping of h_1 on the right-hand side by $p_x : h_1 \mapsto p_x(h_1)$. Thus, we are seeking a fixed point h_1 for the equation $j = p_x(h)$. Set $j' = p_x(h')$. By (6.15),

$$|j - j'| \le K_{g_2} K_{r_3} K_{g_4} K_{r_1} |h - h'|$$

and, by (6.16), p_x is a contraction mapping of R^1 into R^1. So, there truly is a unique fixed point h_1 satisfying (6.25) [94, page 314].

Very similar arguments show that (6.17) and (6.18) uniquely determine w_2, h_3, and w_4 as well. For example, the argument for w_2 starts with

$$w_2 = v_x - r_3(i_x - g_4(v_x - r_1(i_x - g_2(w_2)))). \tag{6.26}$$

Finally, $y = (v_y, i_y)$ is uniquely determined by (6.19) and (6.20) [or by (6.21) and (6.22)]. Thus, our lattice defines a single-valued mapping of R^2 into R^2.

A similar development starting, for example, from (6.19) and (6.20) and the equation

$$h_1 = i_y + g_4(v_y + r_3(i_y + g_2(v_y + r_1(h_1)))) \qquad (6.27)$$

shows that any $y \in R^2$ uniquely determines an $x \in R^2$. Consequently, f is a bijection of R^2 into R^2.

Moreover, for any $x, x' \in R^2$ and the corresponding $h_1, h_1' \in R^1$, we get from (6.25) and restriction 2 of Conditions 6.6-2 that

$$|h_1 - h_1'| = \frac{|i_x - i_{x'}| + K_{g_2}|v_x - v_{x'}| + K_{g_2}K_{r_3}|i_x - i_{x'}| + K_{g_2}K_{r_3}K_{g_4}|v_x - v_{x'}|}{1 - K_{r_1}K_{g_2}K_{r_3}K_{g_4}}$$

which shows that $x \mapsto h_1$ is a continuous mapping. Again, similar arguments show that the mappings $x \mapsto w_2$, $x \mapsto h_3$, and $x \mapsto w_4$ are also continuous. Hence, by (6.19) and (6.20) the mapping $x \mapsto y$ is continuous too. The same development starting with (6.27) shows that $y \mapsto x$ is continuous as well. Thus, f is a homeomorphism.

Furthermore, if $x = 0$, then, by (6.25),

$$|h_1| \leq K_{g_2}K_{r_3}K_{g_4}K_{r_1}|h_1|.$$

In view of (6.16), we can conclude that $h_1 = 0$. Similarly, $w_2 = h_3 = w_4 = 0$, and, by (6.19) and (6.20), $v_y = i_y = 0$. That is, $f(0) = 0$.

Finally, we show that the origin is the only fixed point for f. Assume that $f(x) = x$. By (6.21) and (6.22), $r_1(h_1) = -r_3(h_3)$ and $g_2(w_2) = -g_4(w_4)$. Thus, by restriction 1 of Conditions 6.6-2, h_1 and h_2 are either both zero or have opposite signs. The same may be said of w_2 and w_4. But this in turn implies that, if at least one of the h_1, w_2, h_3, and w_4 is nonzero, then the second equation in either (6.19) or (6.20) is violated. Hence, the only possibility is that $h_1 = w_2 = h_3 = w_4 = 0$. Therefore, $x = 0$.

We have hereby shown that the mapping $f : x \mapsto y$ defined by our lattice satisfies restriction 1 of Conditions 6.6-1.

Before turning to restriction 2 of those conditions, we prove two lemmas.

As before, we adopt the notation $y = f(x)$ and $\eta = f(\xi)$ for any $x, \xi \in R^2$. Moreover, when $\eta = f(\xi)$, we replace h_1, w_2, h_3, and w_4 by χ_1, ω_2, χ_3, and ω_4, respectively.

Lemma 6.6-3: *If $y - \eta \in \bar{Q}_1 \backslash \{0\}$, then $h_1 > \chi_1$, $w_2 > \omega_2$, $h_3 > \chi_3$, and $w_4 > \omega_4$.*

Proof: We will prove only the first conclusion $h_1 > \chi_1$, for the proofs of the others are hardly different; they merely require a permutation in the functions and variables employed below.

Let $m_y(h_1)$ denote the right-hand side of (6.27) and let $m_\eta(\chi_1)$ be the same right-hand side when y is replaced by η and h_1 by χ_1. As was done for p_x, we can prove that m_y and m_η are contraction mappings of R^1 into R^1. Hence, h_1 is the unique fixed point of m_y [see (6.27)], and χ_1 is the unique fixed point of m_η. Now, Lemma 6.1-1 continues to hold when the compact interval I is replaced by R^1 [94, page 314]. Therefore, we may write

$$h_1 = \lim_{n\to\infty} m_y^n(0), \qquad \chi = \lim_{n\to\infty} m_\eta^n(0). \qquad (6.28)$$

Now, if $a, b \in R^1$ and $a \geq b$, then $m_y(a) \geq m_\eta(b)$; this follows from the definition of m_y and m_η, the monotonicities of the r's and g's, and the facts that $v_y \geq v_\eta$ and $i_y \geq i_\eta$. Hence, $h_1 \geq \chi_1$.

We now sharpen this result. Suppose $h_1 = \chi_1$. Thus,

$$i_y + g_4(v_y + r_3(i_y + g_2(v_y + r_1(h_1))))$$
$$= i_\eta + g_4(v_\eta + r_3(i_\eta + g_2(v_\eta + r_1(\chi_1)))).$$

Since $i_y \geq i_\eta$ and g_4 is increasing,

$$v_y + r_3(i_y + g_2(v_y + r_1(h_1))) \leq v_\eta + r_3(i_\eta + g_2(v_\eta + r_1(\chi_1))).$$

Since $v_y \geq v_\eta$ and r_3 is increasing,

$$i_y + g_2(v_y + r_1(h_1)) \leq i_\eta + g_2(v_\eta + r_1(\chi_1)).$$

At this point we can replace the last "$\leq$" by "$<$" because by hypothesis either $i_y > i_\eta$ or $v_y > v_\eta$. (Just repeat the last two steps.) Now, since $i_y \geq i_\eta$ and g_2 is increasing, $v_y + r_1(h_1) < v_\eta + r_1(\chi_1)$, and, since $v_y \geq v_\eta$ and r_1 is increasing, $h_1 < \chi_1$. But this contradicts our supposition. Hence, $h_1 > \chi_1$. ♣

Lemma 6.6-4: *If $x - \xi \in \bar{Q}_4 \backslash \{0\}$, then $h_1 < \chi_1$, $w_2 > \omega_2$, $h_3 < \chi_3$, and $w_4 > \omega_4$.*

This is proven in the same way as was Lemma 6.6-3. For example, to prove $h_1 < \chi_1$ use the mapping p_x defined by the right-hand side of (6.25), and to prove $w_2 > \omega_2$ use the mapping defined by the right-hand side of (6.26). In these cases we have negative signs; nevertheless, the inequalities fall into place in just the right way.

We are now ready to prove restriction 2a of Conditions 6.6-1. Assume that $y = f(x) \in \bar{Q}_1 \backslash \{0\}$. Also, set $\eta = 0$. The latter implies that $\chi_1 = \omega_2 = \chi_3 = \omega_4 = 0$. [For example, $|\chi_1| = |m_0(\chi_1)| \leq K_{g_4} K_{r_3} K_{g_2} K_{r_1} |\chi_1|$, which by (6.16) can only occur if $\chi_1 = 0$.] Substituting these results into Lemma 6.6-3, we conclude that $h_1 > 0$, $w_2 > 0$, $h_3 > 0$, and $w_4 > 0$. We now invoke (6.21) and (6.22) to obtain $v_x > v_y$ and $i_x > i_y$. This verifies restriction 2a of Conditions 6.6-1.

Upon setting $y = 0$ and $\eta = f(\xi) \in \bar{Q}_3 \backslash \{0\}$, we arrive in the same way at restriction 2c of Conditions 6.6-1.

Next, assume that $\xi = 0$. This also implies through (6.16) and (6.26) that $\chi_1 = \omega_2 = \chi_3 = \omega_4 = 0$. So, by Lemma 6.6-4, if $x \in \bar{Q}_4 \backslash \{0\}$, then $h_1 < 0$, $w_2 > 0$, $h_3 < 0$, and $w_4 > 0$. Then, by (6.21) and (6.22) again, $v_x < v_y$ and $i_x > i_y$. This verifies restriction 2d of Conditions 6.6-1. To obtain restriction 2b, start with $x = 0$ and use Lemma 6.6-4 again.

Our next objective is to establish restriction 3a of Conditions 6.6-1. Assume again that $x - \xi \in \bar{Q}_4 \backslash \{0\}$. According to (6.21),

$$v_x - v_\xi = v_y - v_\eta + r_1(h_1) - r_1(\chi_1) + r_3(h_3) - r_3(\chi_3).$$

By Lemma 6.6-4 and the monotonicities of r_1 and r_3,

$$r_1(h_1) - r_1(\chi_1) + r_3(h_3) - r_3(\chi_3) < 0.$$

Hence, $v_x - v_\xi < v_y - v_\eta$. Also, by the hypothesis of restriction 3a, $v_x - v_\xi \geq 0$. Furthermore, according to (6.22),

$$i_x - i_\xi = i_y - i_\eta + g_2(w_2) - g_2(\omega_2) + g_4(w_4) - g_4(\omega_4)$$

and, by Lemma 6.6-4 and the monotonicities of g_2 and g_4, we have $i_x - i_\xi > i_y - i_\eta$. By our hypothesis, $i_x - i_\xi \leq 0$. The four inequalities we have obtained imply that $y - \eta \in Q_4$ and that $\|x - \xi\| < \|y - \eta\|$, which is the conclusion of restriction 3a of Conditions 6.6-1.

Restriction 3b is obtained in the same way from Lemma 6.6-3.

Altogether then, we have shown that any resistive lattice that satisfies Conditions 6.6-2 also satisfies Conditions 6.6-1 and thereby Conditions 6.2-1. Thus, we can transfer our prior conclusions about nonlinear uniform cascades to any uniform cascade $\mathcal{U}$ of identical replications of any lattice satisfying Conditions 6.6-2. So, at every port of $\mathcal{U}$ there are two characteristic immittances; the one corresponding to power flowing to the right (increasing k) – with no power injection at infinity from the right – is the forward characteristic immittance W^f, and the one corresponding to power flowing to the left (decreasing k) – with no power injection at infinity from the left – is the backward characteristic immittance W^b. Actually, we have established even more as a result of our stronger assumption concerning the monotonicity of the lattice parameters; namely, they are monotonic everywhere, rather than in just a neighborhood of the origin. Consequently, restrictions 3 of Conditions 6.2-1 now hold everywhere in R^2, rather than in just a neighborhood of the origin, and therefore the characteristic immittances of $\mathcal{U}$ are monotonic everywhere. In summary, we have

Theorem 6.6-5: *Let $\mathcal{U}$ be a uniform endless cascade of identical replications of a resistive lattice that satisfies Conditions 6.6-2. Then, $\mathcal{U}$*

possesses in the (v, i)-plane a forward characteristic immittance W^f, which is a continuous increasing Jordan curve contained in $\bar{Q}_1 \cup \bar{Q}_3$, passing through the origin, and separating the plane into two open, simply connected sets. It also possesses a backward characteristic immittance W^b, which is a continuous decreasing Jordan curve contained in $\bar{Q}_2 \cup \bar{Q}_4$, passing through the origin, and separating the plane into two open, simply connected sets. Any trajectory having a point in W^f (or in W^b) is contained entirely within W^f (or W^b) and tends to 0 (respectively, ∞) as $k \to \infty$ and to ∞ (respectively, 0) as $k \to -\infty$. Any trajectory that has a point not in $W^f \cup W^b$ never meets or jumps across $W^f \cup W^b$; moreover, it eventually diverges to ∞ in accordance with Theorem 6.5-6.

Let us end our discussion of uniform cascades by briefly mentioning two other features of lattice cascades. The first one concerns lattices whose series resistances r_1 and r_3 and diagonal conductances g_2 and g_4 (see Figure 6.16) have chords with large slopes. Up to now, we only considered the case where those slopes were small enough to satisfy (6.15) and (6.16). When those slopes are large, we can get the same results as before simply by interchanging the port terminals at alternate ports (say, when k is odd).

More specifically, consider the inverse function $g_1 = r_1^{-1}$ of r_1; g_1 is also a continuous increasing function. If the chords of the graph of r_1 have large slopes, then those of g_1 have small slopes. Similar assertions hold for $r_2 = g_2^{-1}$, $g_3 = r_3^{-1}$, and $r_4 = g_4^{-1}$. In fact, with s representing in turn each of r_1, g_2, r_3, and g_4, let us assume that

$$|s(t) - s(t')| \geq K_s |t - t'| \tag{6.29}$$

and

$$K_{r_1} K_{g_2} K_{r_3} K_{g_4} > 1. \tag{6.30}$$

Then, with s^{-1} representing in turn r_1^{-1}, g_2^{-1}, r_3^{-1}, g_4^{-1}, we obtain (6.15) with s replaced by s^{-1} and (6.16) with K_s replaced by $K_{s^{-1}} = K_s^{-1}$. Furthermore, if we switch the output terminals of the lattice of Figure 6.16 but leave the input terminals alone, the series parameters r_1 and r_2 become diagonal parameters and the diagonal parameters g_2 and g_4 become series parameters. For a given cascade let us switch terminals at all the odd-indexed ports but leave the even-indexed ports alone to obtain what we shall call a *swiveled cascade*; accordingly, we also adjust the voltage and current polarities at the odd-indexed ports to make them conform with Figure 6.15. We can conclude that, if a uniform lattice cascade satisfies restriction 1 of Conditions 6.6-2 and also (6.29) and (6.30), then its swiveled cascade satisfies all of Conditions 6.6-2, and conversely. Hence, all our prior conclusions hold for the swiveled cascade. It follows that the trajectories of the original cascade jump back and forth between quadrants – like the trajectories of a second-order difference dynamical system in a neighborhood of a saddle point of the second kind [66], [106].

Our second and last remark concerns the middle range of chord slopes for the lattice parameters, where neither (6.15) and (6.16) nor (6.29) and (6.30) are satisfied. In this case, one or more of the lattices in the cascade may balance and some peculiar behavior may ensue. For example, the nonlinear cascade may be totally unenergized in one part extending infinitely to the left and totally energized in its other part. Also, trajectories may intersect; in fact, an infinity of different trajectories may meet at a nonequilibrium point. Furthermore, there may be certain portions of the voltage-current plane that cannot be penetrated by any trajectory. These strange happenings are discussed in [174, Section VI].

6.7 Nonuniform Cascades: Infinity Imperceptible

The rest of this chapter is devoted to nonuniform cascades, that is, cascades wherein the various stages differ in general. Thus, for a given cascade we no longer have a single forward (or backward) characteristic immittance – the same for every port, but have instead many different forward (or backward) driving-point immittances – varying from port to port. As we shall see in subsequent sections, it is possible now for a voltage-current trajectory to converge to a finite limit other than the origin, a situation that does not occur for a uniform cascade. However, in the present section we shall consider nonuniform cascades for which this situation is disallowed; we shall say that such cascades are *without convergent origin-evading trajectories*. The disallowance is achieved by adding one more assumption (see restriction 4 of Conditions 6.7-1) that mimics the conclusion of Lemma 6.2-4.

Another way we shall assert this disallowance is to say that *infinity is imperceptible*. This terminology is motivated by the fact that there is in this case no need to specify what if anything is connected to 1-nodes of the network. In other words, the local behavior of the cascade is independent of any connections that may be made to the cascade's infinite extremities. (We are disallowing infinite sources.)

With the weaker hypothesis of nonuniformity offset by the addition of restriction 4, the analysis proceeds much as before in Sections 6.2 to 6.5.

The arguments are quite similar, but there are significant differences. Nonetheless, there is so much repetition that another presentation does not seem warranted. So, we shall simply state the definitions, assumptions, and conclusions of the analysis for this case and direct the reader elsewhere [173] for the proofs.

As before, we denote the voltage-current pair at the kth port by $x_k = (v_k, i_k)$, where the usual polarities corresponding to power flowing toward higher k (i.e., to the right) is understood (see Figures 6.6 and 6.7). However, since the stages may now differ, we index the forward and backward mappings as well: $f_k : x_k \mapsto x_{k+1}$ and $b_k : x_{k+1} \mapsto x_k$, respectively. We now impose the following assumptions, which are the same as Conditions 6.2-1 except for the indexing of the forward and backward mappings and the adjoining of restriction 4.

Conditions 6.7-1: *The cascade is endless, and there is no branch coupling between different stages. Moreover, for each k, f_k and b_k satisfy restrictions 1, 2, and 3 of Conditions 6.2-1 (replace f by f_k and b by b_k), as well as the following additional restriction.*

(4)   *Let D be a compact set in R^2 that does not contain the origin. If $D \subset \bar{Q}_2 \cup \bar{Q}_4$ (alternatively, if $D \subset \bar{Q}_1 \cup \bar{Q}_3$), then there exists a positive constant $\alpha \in R^1$, depending on D but not on k, such that $\|f_k(x) - x\| > \alpha$ (respectively, $\|b_k(x) - x\| > \alpha$) for all $x \in D$.*

Restriction 4 imposes the conclusion of Lemma 6.2-4 in a uniform manner with respect to k. However, it restricts the regions where D may appear. Actually, we used that lemma only in this restricted fashion during our analysis of uniform cascades.

As before, a special case is provided by ladder networks, which now are allowed to be nonuniform. Here too, the r's and g's of Figure 6.3 should carry the same indices k as do their arguments. Thus, the forward mapping is given by

$$v_{k+1} = v_k - r_k(i_{k+1}) = v_k - r_k(i_k - g_k(v_k)) \qquad (6.31)$$

$$i_{k+1} = i_k - g_k(v_k) \qquad (6.32)$$

and the backward mapping b_k by

$$v_k = v_{k+1} + r_k(i_{k+1}) \qquad (6.33)$$

$$i_k = i_{k+1} + g_k(v_k) = i_{k+1} + g_k(v_{k+1} + r_k(i_{k+1})) \qquad (6.34)$$

Conditions 6.7-2: *The ladder is endless, and its shunt conductances g_k and series resistances r_k are continuous functions on R^1 into R^1 such* that $g_k(0) = r_k(0) = 0$, $vg_k(v) > 0$ *for all* $v \neq 0$, *and* $ir_k(i) > 0$ *for all* $i \neq 0$. *There is an* $a > 0$ *not depending on* k *such that, on the closed interval* $[-a, a]$, g_k *and* r_k *are increasing functions. Finally, given any numbers* γ *and* δ *with* $0 < \delta < \gamma < \infty$, *there exists a number* $\alpha > 0$ *not depending on* k *such that* $|g_k(v)| > \alpha$ *and* $|r_k(i)| > \alpha$ *whenever* $\delta < |v| < \gamma$ *and* $\delta < |i| < \gamma$ *respectively.*

Lemma 6.7-3: *If a ladder satisfies Conditions 6.7-2, then it also satisfies Conditions 6.7-1.*

Returning to the general case of a nonuniform cascade satisfying Conditions 6.7-1, we define some composite mappings. For $k < n$, set $f_{k,n} = f_{n-1}f_{n-2}\cdots f_{k+1}f_k$ and $b_{k,n} = b_k b_{k+1} \cdots b_{n-2}b_{n-1}$. Thus, $b_{k,n} = f_{k,n}^{-1}$. $f_{k,n}$ is the forward mapping from the kth port to the nth port, and $b_{k,n}$ is the backward mapping from the nth port to the kth port. With $x_k = (v_k, i_k)$ and $x_n = (v_n, i_n)$, we have $x_n = f_{k,n}(x_k)$ and $x_k = b_{k,n}(x_n)$.

As in Section 6.2, $V = \{(v, 0) \in R^2 : v > 0\}$, $I = \{(0, i) \in R^2 : i > 0\}$, and $S = \{(v, i) : 0 < v < a, 0 < i < a\}$, where a is the number used for restriction 3 of Conditions 6.2-1. Within S, for fixed k, and as $n \to \infty$, we have that $b_{k,n}(V)$ converges in the fashion stated in Section 6.4 to a Jordan arc W_k, which is a continuous increasing function of v. Moreover, $b_{k,n}(I)$ tends in a similar way to the same curve W_k.

Lemma 6.7-4: *Let* $k < m$. *If* $x \in W_k$, *then* $f_{k,m}(x) \in W_m$ *and* $\|f_{k,m}(x)\| < \|x\|$. *Conversely, if* $x \in W_m$ *and* $b_{k,m}(x) \in S$, *then* $b_{k,m}(x) \in W_k$ *and* $\|x\| < \|b_{k,m}(x)\|$.

We extend W_k by means of

$$W_{1,k} = \bigcup_{n=k}^{\infty} b_{k,n}(\bar{W}_n),$$

where $b_{k,k}$ denotes the identity mapping. For each k, $W_{1,k}$ is a one-ended Jordan curve in $Q_1 \cup \{0\}$ with its end point at the origin; also, $W_{1,k}$ approaches infinity. It separates Q_1 into two open, simply connected sets.

Upon repeating these (summarized) developments for the other three quadrants, we obtain for each k four curves: $W_{j,k}$ in $Q_j \cup \{0\}$, where $j = 1, 2, 3, 4$. The curve $W_k^f = W_{1,k} \cup W_{3,k}$ lies in $Q_1 \cup \{0\} \cup Q_3$ and is defined to be the *forward driving-point immittance at the kth port.* Similarly, $W_k^b = W_{2,k} \cup W_{4,k}$ lies in $Q_2 \cup \{0\} \cup Q_4$ and is called the *backward driving-point immittance at the kth port.* Thus, for the cascade as a whole, we have in general an infinity of forward and backward

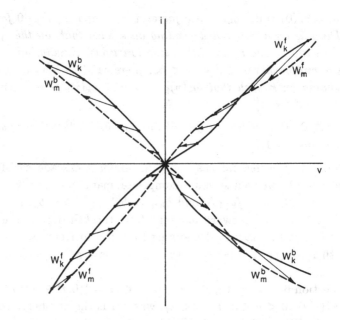

Figure 6.17. Illustration of Equations (6.35) and (6.36), where $k < m$. The solid curves represent W_k^f and W_k^b, and the dashed curves represent W_m^f and W_m^b. The arrows represent $f_{k,m}$.

driving-point immittances, a pair for each port. An additional result arising in this nonuniform case is that, for $k < m$,

$$W_m^f = f_{k,m}(W_k^f), \qquad W_k^f = b_{k,m}(W_m^f) \qquad (6.35)$$

$$W_m^b = f_{k,m}(W_k^b), \qquad W_k^b = b_{k,m}(W_m^b). \qquad (6.36)$$

These relationships are illustrated in Figure 6.17.

If restriction 3 of Conditions 6.7-1 (i.e., of Conditions 6.2-1) holds on all of R^2, then every W_k^f (or W_k^b) is a continuous increasing (respectively, decreasing) function of v for all v.

Let us now examine two sequences, a forward trajectory:

$$\{x, f_{k,k+1}(x), f_{k,k+2}(x), f_{k,k+3}(x), \ldots\} \qquad (6.37)$$

and a backward trajectory:

$$\{x, b_{k-1,k}(x), b_{k-2,k}(x), b_{k-3,k}(x), \ldots\}, \qquad (6.38)$$

starting from the kth port.

Theorem 6.7-5: *If $x \in W_k^f$ and $x \neq 0$, then (6.37) tends to zero and (6.38) tends to infinity; these sequences are contained in Q_1 if $x \in W_{1,k}$ and in Q_3 if $x \in W_{3,k}$. If $x \in W_k^b$ and $x \neq 0$, then (6.37) tends to* infinity and (6.38) tends to zero; these sequences are now contained in Q_2 if $x \in W_{2,k}$ and in Q_4 if $x \in W_{4,k}$.

For each k, $R^2 \backslash (W_k^f \cup W_k^b)$ consists of four open, simply connected sets (depending upon k), each of which contains one of the half axes. We denote these sets by $R_{N,k}$, $R_{E,k}$, $R_{S,k}$, and $R_{W,k}$ in the natural way (see Figure 6.13).

Theorem 6.7-6: *If $x \in R_{E,k} \cup R_{S,k}$ (or if $x \in R_{W,k} \cup R_{N,k}$), then (6.37) tends to infinity, eventually through Q_4 (respectively, Q_2). If $x \in R_{E,k} \cup R_{N,k}$ (or if $x \in R_{W,k} \cup R_{S,k}$), then (6.38) tends to infinity, eventually through Q_1 (respectively, Q_3).*

These two theorems indicate the physical reason for referring to W_k^f and W_k^b as the two driving-point immittances at the kth port of the cascade. Indeed, it is only when $(v_k, i_k) \in W_k^f$ [or when $(v_k, i_k) \in W_k^b$] that (6.37) tends to zero through Q_1 or Q_3 [respectively, (6.38) tends to zero through Q_2 or Q_4] so that the power passing through the kth port toward the right (respectively, toward the left) is equal to the power dissipated in that part of the cascade lying to the right (respectively, to the left) of the kth port. For any other initial point, the sequence diverges to infinity in the manner stated in Theorems 6.7-5 and 6.7-6. This implies that the power dissipated in that part of the cascade is infinite and is supplied primarily by power flowing into the cascade from infinity on the right (respectively, on the left).

Finally, let us remark in passing that the theory for uniform lattice cascades (Section 6.6) can be extended to nonuniform lattice cascades by imposing a uniform lower bound on the lattice parameters [174, Section IV].

6.8 Nonuniform Cascades: Infinity Perceptible

Let us start with an example.

Example 6.8-1. Consider a one-ended linear ladder whose port voltages and currents are indexed as in Figure 6.3 and whose kth stage consists of a shunt conductance g_k with $g_k(v_k) = \gamma_k v_k$ followed by a series resistance r_k with $r_k(i_{k+1}) = \rho_k i_{k+1}$. The symbols γ_k and ρ_k denote positive multiplying constants – and also represent the linear resistance and conductance of the stage. The customary rules for combining resistances in series and in parallel suggest that the driving-point resistance

$R_0 = v_0/i_0$, as observed at the input of the ladder, is given by the *infinite continued fraction*:

$$
R_0 \;=\; \cfrac{1}{\gamma_0 + \cfrac{1}{\rho_0 + \cfrac{1}{\gamma_1 + \cfrac{1}{\rho_1 + \cfrac{1}{\ddots}}}}}
$$

(6.39)

This is in fact true when the continued fraction is convergent, but is not true otherwise.

The convergence of (6.39) is defined in terms of its *odd truncations*:

$$
R_n^o \;=\; \cfrac{1}{\gamma_0 + \cfrac{1}{\ddots + \cfrac{1}{\gamma_n}}}
$$

and its *even truncations*:

$$
R_n^e \;=\; \cfrac{1}{\gamma_0 + \cfrac{1}{\ddots + \cfrac{1}{\rho_n}}}
$$

by letting $n \to \infty$. R_n^o (or R_n^e) is obtained by replacing ρ_n by an open circuit (respectively, γ_{n+1} by a short circuit). It is evident that, for $m < n$, $R_m^o > R_n^o > R_n^e > R_m^e$. Thus, as $n \to \infty$, the R_n^o decrease monotonically and the R_n^e increase monotonically, with one sequence bounding the other. Hence, both sequences converge, but their limits may or may not be the same. A standard result from the theory of continued fractions [140, page 120] is that those two limits are the same when at least one of the two infinite series, $\sum \gamma_n$ and $\sum \rho_n$, diverges; in this case, (6.39) exists as the common limit of its even and odd truncations. In particular, this will truly be the case whenever the linear ladder is uniform.

On the other hand, if both $\sum \gamma_n$ and $\sum \rho_n$ converge, $\lim_{n \to \infty} R_n^o$ and $\lim_{n \to \infty} R_n^e$ will be different and (6.39) will fail to converge; moreover, the ladder will perforce be nonuniform. As we shall see, a driving-point resistance R_0 will exist in this case when we specify what is connected to the ladder at infinity.

(A thorough discussion of resistive ladders is given by Belevitch [18].)

♣

When it is necessary to specify what if anything is connected to 1-nodes of the cascade, we shall say that *infinity is perceptible*. This is the situation we shall examine in the remaining part of this chapter. However, we shall do so in the more general context of a one-ended, nonuniform, nonlinear cascade – either grounded or ungrounded; this will encompass the nonuniform linear ladder of the preceding paragraph as a special case.

In this section we shall show that convergent trajectories that do not approach the origin truly exist. To obtain them, we impose several conditions. The first, Conditions 6.8-2, are similar to – but not the same as – Conditions 6.7-1. For instance, we no longer require that the forward mappings f_k be bijective. Furthermore, the uniform lower bounds on the increments produced by the forward mappings – as stated by restriction 4 of Conditions 6.7-1 – are no longer imposed.

Conditions 6.8-2: *The cascade is one-ended* (as shown in Figure 6.1 or 6.2) *and nonuniform. There is no branch coupling between different stages. Moreover, for each k, the forward mapping f_k satisfies the following restrictions.*

(1) f_k *is a continuous mapping of R^2 into R^2 with $f_k(0) = 0$ and $f_k(x) \neq x$ if $x \neq 0$.*

(2)   *The forward mapping has the following directions:*

    a.   *If $f_k(x) \in Q_1$, then $x - f_k(x) \in Q_1$; if $f_k(x) \in V$, then $x - f_k(x) \in Q_1 \cup I$; if $f_k(x) \in I$, then $x - f_k(x) \in Q_1 \cup V$.*

    b.   *If $x \in Q_2$, then $f_k(x) - x \in Q_2$; if $x \in (-V)$, then $f_k(x) - x \in Q_2 \cup I$; if $x \in I$, then $f_k(x) - x \in Q_2 \cup (-V)$.*

    c.   *If $f_k(x) \in Q_3$, then $x - f_k(x) \in Q_3$; if $f_k(x) \in (-V)$, then $x - f_k(x) \in Q_3 \cup (-I)$; if $f_k(x) \in (-I)$, then $x - f_k(x) \in Q_3 \cup (-V)$.*

    d.   *If $x \in Q_4$, then $f_k(x) - x \in Q_4$; if $x \in V$, then $f_k(x) - x \in Q_4 \cup (-I)$; if $x \in (-I)$, then $f_k(x) - x \in Q_4 \cup V$.*

(3)   *If $x \neq y$ and if $y - x \in \bar{Q}_4$, then $f_k(y) - f_k(x) \in \bar{Q}_4$ and $\|y - x\| < \|f_k(y) - f_k(x)\|$.*

In order to insure the convergence of the trajectories, we have to make the increments produced by the forward mappings f_k decrease in size rapidly enough as $k \to \infty$.

Conditions 6.8-3: *There exists a sequence $\{N_0, N_1, N_2, \ldots\}$ with all $N_k \in R^1$, $N_k > 0$, and $\sum_{k=0}^{\infty} N_k < \infty$ such that, for all $x \in R^2$, we have $\|f_k(x) - x\| \leq N_k \|x\|$.*

(As a different assumption, we could replace the last inequality by

$\|f_k(x) - x\| \leq N_k$. This leads to the same conclusions as those given below – and with a somewhat simpler analysis. Its shortcoming, however, is that it does not encompass linear cascades as a special case. For this alternative development, see [179].)

With Conditions 6.8-2 and 6.8-3 in force, we now show that every voltage-current trajectory converges, whatever be its starting point. This includes trajectories that remain at a distance from the origin. One way of ensuring such a trajectory is to choose x_0 in either Q_2 or Q_4; the subsequent points will be further removed from the origin according to restrictions 2b and 2d of Conditions 6.8-2.

Theorem 6.8-4: *Assume that the cascade satisfies Conditions 6.8-2 and 6.8-3. Then, for every $x_0 \in R^2$, its trajectory $\{x_0, f_0(x_0), f_1(f_0(x_0)), \ldots\}$ converges in R^2.*

Proof: Set $x_{k+1} = f_k(x_k)$ for $k = 0, 1, 2, \ldots$. Since

$$\|f_k(x_k)\| \leq \|f_k(x_k) - x_k\| + \|x_k\| \leq (N_k + 1)\|x_k\|,$$

we may write

$$\|x_{k+1}\| = \|f_k(x_k)\| \leq (1 + N_k)\|x_k\| \leq (1 + N_k)(1 + N_{k-1})\|x_{k-1}\|$$

$$\leq \cdots \leq (1 + N_k)(1 + N_{k-1}) \cdots (1 + N_0)\|x_0\| = \|x_0\| \prod_{j=0}^{k} (1 + N_j).$$

Since $1 + N_j \leq e^{N_j}$,

$$\|x_{k+1}\| \leq \|x_0\| \exp\left(\sum_{j=0}^{k} N_j\right) \leq \|x_0\| \exp\left(\sum_{j=0}^{\infty} N_j\right) < \infty.$$

This shows that the $\|x_{k+1}\|$ for all k comprise a bounded set; that is, the trajectory is bounded. Hence, so too are $\{v_0, v_1, v_2, \ldots\}$ and $\{i_0, i_1, i_2, \ldots\}$, where $x_k = (v_k, i_k)$.

A consequence of restriction 2 of Conditions 6.8-2 is that both $\{v_0, v_1, v_2, \ldots\}$ and $\{i_0, i_1, i_2, \ldots\}$ are eventually monotonic. In fact, one of them will be entirely monotonic whereas the other can change direction at most once, the latter occurring when $\{x_0, x_1, x_2, \ldots\}$ passes from the first or third quadrant into the second or fourth quadrant. It follows that the trajectory converges. ♣

Let us now show that nonuniform ladders are included as a special case of Theorem 6.8-4 when their series resistances r_k and shunt conductances g_k are suitably restricted. As before, the forward mapping for the kth (generally nonlinear) stage is given by (6.31) and (6.32).

Conditions 6.8-5: *For every k, g_k and r_k are (strictly) increasing, continuous mappings of R^1 into R^1 such that $g_k(0) = r_k(0) = 0$.*

As before, some straightforward arguments establish

Lemma 6.8-6: *If the ladder satisfies Conditions 6.8-5, then it also satisfies Conditions 6.8-2.*

Conditions 6.8-7: *There exist two sequences $\{N_{g_0}, N_{g_1}, N_{g_2}, \ldots\}$ and $\{N_{r_0}, N_{r_1}, N_{r_2}, \ldots\}$ such that, for all k, v, and i, we have that N_{g_k} and N_{r_k} are positive constants in R^1, $|g_k(v)| \leq N_{g_k}|v|$, and $|r_k(i)| \leq N_{r_k}|i|$; moreover, $\sum_{k=0}^{\infty} N_{g_k} < \infty$ and $\sum_{k=0}^{\infty} N_{r_k} < \infty$.*

Note that every linear ladder for which (6.39) diverges (i.e., for which $\sum \gamma_n$ and $\sum \rho_n$ both converge) satisfies Conditions 6.8-7.

Lemma 6.8-8: *If a ladder satisfies Conditions 6.8-7, then it also satisfies Conditions 6.8-3.*

Proof: From (6.31) and (6.32), we have

$$|i_{k+1} - i_k| = |g_k(v_k)| \leq N_{g_k}|v_k|$$

and

$$|v_{k+1} - v_k| = |r_k(i_k - g_k(v_k))| \leq N_{r_k}|i_k - g_k(v_k)|$$
$$\leq N_{r_k}|i_k| + N_{r_k}N_{g_k}|v_k|.$$

Since $\|x\| = \|(v, i)\| \leq |v| + |i|$, $|v| \leq \|x\|$, and $|i| \leq \|x\|$, we also have

$$\|f_k(x_k) - x_k\| = \|x_{k+1} - x_k\| \leq |v_{k+1} - v_k| + |i_{k+1} - i_k|$$
$$\leq (N_{g_k} + N_{r_k} + N_{g_k}N_{r_k})\|x_k\| = N_k\|x_k\|$$

where $N_k = N_{g_k} + N_{r_k} + N_{g_k}N_{r_k}$. Finally, since $\sum_{k=0}^{\infty} N_{g_k} < \infty$, the N_{g_k} are all bounded by some constant C. Therefore, with $\sum$ denoting $\sum_{k=0}^{\infty}$, we have $\sum N_{g_k}N_{r_k} < C \sum N_{r_k} < \infty$. Hence, $\sum N_k = \sum N_{g_k} + \sum N_{r_k} + \sum N_{g_k}N_{r_k} < \infty$. ♣

By Lemmas 6.8-6 and 6.8-8 and Theorem 6.8-4, we can state

Corollary 6.8-9: *Every trajectory of a one-ended nonuniform ladder satisfying Conditions 6.8-5 and 6.8-7 converges.*

6.9 Input–Output Mappings

Because all trajectories converge under Conditions 6.8-2 and 6.8-3, the one-ended cascades of Figures 6.1 and 6.2 can be viewed as input–output systems in the following way.

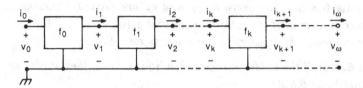

Figure 6.18. An input–output system obtained from a one-ended grounded cascade that satisfies Conditions 6.8-2 and 6.8-3. The output terminal-pair consists of a 1-node and the ground node.

Consider first any three-terminal stage of Figure 6.1. Remember that, by assumption, it is a finite network. No external terminal can be shorted to any other external terminal if Conditions 6.8-2 are to be satisfied. Also, those conditions imply that there is a path from the upper input node to the upper output node that does not pass through ground. Thus, every one-ended path in that grounded cascade must proceed along all but a finite number of upper port nodes and can meet the ground node at most once. Thus, every 0-tip is not disconnectable from every other 0-tip. Define a 1-node as the union of all the 0-tips. We shall assign to that 1-node the index ω, the smallest transfinite ordinal. Finally, an output pair of terminals can be taken to be that 1-node and the ground node. The ground node is of course an infinite 0-node. This 1-network is illustrated in Figure 6.18, where now each stage is labeled with its forward mapping. By virtue of Theorem 6.8-4, any voltage-current pair $x_0 = (v_0, i_0) \in R^2$ at the cascade's input uniquely determines an output voltage-current pair $x_\omega = (v_\omega, i_\omega) \in R^2$ as the limit of the corresponding trajectory. Of course, i_ω can be nonzero only if some load is connected to the output. We shall explore the effects of various output loads in Section 6.11.

Now consider any two-port in the ungrounded cascade of Figure 6.2. That too is understood to be a finite network. One input terminal may be internally shorted to one output terminal to yield in effect a three-terminal network. For any other combination of internal shorts, Conditions 6.8-2 will again be violated. Thus, the possibility arises that, after some port, we have effectively a grounded cascade, but then an output port can be defined as in the preceding paragraph. So, assume this is not the case. Assume furthermore that we can choose a one-ended path that passes along the upper port nodes – but avoids all the lower port nodes – to obtain a representative of a 0-tip. We take the upper output 1-node to be the union of all such 0-tips. Under a similar assumption, a lower output 1-node can be defined in the same way except that each of its embraced 0-tips has a representative that passes through the lower port nodes and avoids all the upper port nodes. Once again, we index the upper port 1-node by ω. Also, by Theorem 6.8-4 again, given any input $x_0 = (v_0, i_0) \in R^2$, we have an output $x_\omega = (v_\omega, i_\omega) \in R^2$ as the limit of the corresponding trajectory.

In both cases, we shall refer to the 0-network consisting of just the given one-ended cascade as an $\vec{\omega}$-*cascade* and to the 1-network obtained by appending the output terminals as an ω-*cascade*. For the latter, we can define $f^\omega : x_0 \mapsto x_\omega$ as the *input–output mapping* or synonymously as the *forward mapping* of the ω-cascade. In fact, f^ω is the pointwise limit of the sequence $\{f_{0,k}\}_{k=1}^\infty$ of composite forward mappings $f_{0,k} = f_{k-1} \cdots f_0$.

Theorem 6.9-1: *If an $\vec{\omega}$-cascade satisfies Conditions 6.8-2 and 6.8-3, then the corresponding ω-cascade has a forward mapping f^ω that satisfies restrictions 1, 2, and 3 of Conditions 6.8-2. Moreover, there exists a positive constant $N^\omega \in R^1$ such that $\|f^\omega(x) - x\| \leq N^\omega \|x\|$ for all $x \in R^2$.*

Proof: Theorem 6.8-4 ensures that f^ω maps all of R^2 into R^2. Since $f_k(0) = 0$ for each k, $f^\omega(0) = 0$ too. By the monotonicities pointed out in the proof of Theorem 6.8-4, $f^\omega(x) \neq x$ if $x \neq 0$. Moreover, restrictions 2 and 3 of Conditions 6.8-2 transfer from the individual mappings f_k onto the composite mappings $f_{0,k}$ and then onto f^ω.

To prove the continuity of f^ω, fix $x \in R^2$. With η denoting a positive number, let $\Omega(x, \eta) = \{y \in R^2 : \|y - x\| < \eta\}$. Set $y_k = f_{0,k}(y) = f_{k-1} \cdots f_0(y)$ for $k = 1, 2, 3, \ldots$ and set $y_0 = y$. We may write

$$\|f_k(y_k)\| \leq \|y_k\| + \|f_k(y_k) - y_k\| \leq (1 + N_k)\|y_k\|$$

according to Conditions 6.8-3. So,

$$\begin{aligned} \|y_{k+1}\| &= \|f_k(y_k)\| \leq (1 + N_k)\|y_k\| \\ &\leq (1 + N_k)(1 + N_{k-1})\|y_{k-1}\| \\ &\leq \cdots \leq \|y\| \prod_{j=0}^{k}(1 + N_j). \end{aligned}$$

Since $1 + N_j \leq e^{N_j}$,

$$\prod_{j=0}^{k}(1 + N_j) \leq \exp\left(\sum_{j=0}^{k} N_j\right).$$

According to Conditions 6.8-3 again, the quantity on the right-hand side is bounded for all k by $B = \exp(\sum_{j=0}^\infty N_j) < \infty$. Thus, $\|y_{k+1}\| \leq \|y\|B$.

Set $y_\omega = f^\omega(y)$. Since every trajectory converges in R^2, we have for every K

$$\|y_\omega - y_K\| \leq \|y_{K+1} - y_K\| + \|y_{K+2} - y_{K+1}\| + \|y_{K+3} - y_{K+2}\| + \cdots$$

$$\leq N_K\|y_K\| + N_{K+1}\|y_{K+1}\| + N_{K+2}\|y_{K+2}\| + \cdots$$

$$\leq \|y\|B\sum_{k=K}^{\infty} N_k \leq (\|x\| + \eta)B\sum_{k=K}^{\infty} N_k$$

whenever $y \in \Omega(x,\eta)$. Since $\sum_{k=0}^{\infty} N_k < \infty$, given any $\epsilon > 0$, we can choose K so large and η so small that $\|y_\omega - y_K\| < \epsilon/3$ for all $y \in \Omega(x,\eta)$. Fix K and η this way.

Now, $y_K = f_{0,K}(y)$. Set $x_K = f_{0,K}(x)$. Since $f_{0,K}$ is a continuous mapping, η_0 can be chosen so small that $0 < \eta_0 < \eta$ and $\|y_K - x_K\| < \epsilon/3$ for every $y \in \Omega(x,\eta_0)$. Hence, for all such y and with $x_\omega = f^\omega(x)$, we have

$$\|y_\omega - x_\omega\| \leq \|y_\omega - y_K\| + \|y_K - x_K\| + \|x_K - x_\omega\|$$

$$\leq \frac{\epsilon}{3} + \frac{\epsilon}{3} + \frac{\epsilon}{3} = \epsilon$$

for every $y \in \Omega(x,\eta_0)$. This proves the continuity of f^ω. Thus, f^ω satisfies all three restrictions of Conditions 6.8-2.

Finally, we have to establish the asserted bound on $f^\omega(x) - x$. For any $x \in R^2$ and as $k \to \infty$,

$$\|f^\omega(x) - x\| \leftarrow \|f_{k-1}\cdots f_0(x) - x\|$$

$$\leq (1 + N_{k-1})\cdots(1 + N_0)\|x\| + \|x\|$$

$$\rightarrow \|x\|\left(1 + \prod_{j=0}^{\infty}(1 + N_j)\right)$$

$$\leq \|x\|\left(1 + \exp\left(\sum_{j=0}^{\infty} N_j\right)\right) = N^\omega\|x\|.$$

This completes the proof. ♣

6.10 ω^p-Cascades

Two ω-cascades, whose corresponding $\bar{\omega}$-cascades satisfy Conditions 6.8-2 and 6.8-3, can themselves be connected in cascade with the output of one connected to the input of the other. The overall structure will be an input–output system too, and its forward mapping will again satisfy the three restrictions of Conditions 6.8-2 as well as a bound like that of Conditions 6.8-3. Clearly, this can be repeated any finite number of

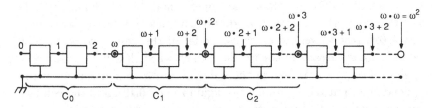

Figure 6.19. A grounded ω^2-cascade of finite three-terminal networks. Its output terminal-pair consists of a 2-node and the ground node.

times – and indeed infinitely often to obtain an $\bar{\omega}$-cascade of ω-cascades. The last structure will in turn define an input–output mapping so long as the constants N decrease rapidly enough.

To be more specific, assume that we have an $\bar{\omega}$-cascade of ω-cascades. That is, for each $k = 0, 1, 2, \ldots$, we have an ω-cascade C_k, whose corresponding $\bar{\omega}$-cascade satisfies Conditions 6.8-2 and 6.8-3. The output of C_k is connected to the input of C_{k+1}. This is illustrated in Figure 6.19 for the grounded case. As always, no branch coupling between stages is allowed. Let f_k^ω be the forward mapping of C_k. By Theorem 6.9-1, f_k^ω satisfies the three restrictions of Conditions 6.8-2, and moreover $\|f_k^\omega(x) - x\| \leq N_k^\omega \|x\|$ for all $x \in R^2$, where N_k^ω is a constant independent of x. (The proof of Theorem 6.9-1 had $N_k^\omega > 1$; nonetheless, $N_k^\omega < 1$ is possible for suitably restricted cascades.) If $\sum_{k=0}^\infty N_k^\omega < \infty$, then the hypotheses and proofs of Theorems 6.8-4 and 6.9-1 continue to hold when the finite two-ports of those theorems are replaced by the present ω-cascades.

Now, we may transfinitely index the ports of this $\bar{\omega}$-cascade of ω-cascades C_k in the natural way: $0, 1, 2, \ldots$ for C_0; $\omega, \omega + 1, \omega + 2, \ldots$ for C_1; $\omega \cdot 2, \omega \cdot 2 + 1, \omega \cdot 2 + 2, \ldots$ for C_3; and so forth. ($\omega \cdot 2$ denotes $\omega + \omega$.) Letting $x_q = (v_q, i_q)$ be the voltage-current pair at the qth port, we can conclude that the *transfinite trajectory*:

$$\{x_0, x_1, x_2, \ldots, x_\omega, x_{\omega+1}, x_{\omega+2}, \ldots, x_{\omega \cdot 2}, x_{\omega \cdot 2 + 1}, x_{\omega \cdot 2 + 2}, \ldots\} \quad (6.40)$$

converges – in the sense that the ordinary sequence $\{x_\omega, x_{\omega \cdot 2}, x_{\omega \cdot 3}, \ldots\}$ of port voltages at the outputs of the C_k converges. Let us denote the limit of the last sequence by x_{ω^2}.

Furthermore, just as an output pair of terminals could be appended to an $\bar{\omega}$-cascade of two-ports, so too can an output pair of terminals be appended to this $\bar{\omega}$-cascade of ω-cascades by using two 2-nodes in the ungrounded case (or a 2-node along with the ground node in the grounded case). For example, the upper 2-node is the union of all 1-tips having representatives that do not embrace any of the lower port nodes (or, respectively, the ground node) of the finite two-ports. The result is an ω-cascade of ω-cascades, which we shall call an ω^2-*cascade of finite two-ports*. We assign the index ω^2 to the upper 2-node of the appended output terminal-pair (see again Figure 6.19), and the voltage-current pair x_{ω^2} to that output terminal-pair. This defines in turn a forward mapping $f^{\omega^2} : x_0 \mapsto x_{\omega^2}$, which can also be viewed as the pointwise limit as $k \to \infty$ of the composite mappings $f_k^\omega f_{k-1}^\omega \cdots f_0^\omega$. Also, by Theorem 6.9-1 again, there is a constant $N^{\omega^2} \in R^1$ such that $\|f^{\omega^2}(x) - x\| < N^{\omega^2}\|x\|$ for all $x \in R^2$.

This argument can be repeated for an $\vec{\omega}$-cascade of ω^2-cascades to obtain an ω^3-*cascade of finite two-ports* – after an output terminal-pair of a 3-node coupled with either another 3-node or the ground node is appended. Continuing in this fashion, we can get, for any natural number p, an $\vec{\omega}$-cascade of ω^{p-1}-cascades, which we also refer to as an $\vec{\omega^p}$-*cascade of finite two-ports*. Upon appending an output pair of terminals as indicated above, we obtain an ω^p-*cascade of finite two-ports*. All this requires, however, that a hierarchy of bounds be placed on the constants N. In particular, let q be any natural number with $q < p$. Within the ω^p-cascade consider any one of the $\vec{\omega}$-cascades of ω^q-cascades of finite two-ports, where for $q = 0$ the ω^q-cascades are themselves finite two-ports. We number consecutively those ω^q-cascades by the natural numbers $k = 0, 1, 2, \ldots$. Let $f_k^{\omega^q}$ be the forward mapping of the kth ω^q-cascade. By the proofs of Theorems 6.8-4 and 6.9-1 it follows recursively that, if each $\vec{\omega}$-cascade of finite two-ports satisfies Conditions 6.8-2 and 6.8-3, then, for each $q = 0, 1, 2, \ldots, p-1$, for each natural number k, and for all $x \in R^2$, we have that $f_k^{\omega^q}$ satisfies the three restrictions of Conditions 6.8-2 and the bound $\|f_k^{\omega^q}(x) - x\| \leq N_k^{\omega^q}\|x\|$, where $N_k^{\omega^q}$ is a positive constant – if in addition the following requirement is also fulfilled.

Conditions 6.10-1: *For every $\vec{\omega}$-cascade of ω^q-cascades $(0 \leq q \leq p-1)$ in a given ω^p-cascade, the just-defined $N_k^{\omega^q}$ satisfy $\sum_{k=0}^{\omega^q} N_k^{\omega^q} < \infty$.*

Moreover, one more application of this argument shows that the forward mapping f^{ω^p} of the overall ω^p-cascade also has the same properties. In fact, we obtain the convergence of the transfinite trajectory (6.40) extended to all the ordinals less than ω^p:

$$\{x_0, \ldots, x_{\omega^{p-1}}, \ldots, x_{\omega^{p-1} \cdot 2}, \ldots, x_{\omega^{p-1} \cdot 3}, \ldots\} \qquad (6.41)$$

However, that convergence has so far only been defined as a hierarchy of convergences, namely, the convergence of the port voltage-current pairs for each $\tilde{\omega}$-cascade of finite two-ports, then the convergence of the output-port voltage-current pairs for each $\tilde{\omega}$-cascade of ω-cascades, and so forth up to the convergence of the output-port voltage-current pairs for the $\tilde{\omega}$-cascade of ω^{p-1}-cascades.

Actually, this can be reinterpreted as the convergence in R^2 of an ordinary series if we examine the forward-mapping increments $f_k(x) - x$ for all the finite two-ports, where k traverses the finite and transfinite ordinals up to, but not including, ω^p. Since ω^p is a countable ordinal, we can renumber those increments using only the natural numbers n to index all of them. Then, the input-to-output increment $f^{\omega^p}(x_0) - x_0$ for the overall ω^p-cascade can be written as the ordinary series

$$f^{\omega^p}(x_0) - x_0 = \sum_{n=0}^{\infty} (f_n(x_n) - x_n) \qquad (6.42)$$

where x_n is the renumbered input to the nth finite two-port. To show that (6.42) converges in R^2 in the usual sense, let us go back to the original transfinite sequence (6.41) and consider the increments $f_i^{\omega^{p-1}}(y_i) - y_i$ in the final $\tilde{\omega}$-cascade of ω^{p-1}-cascades, where now $y_i = x_{\omega^{p-1} \cdot i}$ denotes the input to the ith ω^{p-1}-cascade. Given any $\epsilon > 0$, we can choose the integer K^{ω^p} so large that

$$\left\| f^{\omega^p}(x_0) - x_0 - \left(\sum_{i=0}^{K^{\omega^p}} f_i^{\omega^{p-1}}(y_i) - y_i \right) \right\| < \frac{\epsilon}{2} .$$

This eliminates all but a finite number $K^{\omega^p} + 1$ of the ω^{p-1}-cascades. Then, for each of the noneliminated ω^{p-1}-cascades, we can choose the integer $K^{\omega^{p-1}}$ so large that

$$\left\| f_i^{\omega^{p-1}}(y_i) - y_i - \left(\sum_{j=1}^{K^{\omega^{p-1}}} f_j^{\omega^{p-2}}(z_j) - z_j \right) \right\| < \frac{\epsilon}{4(K^{\omega^p} + 1)} \qquad (6.43)$$

where $f_j^{\omega^{p-2}}$ is the forward mapping of the jth ω^{p-2}-cascade within the considered ω^{p-1}-cascade and z_j denotes the input to that ω^{p-2}-cascade. This $K^{\omega^{p-1}}$ can be chosen to satisfy (6.43) for every one of the ω^{p-1}-cascades with indices no larger than K^{ω^p}. Moreover, this eliminates all but a finite number of the ω^{p-2}-cascades. That number is $(K^{\omega^p} + 1)(K^{\omega^{p-1}} + 1)$. Proceeding down the hierarchy in this fashion, we can finally choose an integer K^{ω} so large that, for every one of the finitely many, noneliminated $\vec{\omega}$-cascades of finite two-ports, we have

$$\left\| f^\omega(\xi_i) - \xi_i - \left(\sum_{j=0}^{K^\omega} f_j(\zeta_j) - \zeta_j \right) \right\| < \frac{\epsilon}{2^p (K^{\omega^p} + 1) \cdots (K^{\omega^2} + 1)}, \quad (6.44)$$

where again K^ω can be chosen to satisfy (6.44) for every one of those finitely many $\vec{\omega}$-cascades of finite two-ports. It follows from all this that there is a finite number of increments between consecutive terms in (6.40) – extended as in (6.41) – whose sum is within an ϵ-neighborhood of $x_{\omega^p} - x_0$. We can conclude that the ordinary series (6.42) with the renumbered port voltage-current pairs converges in R^2 in the usual sense.

Actually, we need not resort to the stated renumbering. We can write

$$f^{\omega^p}(x_0) - x_0 = \sum_{k \in A} (f_k(x_k) - x_k) \quad (6.45)$$

where now A is the set of all finite and transfinite ordinals up to (but not including) ω^p. From what we have shown, we can say that the right-hand side of (6.45) is *summable in R^2* in the sense that, given any $\epsilon > 0$, there is a finite subset M of A such that

$$\left\| f^{\omega^p}(x_0) - x_0 - \sum_{k \in M} (f_k(x_k) - x_k) \right\| < \epsilon.$$

In summary, we can state

Theorem 6.10-2: *For any natural number $p \geq 1$, consider an ω^p-cascade all of whose $\vec{\omega}$-cascades of finite two-ports satisfy Conditions 6.8-2 and 6.8-3. Assume furthermore that Conditions 6.10-1 are satisfied. Then, for any input voltage-current pair x_0, the transfinite trajectory (6.40) of port voltage-current pairs along the corresponding $\vec{\omega^p}$-cascade of finite two-ports converges – in the sense that the right-hand side of (6.45) is summable in R^2 – to a limit $x_{\omega^p} \in R^2$. Moreover, the forward mapping $f^{\omega^p} : x_0 \mapsto x_{\omega^p}$ satisfies the three restrictions of Conditions 6.8-2, and there is a positive constant N^{ω^p} such that $\|f^{\omega^p}(x_0) - x_0\| \leq N^{\omega^p} \|x_0\|$ for all $x_0 \in R^2$.*

Finally, let us indicate briefly how our ideas can be extended to ω^ω-cascades, and beyond. Essentially, the procedure used to define $\vec{\omega}$-tips and ω-nodes (Section 5.2) can be used here as well. We start by continuing indefinitely the process of constructing an ω^p-cascade from an $\vec{\omega}$-cascade of ω^{p-1}-cascades. This yields a cascade containing ω^p-cascades for every natural number p; we call this an $\omega^{\vec{\omega}}$-cascade. Assume that

Conditions 6.8-2 and 6.8-3 are satisfied by all the $\vec{\omega}$-cascades of finite two-ports within the $\omega^{\vec{\omega}}$-cascade and that Conditions 6.10-1 are satisfied for all natural numbers q. Now, choose any increasing sequence of natural numbers $\{p_0, p_1, p_2, \ldots\}$ such that $p_k \to \infty$ as $k \to \infty$. Set $\kappa = \omega^{p_k}$ and consider the sequence of ports with indices κ and port voltage-current pairs x_κ. Then, for $\kappa' = \omega^{p_{k+1}}$, $f_k^{\kappa'} : x_\kappa \mapsto x_{\kappa'}$ is the forward mapping for a κ'-cascade. (That κ and κ' depend upon k should be kept in mind.) As before, $f_k^{\kappa'}$ will satisfy the three restrictions of Conditions 6.8-2 and the bound $\|f_k^{\kappa'}(x_k) - x_k\| < N_k^{\kappa'}\|x_k\|$ for all $x_k \in R^2$, where $N_k^{\kappa'}$ is a constant. Finally, assume that $\sum_{k=0}^{\infty} N_k^{\kappa'} < \infty$. It then follows much as before that the sequence of x_κ converges in R^2 to, say, x_{ω^ω}. It can be shown that x_{ω^ω} is independent of the choices of the p_k. Assign x_{ω^ω} to an output terminal-pair appended to the $\omega^{\vec{\omega}}$-cascade. At least one of those output terminals will be an ω-node; the other will be either a ground node or another ω-node. This yields an ω^ω-cascade with the forward mapping $f^{\omega^\omega} : x_0 \mapsto x_{\omega^\omega}$; f^{ω^ω} will also satisfy conclusions like those of Theorem 6.10-2.

These procedures can be continued to obtain cascades of even higher ranks by using the technique of the last paragraph for a limit-ordinal rank and the technique of the earlier paragraphs for a successor-ordinal rank.

6.11 Loaded Cascades

For the sake of definiteness we shall consider an ω^p-cascade satisfying the hypothesis of Theorem 6.10-2; nevertheless, our remarks will also apply to ω^ω-cascades and to cascades of still higher ranks.

So far, we have shown that every trajectory converges. However, just which trajectory is at hand depends on how the cascade is loaded on its input and output. The kinds of loads we shall permit are power sources or power dissipators and appear in two general forms: The *Thevenin's load* and the *Norton's load.*

The Thevenin's load is shown in Figure 6.20(a). It consists of a pure voltage source e in series with a resistance $r : i \mapsto r(i)$. The value e of the voltage source may be any real number including zero. Moreover, r may be the zero mapping (i.e., it may be absent), but, if it differs from the zero mapping, it is assumed to satisfy Conditions 6.8-5. Thus, r is a mapping of R^1 into R^1, but not necessarily onto R^1; its range is an interval in R^1. A typical plot of current versus voltage for the polarities indicated in Figure 6.20(a) is shown in Figure 6.20(b).

The Norton's load is shown in Figure 6.21(a) and consists of a pure

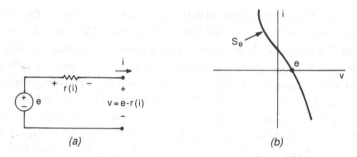

(a)   (b)

Figure 6.20. (a) A Thevenin's load. (b) Its typical voltage-current characteristic S_e.

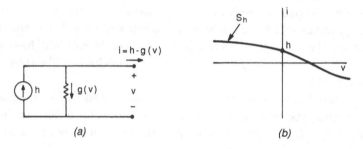

(a)   (b)

Figure 6.21. (a) A Norton's load. (b) Its typical voltage-current characteristic S_h.

current source h in parallel with a conductance $g : v \mapsto g(v)$. Here too, h may be any real number, and g may or may not be present. If g is present, it too is assumed to satisfy Conditions 6.8-5. Moreover, the range of g is an interval in R^1, which need not be all of R^1. A typical voltage-current characteristic for the polarities assumed in Figure 6.21(a) is shown in Figure 6.21(b).

Now, consider the loaded ω^p-cascade shown in Figure 6.22. For the sake of definiteness we have shown a Thevenin's load on the input and a Norton's load on the output, but either kind of load is allowed on the input and on the output. The voltage-current characteristic S_a for the input load can be written in parametric form as follows. For a Thevenin's load, we take i to be a parameter varying throughout R^1 and represent any point on S_a by $(v(i), i)$, where $v(i) = e_a - r_a(i)$. For a Norton's load, we take v to be a parameter varying throughout R^1 and represent any point on S_a by $(v, i(v))$, where $i(v) = h_a - g_a(v)$. Since r_a and g_a satisfy Conditions 6.8-5 or are the zero mapping, it follows that S_a is a

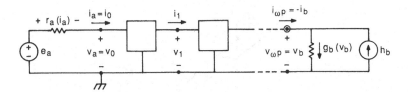

Figure 6.22. A loaded ω^p-cascade.

Jordan curve. Moreover, it is endless in the sense that, as the parameter i or v tends to either $+\infty$ or $-\infty$, the corresponding point on S_a tends to infinity within R^2. Furthermore, the chords of S_a never have positive slopes; that is, if $v_x < v_y$ or $i_x > i_y$ and if $x = (v_x, i_x)$ and $y = (v_y, i_y)$ are the corresponding points on S_a, then $y - x \in \bar{Q}_4$. In the following, we shall write $x \prec y$ to indicate that two points x and y on a voltage-current characteristic, whose chords all have nonpositive slopes, satisfy $v_x < v_y$ or $i_x > i_y$.

Let us examine how the ω^p-cascade transforms S_a into another voltage-current characteristic $f^{\omega^p}(S_a)$ on its output. According to Theorem 6.10-2, f^{ω^p} is a continuous mapping on R^2. Now, even though f^{ω^p} is not in general injective, its restriction to S_a is a one-to-one mapping of S_a onto $f^{\omega^p}(S_a)$. This is an immediate consequence of restriction 3 of Conditions 6.8-2. Hence, $f^{\omega^p}(S_a)$ is a Jordan curve.

Furthermore, the chords of $f^{\omega^p}(S_a)$ have nonpositive slopes. Indeed, choose any two points w and z on $f^{\omega^p}(S_a)$ and let x and y be respectively their unique preimages on S_a according to f^{ω^p}. As above, if $x \prec y$, then $y - x \in \bar{Q}_4$. Therefore, by restriction 3 of Conditions 6.8-2, $z - w = f^{\omega^p}(y) - f^{\omega^p}(x) \in \bar{Q}_4$ too.

Finally, $f^{\omega^p}(S_a)$ is endless. Indeed, fix x on S_a and let y tend to infinity on S_a in either direction. If $f^{\omega^p}(y)$ does not tend to infinity, then restriction 3 of Conditions 6.8-2 will eventually be violated.

All this shows that, under the hypothesis of Theorem 6.10-2, $T = f^{\omega^p}(S_a)$ is an endless Jordan curve, which approaches infinity in both ways and whose chords have nonpositive slopes. This is illustrated in Figure 6.23.

Now, let us assume that the output load is either a Thevenin's load with resistance or a Norton's load with conductance. Consequently, its voltage-current characteristic $S_b = \{(v_b, i_b)\}$ is also an endless Jordan curve, but, by virtue of Conditions 6.8-5, its chords have negative slopes (no chords with zero or infinite slopes). In view of the polarities of i_b and i_{ω^p} (see Figure 6.22), we have $i_b = -i_{\omega^p}$. Thus, with respect to

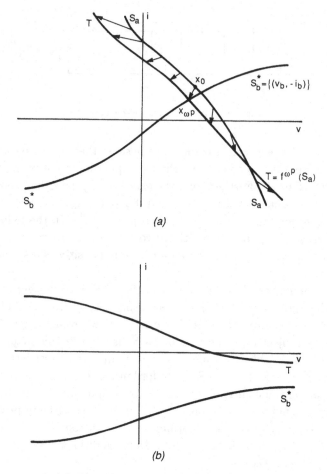

(a)

(b)

Figure 6.23. (a) The construction of the unique input x_0 and output x_{ω^p} for a loaded ω^p-cascade in a case where a trajectory exists. (b) A case where no trajectory exists.

the axes of Figure 6.23, the output load is represented by the voltage-current characteristic $S_b^* = \{(v_b, -i_b)\}$, which is an endless Jordan curve approaching infinity in both ways, whose chords all have positive slopes, as is also indicated in Figure 6.23.

In view of the properties of T and S_b^*, there is either a unique intersection point between T and S_b^* or none at all, depending on whether T and S_b^* meet or miss each other. An example of the first case is shown in Figure 6.23(a) and an example of the second in Figure 6.23(b). Consider the case where T and S_b^* do meet at, say, x_{ω^p}. Since the restriction of f^{ω^p} to S_a is injective, there is a unique point $x_0 = (v_0, i_0)$ on S_a that is mapped by f^{ω^p} onto x_{ω^p}. In fact, a unique transfinite trajectory $\{x_0, x_1, \ldots, x_{\omega^p}\}$ is determined by the assumed input and output loads. In summary, we may state the following.

Theorem 6.11-1: *Assume that an ω^p-cascade satisfies the hypothesis of Theorem 6.10-2, is loaded on its input by either a Thevenin's load (with or without resistance) or a Norton's load (with or without conductance) having the voltage-current characteristic S_a, and is loaded on its output by either a Thevenin's load with resistance or a Norton's load with conductance having the voltage-current characteristic S_b. Then, the ω^p-cascade has either a unique voltage-current trajectory $\{x_0, x_1, \ldots, x_{\omega^p}\}$ or none at all. When the trajectory exists, its initial point x_0 is the unique point on S_a that is mapped by f^{ω^p} onto the unique intersection point between $T = f^{\omega^p}(S_a)$ and S_b^*.*

Corollary 6.11-2: *Under the hypothesis of Theorem 6.11-1, the total power absorbed in all the stages of the cascade is finite whenever a trajectory exists.*

Proof: The power absorbed in the kth stage is $v_k i_k - v_{k+1} i_{k+1}$. This is in fact a nonnegative quantity because of restriction 2 of Conditions 6.8-2. Moreover, it is no larger than

$$|v_k|\,|i_k - i_{k+1}| + |v_k - v_{k+1}|\,|i_{k+1}|.$$

Since the trajectory is bounded, there is a constant M such that $|v_k| < M$ and $|i_{k+1}| < M$ for all k. Summing the absorbed powers over all stages, we find the sum to be bounded by

$$M \sum |i_k - i_{k+1}| + M \sum |v_k - v_{k+1}|.$$

By the aforementioned monotonicities in the currents and voltages and by the boundedness of the trajectory again, the last quantity is finite.

♣

Various sufficient conditions can be stated which insure that T and S_b^* intersect. For example, assume that S_a meets both the second and fourth quadrants and that S_b does the same. Under the hypothesis of Theorem 6.10-2, f^{ω^p} satisfies restrictions 2b and 2d of Conditions 6.8-2, which implies that $T = f^{\omega^p}(S_a)$ also meets the second and fourth quadrants. On the other hand, S_b^* meets the first and third quadrants. This implies that T and S_b^* must meet because they are endless Jordan curves approaching infinity as indicated above.

As another example, assume there is a pure voltage source on the input and a Norton's load with conductance on the output. Then, S_a is a vertical line, and, by Theorem 6.10-2 and restriction 3 of Conditions 6.8-2, the projection of $T = f^{\omega^p}(S_a)$ onto the i-axis covers all of the i-axis. On the other hand, the projection of S_b^* onto the v-axis covers all of the v-axis. It again follows that T and S_b^* must meet.

Let us close by noting that this determination of a voltage-current trajectory can be interpreted as a generalization of Thevenin's theorem or Norton's theorem whenever $T \cap S_b^* \neq \emptyset$. For instance, given an ω^p-cascade satisfying the hypothesis of Theorem 6.10-2 and a Norton's load with conductance (or a Thevenin's load with resistance) on the output, we can let the input load be a pure voltage source e_a (respectively, a pure current source h_a) and then let e_a (or h_a) vary throughout R^1 to get an input curve W of initial voltage-current pairs (v_0, i_0) . It is in this way that the output load S_b can be "seen" as the curve W by an observer at the input even though S_b is at the end of a transfinite cascade.

Chapter 7

Grids

Perhaps the most important infinite electrical networks with respect to physical phenomena – putting aside the finite networks – are the infinite grids. This is because finite-difference approximations of various partial differential equations have realizations as electrical networks whose nodes are located at the sample points of the approximation. Those sample points are distributed in accordance with increments in each of the coordinates, hence the gridlike structure. Moreover, if the phenomenon exists throughout an infinite domain, it is natural (but, to be sure, not always necessary) to choose an infinity of sample points. In this way, one is led to infinite electrical grids as models for the so-called "exterior problems" of certain partial differential equations. Two cases of this were presented in Section 1.7, and more will be discussed in the next chapter. The grids we examine are of two general types: the grounded grids, wherein a resistor connects each node to a common ground node, and the ungrounded grids, wherein those grounding resistors are entirely absent. Grounded grids are readily analyzed, but ungrounded grids are more problematic because of a singularity in a certain function that characterizes the network.

This chapter is devoted to rectangular grids, the natural finite-difference model for Cartesian coordinates, but the analysis can be extended to other coordinate systems such as cylindrical and spherical ones [180], [183], [184], [188]. The uniform n-dimensional rectangular grid is examined in Sections 7.2 and 7.3, and a complete solution for the voltage-current regime is obtained for every n in the grounded case and for $n \geq 3$ in the ungrounded case. The ungrounded case for $n = 1$ is trivial. For $n = 2$, the ungrounded grid has an additional difficulty; a more limited conclusion is obtained for it in Section 7.3, and a complete solution is finally attained in Section 7.6.

The rest of this chapter is devoted to rectangular grids having non-uniformity with respect to one of the coordinates. They model layered media such as a stratified earth, a not uncommon physical configuration. The Fourier-series transformation is used to convert the grid into a ladder network of operators on L_2, the classical space of quadratically integrable functions defined on the unit circle or on a Cartesian product of unit circles. Upon borrowing results from the preceding chapter, one may analyze the ladder and then apply the finite Fourier transformation, which is the inverse of the Fourier-series transformation, to obtain an exact solution for the grid. We end by examining solutions for grids that extend beyond infinity in one direction; this entails the matching of boundary conditions at the transfinite operator-ladder's output nodes.

The Fourier transformation, which is used for both the uniform and one-dimensionally nonuniform cases, requires of course that the grid be linear. There is virtually no information about a general method for solving a nonlinear grid exactly. An approximate voltage-current regime for the latter might be computed iteratively by means of the standard technique of approaching the nonlinear grid with a sequence of linear ones.

Also, how to obtain exactly the voltage-current regime for a linear infinite grid having nonuniformity along more than one coordinate is by-and-large unknown. One obvious but approximate approach is to choose an expanding sequence of finite subnetworks, which fill out the infinite grid, and then invoke an approximation theorem such as Flanders' Theorem 3.1-8 or the result in [180]. Another iterative approach has been devised for linear rectangular grounded grids whose grounding resistors may vary in all coordinates but whose other resistors are all the same [186]. A more important situation for physical applications arises when a finite nonuniform subnetwork is embedded in a uniform or layered grid. Our results concerning one-dimensional nonuniformity can be used to contract the domain of analysis down to a layer containing the subnetwork; this technique is discussed in Section 8.2.

7.1 Laurent Operators and the Fourier-Series Transformation

We will need some standard results concerning a certain kind of operator, called a "Laurent operator," acting on an n-dimensional version of Hilbert's coordinate space. We gather them in this section.

Given any two Banach spaces $\mathcal{A}$ and $\mathcal{B}$, both real or both complex, the set of bounded linear mappings of $\mathcal{A}$ into $\mathcal{B}$ will be denoted by $[\mathcal{A};\mathcal{B}]$. $[\mathcal{A};\mathcal{B}]$ is itself a linear space when the addition of operators and the multiplication of them by a real or respectively complex number is defined pointwise. With the norm of any $\mathbf{f} \in [\mathcal{A};\mathcal{B}]$ being defined by

$$\|f\| = \sup\{\|f(x)\| : \|x\| = 1, \, x \in \mathcal{A}\},$$

$[\mathcal{A};\mathcal{B}]$ is a Banach space too [94, page 413]. To simplify notation we shall denote $f(x)$ simply by fx. An inequality we shall need is $\|fx\| \le \|f\|\,\|x\|$ for any $x \in A$.

If $\mathcal{H}$ is a complex Hilbert space with the inner product $(\cdot, \cdot)$, then the *numerical range* $W(f)$ of any $f \in [\mathcal{H};\mathcal{H}]$ is defined by

$$W(f) = \{(fx, x) : x \in \mathcal{H}, \|x\| = 1\}.$$

f is called *positive* if $(fx, x) \ge 0$ for all $x \in \mathcal{H}$, and f is also invertible if and only if $W(f)$ is bounded away from the origin.

By an *n-dimensional integer* or *lattice point* we mean a vector $p = (p_1, \ldots, p_n)$ of n integers. The set of all n-dimensional integers will be denoted by Z^n. A norm we shall use for them is $\|p\|_1 = |p_1| + \cdots + |p_n|$.

Next, l_2^n will denote the complex Hilbert space of all vectors $\mathbf{x} = \{x_p : p \in Z^n\}$ of complex numbers x_p satisfying $\sum |x_p|^2 < \infty$, where it is understood that the summation is over all indices $p \in Z^n$. The inner product for any $\mathbf{x}, \mathbf{y} \in l_2^n$ is $(\mathbf{x}, \mathbf{y}) = \sum x_p \bar{y}_p$, where the overbar signifies the complex conjugate. Any $\mathbf{f} \in [l_2^n; l_2^n]$ has a unique matrix representation $[f_{p,q}]$ of complex numbers $f_{p,q}$ such that, for any $\mathbf{x} \in l_2^n$, the pth component of $\mathbf{y} = \mathbf{f}\mathbf{x}$ is given by

$$y_p = \sum_{q \in Z^n} f_{p,q} x_q. \tag{7.1}$$

In fact, $[f_{p,q}]$ can be viewed as a $2n$-dimensional infinite array of the complex numbers $f_{p,q}$. Not all such arrays will represent some member of $[l_2^n; l_2^n]$ [70, page 119], but those we encounter will do so. Throughout the following, we shall employ the same symbol to denote an operator $\mathbf{f} \in [l_2^n; l_2^n]$ and its matrix representation $\mathbf{f} = [f_{p,q}]$. This should not lead to any confusion in the formulas we shall be using.

An $\mathbf{f} \in [l_2^n; l_2^n]$ will be called a *Laurent operator* if $f_{p,q} = f_{p+m,q+m}$ for every $m, p, q \in Z^n$, and the matrix $[f_{p,q}]$ is then called a *Laurent matrix*. On the other hand, the *shift operator* σ_m is defined on l_2^n as follows. Given any $m \in Z^n$ and any $\mathbf{x} = \{x_p : p \in Z^n\} \in l_2^n$, σ_m maps $\mathbf{x}$ into $\sigma_m \mathbf{x} = \{x_{p+m} : p \in Z^n\} \in l_2^n$. It follows that every Laurent operator $\mathbf{f}$ commutes with σ_m.

A convenient operational calculus for these Laurent operators arises from the Fourier-series transformation defined on an n-dimensional version of the classical L_2 space [157, page 221]. Let Θ^n denote the Cartesian product of n replicates of the unit circle; thus, any $\theta \in \Theta^n$ has the form $\theta = (\theta_1, \ldots, \theta_n)$, where $-\pi \le \theta_\nu \le \pi$ and $\nu = 1, \ldots, n$. L_2^n will denote the Hilbert space of (equivalence classes of) complex-valued, quadratically integrable functions on Θ^n. The inner product of any $X, Y \in L_2^n$ is

$$(X, Y) = \frac{1}{(2\pi)^n} \int_{\Theta^n} X(\theta) \bar{Y}(\theta) \, d\theta,$$

where again the overbar signifies the complex conjugate. The right-hand side denotes a multiple integration over the n components of θ between $-\pi$ and π with $d\theta = d\theta_1 \cdots d\theta_n$. The norm of any $X \in L_2^n$ is

$$\|X\| = \left[\frac{1}{(2\pi)^n} \int_{\Theta^n} |X(\theta)|^2 \, d\theta \right]^{1/2}.$$

Given any $\mathbf{x} = \{x_p : p \in Z^n\} \in l_2^n$, set

$$X(\theta) = \sum_{p \in Z^n} x_p e^{\iota p \cdot \theta}, \quad \theta \in \Theta^n, \tag{7.2}$$

where $\iota = \sqrt{-1}$ and $p \cdot \theta = p_1 \theta_1 + \cdots + p_n \theta_n$. It is a fact that $\mathbf{x} \in l_2^n$ if and only if $X \in L_2^n$. Moreover, $\mathbf{x}$ is determined from X by

$$x_p = \frac{1}{(2\pi)^n} \int_{\Theta^n} X(\theta) e^{-\iota p \cdot \theta} \, d\theta, \quad p \in Z^n. \tag{7.3}$$

(7.2) defines the *n-dimensional Fourier-series transformation* $\mathcal{F} : \mathbf{x} \mapsto X$, and (7.3) specifies its *inverse* $\mathcal{F}^{-1} : X \mapsto \mathbf{x}$. ($\mathcal{F}^{-1}$ is also called the "n-dimensional finite Fourier transformation.") It is also a standard result that $\mathcal{F}$ is a "topological linear isomorphism" from l_2^n onto L_2^n; this means that $\mathcal{F}$ is a linear bijection that maps convergent sequences in l_2^n into convergent sequences in L_2^n – and conversely. In fact, $\mathcal{F}$ preserves

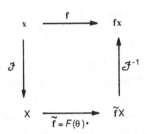

Figure 7.1. The transformation through $\mathcal{F}$ of the Laurent operator $\mathbf{f} \in [l_2^n; l_2^n]$ to obtain the multiplicative mapping $\tilde{\mathbf{f}} = F(\theta)\cdot$ on L_2^n. The arrows illustrate the equation $\mathbf{f} = \mathcal{F}^{-1}\tilde{\mathbf{f}}\mathcal{F}$.

the norms: $\|\mathbf{x}\| = \|X\|$. A more general and detailed discussion of these ideas is given in [156, Chapter 9].

The mapping of any $\mathbf{x} \in l_2^n$ into $\mathbf{fx} \in l_2^n$ by any $\mathbf{f} \in [l_2^n; l_2^n]$ is transformed through $\mathcal{F}$ into the mapping of $X = \mathcal{F}\mathbf{x}$ into $\mathcal{F}\mathbf{fx}$. Since $\mathbf{x} = \mathcal{F}^{-1}X$, we have that the transformed mapping is $X \mapsto \mathcal{F}\mathbf{f}\mathcal{F}^{-1}X$. Set $\tilde{\mathbf{f}} = \mathcal{F}\mathbf{f}\mathcal{F}^{-1}$. Since $\mathcal{F}$ is a topological linear isomorphism, $\tilde{\mathbf{f}}$ is a member of $[L_2^n; L_2^n]$. Moreover, $\mathbf{f} = \mathcal{F}^{-1}\tilde{\mathbf{f}}\mathcal{F}$. This last equation is illustrated in Figure 7.1. If there is a Lebesgue measurable function $F: \Theta^n \rightsquigarrow C^1$, where C^1 denotes the complex plane, such that $(\tilde{\mathbf{f}}X)(\theta) = F(\theta)X(\theta)$ for every $X \in L_2^n$, then $\tilde{\mathbf{f}}$ is called *multiplication by $F(\theta)$*.

When $\mathbf{f}$ is Laurent, we have in particular that $f_{p,q} = f_{0,q-p}$, where 0 now denotes the origin of R^n. We shall refer to the n-dimensional vector $\{f_{0,q}: q \in Z^n\}$ as the *principal array* of $\mathbf{f}$. Set

$$F(\theta) = \sum_{q \in Z^n} f_{0,q} e^{-\iota q \cdot \theta}. \tag{7.4}$$

Here is all we shall need to know about Laurent operators [29].

Lemma 7.1-1: $\mathbf{f}$ *is a Laurent operator in* $[l_2^n; l_2^n]$ *if and only if* $\tilde{\mathbf{f}}$ *is multiplication by* $F(\theta)$, *with* F *being an essentially bounded function on* Θ^n. *Moreover,*

$$\|\mathbf{f}\| = \text{ess sup } \{|F(\theta)|: \theta \in \Theta^n\}.$$

In addition, $\mathbf{f}$ *is positive and invertible if and only if* F *is real-valued and*

$$\text{ess inf } \{F(\theta): \theta \in \Theta^n\} > 0.$$

When all this holds, the transform $\mathcal{F}\mathbf{f}^{-1}\mathcal{F}^{-1}$ *of the inverse* $\mathbf{f}^{-1}$ *of* $\mathbf{f}$ *is multiplication by* $(F(\theta))^{-1}$, *and the numerical range of* $\mathbf{f}$ *is contained in the closed interval between the essential supremum and the essential infimum of* $\{F(\theta): \theta \in \Theta^n\}$.

The multiplicative character of $\tilde{\mathbf{f}}$ is due to the fact that $\mathbf{f}$ commutes with the shifting operator σ_m. This yields an operational calculus for Laurent operators, which is closely related to the "two-sided" (or synonymously the "bilateral") $\mathcal{Z}$ transformation used on digital signals in linear time-invariant systems [13, page 34], [105, Chapter 10]; instead of working with analytic functions on annuli of convergence when using the $\mathcal{Z}$ transformation, we employ here quadratically integrable functions on Θ^n.

As always, we will be examining in the following sections resistive networks whose voltages and currents are real-valued. Hence, we will be dealing with the real Hilbert space l_{2r}^n, which is defined exactly as is l_2^n except that complex quantities are replaced by real ones. Also, the mappings generated by the resistive network will be in $[l_{2r}^n; l_{2r}^n]$. Actually, l_{2r}^n is a subset of l_2^n, and l_2^n is the "complexification of l_{2r}^n." [157, pages 137-138]. Any $\mathbf{f} \in [l_{2r}^n; l_{2r}^n]$ has a unique natural extension onto l_2^n obtained by linearly extending $\mathbf{f}$ onto the real and imaginary parts of any $\mathbf{x} \in l_2^n$. We call such an $\mathbf{f}$ a *real member* of $[l_2^n; l_2^n]$. It is characterized by the fact that the real and imaginary parts of $F(\theta)$ are even and odd functions of θ respectively.

7.2 n-Dimensional Rectangular Grids

We shall first define an arbitrary n-dimensional rectangular grid and then specialize to its uniform version. As before, we let Z^n be the set of all n-dimensional integers $p = (p_1, \ldots, p_n)$ and set $\|p\|_1 = |p_1| + \cdots + |p_n|$. In the following, the subscripts p and q will be members of Z^n, but the subscript g will not.

An *n-dimensional rectangular grid* or simply a *grid* is an infinite electrical network having a node n_p for each $p \in Z^n$, possibly one more node n_g, a branch incident to nodes n_p and n_q if and only if $\|p - q\|_1 = 1$, and possibly a branch connecting node n_p to n_g for various p. This is a 0-network; no 1-nodes have been specified. The nodes n_p will be called *floating nodes*, and n_g will be called *ground*. A branch incident to two floating nodes will also be said to be *floating*, as will its parameters. If the unit increment in the indices for the two nodes of a floating branch occurs in the νth component, we shall say that the branch *lies parallel to the νth coordinate*. It will always be understood that a floating branch has a positive resistance. On the other hand, a branch connecting n_p to n_g may be a pure source, but, if it does possess a positive resistor, that resistor will be called a *grounding resistance* or a *grounding conductance*, and the branch itself will be called a *grounding branch*. The grid itself will be called *grounded* if every possible grounding branch (a positive resistor required) exists. If there is no ground (and thereby no grounding branches and no pure current sources connected to ground), the grid is called *ungrounded*. Thus, in a grounded grid, n_g is an infinite node, and the degree of every floating node is $2n + 1$. In an ungrounded grid, the degree of every node is $2n$. (Later on when we examine "ungrounded semi-infinite grids" in Sections 7.4 and 7.6, pure current sources connecting an appended ground node to floating nodes will be introduced.)

The one-dimensional rectangular grid without any sources is simply an endless, purely resistive path in the ungrounded case and an endless, purely resistive ladder in the grounded case. The two-dimensional rectangular grid is called the *square grid* and is illustrated in Figure 5.1 for the ungrounded case. In higher dimensions we have *cubic grids, quartic grids, quintic grids*, and so forth. See Theorem 3.5-6 for one set of conditions ensuring a unique voltage-current regime in any grid.

An n-dimensional rectangular grid is called *uniform* if for each coordinate the resistors in the branches parallel to that coordinate all have the same value and – in the grounded case – all the resistors in the grounding branches have a common value too. As for the sources, we place no restrictions on their values other than the requirement that the total available power be finite. Since for a uniform grid there is only a finite number of different resistance values, this means that the sum of the squares of all source values is required to be finite, that is, those source values are quadratically summable. Thus, if all branches are in the Thevenin form (or Norton form), we require that $\sum e_j^2 < \infty$ (respectively, $\sum h_j^2 < \infty$).

(It may be worth mentioning at this point that our analysis of uniform grids can be extended to more general types of uniform grids, ones wherein branches having the same resistance value connect nodes n_p and n_{p+m}, where $m \in Z^n$ is fixed and p traverses Z^n – so long as the floating nodes are of finite degree [170]. In the grounded case, this requires no more than an extension of our notation; in the ungrounded case, it needs to be assumed in addition that this more general type of grid is 0-connected.)

Up to and including Section 7.6, we shall assume that all branches are in the Norton form. Moreover, we shall make use of a *nodal current-source vector* $\mathbf{x}$, which is defined as follows. Let x_p be the sum of all the branch current sources $\pm h_{j_i}$ incident to the floating node n_p, with a plus (minus) sign appended to a branch-current-source value h_{j_i} when the source is oriented toward (away from) n_p. There are at most $2n + 1$ such current sources. We set $\mathbf{x} = \{x_p : p \in Z^n\}$. This is a known vector

for any specified grid. Note that the sum of the branch current sources
incident to the ground does not appear as a component of **x**.

Lemma 7.2-1: *In a uniform grid (grounded or ungrounded) whose branch-
current-source values are quadratically summable, the nodal current-source
vector* **x** *is a member of* l_{2r}^n.

Proof: The sum of all the branch current sources $\pm h_{j_i}$ feeding a given
floating node is bounded according to

$$\left(\sum_{i=1}^{2n+1} \pm h_{j_i}\right)^2 \leq (2n+1)\sum h_{j_i}^2.$$

Now, sum the right-hand side over all floating nodes. Each branch current source appears exactly once in the latter sum if that branch is a
grounding branch and exactly twice if it is a floating branch. This leads
to the conclusion. ♣

Let us now restrict our attention to the grounded case. Let a uniform grounded grid have the voltage-current regime dictated by Theorem
3.5-6 and Corollary 3.5-7. Assign to each grounding branch the same index $p \in Z^n$ as that of the floating node n_p to which it is incident. Let u_p
be the voltage of the pth grounding branch oriented toward the ground
node n_g. Also, assign a node voltage of zero to n_g, and assign u_p as
the node voltage of n_p. This defines $\mathbf{u} = \{u_p : p \in Z^n\}$ as the vector of
floating-node voltages, which we shall refer to simply as the *node-voltage
vector*. By Corollary 3.5-7 and the fact that the u_p are grounding-branch
voltages, we have that $\sum u_p^2 < \infty$, that is, $\mathbf{u} \in l_{2r}^n$.

As for the uniform ungrounded grid, we start with a grounded uniform
grid with all branches in Norton form and with no current sources in its
grounding branches. Let us examine what may happen as the common
value g_g of all the grounding conductances tends to zero. In particular,
let us assume that the node-voltage vector **u** of the grounded grid converges nodewise to a node-voltage vector $\mathbf{w} = \{w_p : p \in Z^n\}$ and that
$\sum w_p^2 < \infty$. Now, **w** in turn determines a voltage-current regime in the
resulting ungrounded grid because every branch voltage is the difference
between the voltages at the branches nodes. We now argue that the
voltage-current regime determined by this passage to the limit from the
grounded grid to the ungrounded grid is identical to the one dictated by
Corollary 3.5-7 for the ungrounded grid.

To see this, first note that any choice of node voltages determines
branch voltages that automatically satisfy Kirchhoff's voltage law around
all 0-loops. Moreover, since every floating node remains of finite degree,

Kirchhoff's current law will continue to be satisfied at those nodes in the limit as g_g tends to zero. Hence, in order to invoke Corollary 3.5-7, we need only note that the branch voltages determined by $\mathbf{w} \in l_{2r}^n$ are quadratically summable. Indeed, since each branch voltage is the difference between its nodes' voltages, all the voltages on the branches lying parallel to the first coordinate must be quadratically summable, and so too for every one of the n coordinates. We will have need of this result later on, so let us restate it as

Lemma 7.2-2: *Given any uniform grounded grid with its current sources appearing only in the floating branches and with a voltage-current regime dictated by Corollary 3.5-7, let the common value g_g of its grounding conductances tend to zero with all the floating conductances and current sources held fixed. Assume that that grid's node-voltage vector converges nodewise to a $\mathbf{w} \in l_{2r}^n$. Then, the voltage-current regime determined by the node-voltage vector $\mathbf{w}$ for the resulting ungrounded grid is the same as the regime dictated by Corollary 3.5-7.*

7.3 A Nodal Analysis for Uniform Grids

We continue to maintain our tacit assumption that all branches are in the Norton form. When using node voltages for 0-networks, Kirchhoff's voltage law is automatically satisfied. A nodal analysis applies just Kirchhoff's current law and Ohm's law in terms of node voltages.

Consider a uniform grounded grid, and assign a zero node voltage to its ground. Let $\mathbf{u} = \{u_p : p \in Z^n\}$ be an unknown node-voltage vector; as was indicated in the preceding section, $\mathbf{u}$ will be a member of l_{2r}^n if and only if all the branch voltages are quadratically summable. Moreover, let $\mathbf{x} = \{x_p : p \in Z^n\}$ be a known nodal current-source vector; by Lemma 7.2-1, $\mathbf{x}$ is in l_{2r}^n too whenever the branch current sources are quadratically summable. Furthermore, let g_g be the common conductance of all the grounding branches and g_ν be the common conductance of all the branches lying parallel to the νth coordinate.

Upon applying Kirchhoff's current law at the pth floating node, we obtain

$$\sum_{q \in Z^n} g_{p,q} u_q = x_p. \qquad (7.5)$$

The conductance coefficients $g_{p,q}$ are given by $g_{p,p} = g_g + 2 \sum_{\nu=1}^n g_\nu$, by $g_{p,q} = -g_\nu$ whenever p and q differ by exactly 1 in one of their components – the νth component – and are otherwise equal, and by

$g_{p,q} = 0$ if p and q differ in any other way. To encompass all p, (7.5) may be written in matrix form: $[g_{p,q}][u_q] = [x_p]$, or more concisely as

$$\mathbf{gu} = \mathbf{x}, \tag{7.6}$$

where $\mathbf{u} = [u_q]$ and $\mathbf{x} = [x_p]$ are as before and $\mathbf{g} = [g_{p,q}]$ is a $2n$-dimensional array that satisfies $g_{p,q} = g_{p+m,q+m}$ for all $p, q, m \in Z^n$. We wish to show that $\mathbf{g} \in [l_2^n; l_2^n]$. The linearity of $\mathbf{g}$ being clear, consider its boundedness. For any $z \in l_2^n$,

$$\|\mathbf{gz}\|^2 = \sum_{p \in Z^n} \left| \sum_{q \in Z^n} g_{p,q} z_q \right|^2 \le M \sum_{p \in Z^n} \left(\sum_{p,q} |z_q| \right)^2,$$

where $M = (g_g + 2 \sum_{\nu=1}^n g_\nu)^2$ and, for each p, $\sum_{p,q}$ denotes a summation over all q for which $\|p - q\|_1 \le 1$. There are $2n + 1$ such q. Hence, the right-hand side is bounded by $M(2n + 1) \sum_{p \in Z^n} \sum_{p,q} |z_q|^2$. Each z_q appears in this last expression $2n + 1$ times. So, altogether we have

$$\|\mathbf{gz}\|^2 \le M(2n + 1)^2 \sum_{p \in Z^n} |z_p|^2 = M(2n + 1)^2 \|z\|^2.$$

This proves the boundedness of $\mathbf{g}$, as desired. So truly, $\mathbf{g}$ is a Laurent operator in $[l_2^n; l_2^n]$; in fact, $\mathbf{g} \in [l_{2r}^n; l_{2r}^n]$.

Let us now apply the Fourier-series transformation $\mathcal{F}$ to (7.6) to obtain $G(\theta)U(\theta) = X(\theta)$, where $X \in L_2^n$ is given by (7.2), U is being sought as a member of L_2^n, and $G(\theta)$ is obtained from the principal array of $[g_{p,q}]$ in accordance with (7.4). By an easy manipulation,

$$G(\theta) = g_g + 2 \sum_{\nu=1}^n g_\nu (1 - \cos \theta_\nu). \tag{7.7}$$

Since $0 < g_g \le G(\theta) \le g_g + 4 \sum g_\nu$, it follows from Lemma 7.1-1 that $\mathbf{g}$ is positive and invertible and that $U \in L_2^n$ is given by $U(\theta) = X(\theta)/G(\theta)$. Hence, upon applying $\mathcal{F}^{-1}$ [see (7.3)], we obtain the pth node voltage as

$$u_p = \frac{1}{(2\pi)^n} \int_{\Theta^n} \frac{X(\theta)}{G(\theta)} e^{-\iota p \cdot \theta} \, d\theta. \tag{7.8}$$

This determines in turn all the branch voltages and branch currents. Thus, the $2n$-dimensional uniform grounded grid is completely solved by (7.8). Indeed, $\mathbf{u} = [u_p] \in l_2^n$ because $X(\theta)/G(\theta) \in L_2^n$. Hence, the branch voltages are quadratically summable. Moreover, every u_p is real because the real part (imaginary part) of $X(\theta)/G(\theta)$ is an even (respectively, odd) function of θ whenever the branch current sources and thereby the x_p are real. Furthermore, Kirchhoff's voltage law around 0-loops is automatically satisfied, and Kirchhoff's current law is satisfied at all floating nodes by virtue of the nodal equations (7.6). So, Corollary 3.5-7 justifies our assertion. We have obtained

Theorem 7.3-1: *Given a uniform n-dimensional grounded grid whose branch current sources are quadratically summable, let* $\mathbf{x} \in l^n_{2r}$ *be the corresponding nodal current-source vector and let its node-voltage vector* $\mathbf{u} = [u_p]$ *be given by (7.8). Then, the voltage-current regime determined by* $\mathbf{u}$ *is the same as that dictated by Corollary 3.5-7.*

We turn now to uniform ungrounded grids. Let $G_0(\theta)$ denote $G(\theta)$ when $g_g = 0$:

$$G_0(\theta) = 2 \sum_{\nu=1}^{n} g_\nu (1 - \cos \theta_\nu). \qquad (7.9)$$

In this case, there is a difficulty with (7.8) as a solution for the node voltage u_p when $G(\theta)$ is replaced by $G_0(\theta)$; $1/G_0(\theta)$ tends to infinity as $\|\theta\| \to 0$. A restriction may have to be imposed on $\mathbf{x}$ to insure that a zero in $X(\theta)$ renders $|X(\theta)/G_0(\theta)|^2$ integrable. As will be shown later on, a sufficient condition for this purpose, when the grid's dimension is no less than 3, is that there be just finitely many nonzero branch current sources. The following is obvious.

Lemma 7.3-2: *If an ungrounded grid has only a finite number of nonzero branch current sources, then the nodal current-source vector* $\mathbf{x} = [x_p]$ *has only a finite number of nonzero components and* $\sum x_p = 0$.

We shall use the procedure indicated in Lemma 7.2-2 to obtain the node-voltage vector $\mathbf{w} = [w_p]$ of the ungrounded grid; that is, we append grounding conductances, all of the same value $g_g > 0$, to all the floating nodes and then examine (7.8) as $g_g \to 0$. Since there is only a finite number of nonzero x_p, $X(\theta)$ is a continuous function for all $\theta \in \Theta^n$. In view of (7.7) and (7.9), $X(\theta)/G(\theta)$ is continuous for all $\theta \in \Theta^n$, and $X(\theta)/G_0(\theta)$ is continuous everywhere except possibly at $\theta = 0$.

We now argue that, for an n-dimensional ungrounded grid with $n \geq 3$ and with only a finite number of nonzero branch current sources, $|X(\theta)/G_0(\theta)|^2$ is integrable on Θ^n. We need only examine the asymptotic behaviors of X and G_0 near the origin. From (7.9), we have $G_0(\theta) \sim \sum_{\nu=1}^{n} g_\nu \theta_\nu^2$ as $\|\theta\| \to 0$. As for $X(\theta)$, since $\sum x_p = 0$ according to Lemma 7.3-2, we may write

$$X(\theta) = \sum x_p (e^{\iota p \cdot \theta} - 1)$$

$$\sim \sum x_p \iota p \cdot \theta \qquad \text{as } \|\theta\| \to 0$$

$$= M_1 \theta_1 + \cdots + M_n \theta_n$$

where

$$M_\nu = \iota \sum_{p_1} \cdots \sum_{p_n} p_\nu x_{(p_1,\ldots,p_n)}, \qquad \nu = 1, \ldots, n.$$

Hence, as $\|\theta\| \to 0$,

$$\left| \frac{X(\theta)}{G_0(\theta)} \right|^2 \sim \left| \frac{M_1\theta_1 + \cdots + M_n\theta_n}{g_1\theta_1^2 + \cdots + g_n\theta_n^2} \right|^2$$

$$\leq n \sum_{\nu=1}^{n} \frac{|M_\nu|^2\theta_\nu^2}{(g_1\theta_1^2 + \cdots + g_n\theta_n^2)^2}.$$

Now, each term in this last sum is integrable over a neighborhood of the origin if $n \geq 3$. This can be seen for, say, $\nu = 1$ by making the change of variables: $\theta_1 = r\cos\psi$, $\theta_2 = r\sin\psi$, and $\theta_\nu = \theta_\nu$ for $\nu = 3, \ldots, 4$ to factor out $\int_{-\pi}^{\pi} (\cos\psi)^2 \, d\psi$, and then making another change of variables: $r = \rho\cos\alpha$, $\theta_3 = \rho\sin\alpha$, and $\theta_\nu = \theta_\nu$ for $\nu = 4, \ldots, n$ (if $\nu > 3$) to factor out $\int_{-\pi/2}^{\pi/2} (\cos\alpha)^3 \, d\alpha$. The remaining integrand is the constant 1. So truly, $|X(\theta)/G_0(\theta)|^2$ is integrable over Θ^n when $n \geq 3$. It follows that $|X(\theta)/G_0(\theta)|$ is also integrable over Θ^n.

These integrability results allow us to use Lemma 7.2.-2 to solve an n-dimensional ungrounded grid when $n \geq 3$.

Theorem 7.3-3: *Given a uniform n-dimensional ungrounded grid with $n \geq 3$ and with only a finite number of branch current sources, let $\mathbf{x}$ be the nodal-current-source vector and let the node voltages be determined by*

$$w_p = \frac{1}{(2\pi)^n} \int_{\Theta^n} \frac{X(\theta)}{G_0(\theta)} e^{-\iota p \cdot \theta} \, d\theta. \qquad (7.10)$$

Then, the voltage-current regime determined by the node voltages (7.10) is the same as that dictated by Corollary 3.5-7.

Proof: By Lemma 7.3-2, there is only a finite number of nonzero x_p, and therefore X is a continuous function. By the integrability of $|X(\theta)/G_0(\theta)|$, u_p exists for every p. Now, consider the corresponding uniform grounded grid obtained by appending the positive conductance g_g to every floating node. The latter grid has node voltages in accordance with Theorem 7.3-1. Upon letting $g_g \to 0$, we can assert that the pth such node voltage converges to w_p. Indeed, by (7.7) and (7.9), $|X(\theta)/G_0(\theta)| > |X(\theta)/G(\theta)|$. Hence, we may use Lebesgue's theorem of dominated convergence [48, page 151] to take the limit under the integral sign in order to obtain (7.10) from (7.8). w_p is real because u_p is real. Moreover, since the function $\theta \mapsto X(\theta)/G_0(\theta)$ is a member of L_2^n, we have $\mathbf{w} = \{w_p : p \in Z^n\} \in l_{2r}^n$. Lemma 7.2-2 now completes the proof. ♣

Let us now discuss the cases where $n = 1$ and $n = 2$. The one-dimensional case of an ungrounded grid is just an endless path. Hence, its solution is trivial: The current from any branch's current source flows through that branch's conductance alone. (Remember that we have a 0-network – there are no connections at infinity.)

There is a difficulty in the two-dimensional case. Integrability on Θ^2 can be shown for $|X(\theta)/G_0(\theta)|$ much as before, but not for $|X(\theta)/G_0(\theta)|^2$. The latter function has a nonintegrable singularity at the origin. This means that Lebesgue's theorem of dominated convergence can be used to prove the existence of the node voltages w_p as the nodewise limits of the node voltages u_p. However, we cannot conclude that $\mathbf{w} = \{w_p : p \in Z^2\}$ is a member of l_{2r}^2 and therefore cannot invoke Lemma 7.2-2. Anyway, we can state the following limited result.

Theorem 7.3-4: *Given a uniform two-dimensional ungrounded grid with only a finite number of branch current sources, the node voltages, given by (7.10) with $n = 2$, are the nodewise limits of the node voltages in the corresponding uniform grounded grid as the grounding-conductance value g_g is taken to zero.*

Later on, we will show through a different analysis that the voltage-current regime arising from these node voltages in this two-dimensional case is the same as that of Theorem 3.5-6; see Theorem 7.6-5.

Example 7.3-5. Let us compute various driving-point resistances for the purely resistive, uniform, n-dimensional, ungrounded grid [171]. Such a resistance can be calculated as the difference between two node voltages induced by connecting a 1 A pure current source to those two nodes. Those nodes need not be adjacent.

Let 1 A be extracted from node n_p and injected into node n_q. Then, $X(\theta) = e^{\iota q \cdot \theta} - e^{\iota p \cdot \theta}$, and, by (7.10), the driving-point resistance between those two nodes is

$$\frac{1}{(2\pi)^n} \int_{\Theta^n} \frac{1}{G_0(\theta)} (e^{\iota q \cdot \theta} - e^{\iota p \cdot \theta})(e^{-\iota q \cdot \theta} - e^{-\iota p \cdot \theta}) \, d\theta$$

$$= \frac{1}{(2\pi)^n} \int_{\Theta^n} \frac{2 - 2\cos((q - p) \cdot \theta)}{G_0(\theta)} \, d\theta. \tag{7.11}$$

More particularly, assume now that all resistors are 1 Ω and that n_p and n_q are adjacent, that is, let the driving-point resistance be measured across a single branch, which we take to be parallel to the first coordinate. Then, (7.11) becomes

$$\frac{1}{(2\pi)^n} \int_{-\pi}^{\pi} \cdots \int_{-\pi}^{\pi} \frac{1 - \cos\theta_1}{n - \cos\theta_1 - \cdots - \cos\theta_n} \, d\theta_1 \cdots d\theta_n. \qquad (7.12)$$

Several numerical integrations suggest that (7.12) is equal to $1/n$.

As other examples for grids of $1\ \Omega$ resistors, we have the driving-point resistance across the diagonal of one face of a square grid:

$$\frac{1}{(2\pi)^2} \int_{-\pi}^{\pi} \int_{-\pi}^{\pi} \frac{1 - \cos(\theta_1 + \theta_2)}{2 - \cos\theta_1 - \cos\theta_2} \, d\theta_1 d\theta_2 = 0.6366 \approx \frac{2}{\pi}, \qquad (7.13)$$

across the diagonal of a square face in a cubic grid:

$$\frac{1}{(8\pi)^3} \int_{-\pi}^{\pi} \int_{-\pi}^{\pi} \int_{-\pi}^{\pi} \frac{1 - \cos(\theta_1 + \theta_2)}{3 - \cos\theta_1 - \cos\theta_2 - \cos\theta_3} \, d\theta_1 d\theta_2 d\theta_3 = 0.3951,$$
$$\qquad (7.14)$$

and across the diagonal of one cube in a cubic grid:

$$\frac{1}{(8\pi)^3} \int_{-\pi}^{\pi} \int_{-\pi}^{\pi} \int_{-\pi}^{\pi} \frac{1 - \cos(\theta_1 + \theta_2 + \theta_3)}{3 - \cos\theta_1 - \cos\theta_2 - \cos\theta_3} \, d\theta_1 d\theta_2 d\theta_3 = 0.4183.$$
$$\qquad (7.15)$$

The values of these integrals were obtained through numerical integrations.

The determination of various driving-point resistances of a uniform, rectangular, n-dimensional grid is a classical problem; see [43], [52], [129], and the references therein. The present analysis by means of the n-dimensional Fourier-series transformation completely solves the problem. Moreover, some integrals like (7.12) have been evaluated in terms of elliptic integrals and the gamma function [59], [141]; see also the correction to [59] pointed out in [43, page 126]. ♣

Example 7.3-6. The purpose of this example is to compare our value for (7.12) with a value obtained from another way of calculating driving-point resistances in uniform grids. The latter is based on a formula for finite networks, known as *Foster's averaging formula* [54], [55]; see also [53] and [143, pages 170–176]. It states that, for a purely resistive, finite network of B branches and N nodes with no self-loops,

$$\sum_{j=1}^{B} \frac{R_j}{r_j} = N - 1, \qquad (7.16)$$

where r_j is the jth branch resistance and R_j is the driving-point resistance measured across the jth branch. If all the r_j are $1\ \Omega$ and if the network is symmetrical with respect to its branches (i.e., the graph of

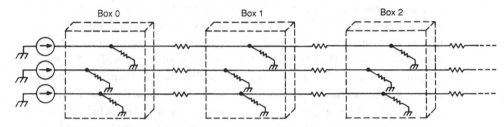

Figure 7.2. A semi-infinite, $(n + 1)$-dimensional, rectangular grid. Each box denotes schematically an n-dimensional uniform grid, which may be either grounded or ungrounded. Pure current sources – external to the boxes – feed current from ground to some or all of the nodes in box 0.

the network is topologically the same when viewed from any branch), then (7.16) becomes

$$R_j = \frac{2(N-1)}{Nd},\qquad (7.17)$$

where d is the common degree of the nodes. Let us assume that this formula remains valid in the limit as $N \to \infty$. After all, any ungrounded grid of 1 Ω resistors can be viewed as an extreme case for Foster's formula (7.17). For that grid, $d = 2n$ and (7.17) tends toward $1/n$. This agrees with our evaluation of (7.12). Actually, Foster's formula does have a rigorous justification for infinite networks [129, Section 3].

It is noteworthy that, for an infinite hexagonal grid of 1 Ω resistors, d equals 3, and therefore Foster's formula (7.17) yields the limiting value of 2/3 for the driving-point resistance across any branch. This result is unavailable from our theory, which is restricted to uniform grids whose branches lie in a rectangular configuration. ♣

7.4 One-dimensional Nonuniformity

A grid whose resistance values vary in one and only one of the coordinate directions is our next subject. It is more convenient now to study an $(n + 1)$-dimensional, rectangular grid, which is *semi-infinite* in the sense that the grid is one-way infinite in the direction of the nonuniformity. This is illustrated schematically in Figure 7.2. Each box therein represents an imaginary surface that encloses a uniform n-dimensional rectangular grid, which may be either grounded or ungrounded. We have indicated the grounded case with just three nodes and have not drawn the floating branches within the box. The boxes are indexed by $k = 0, 1, 2, \ldots$. The grid within any box will be called a *boxed* grid, and all its elements will be called *boxed* elements. Here too, the nodes within any box are indexed by $p \in Z^n$; that is, $n_{k;p}$ is the pth node in box k. Also, $g_{k;\nu}$ will denote the common positive conductance for the branches in box k that lie parallel to the νth coordinate and, if the grid is grounded, g_{gk} will be the common positive conductance to ground in box k. Furthermore, the branches between consecutive boxes will be called *unboxed* branches, and their resistances will be called *unboxed* resistors. Between box k and box $k+1$, the unboxed resistors all have the same positive value r_k. The incident nodes of any unboxed branch have the same index $p \in Z^n$. As always, all conductance or resistance values are positive, but now they may vary as k varies. Finally, we shall take it that the only sources are pure current sources connected between ground and some or all of the nodes of box 0, every source being incident to ground and oriented as shown in Figure 7.2 (i.e., the current-source value is positive if the current enters the floating node). We view those sources as being outside the box. $\mathbf{h} = \{h_p : p \in Z^n\}$ denotes the vector of those current sources. (In Section 7.7, we will specify certain 1-nodes at infinity and will connect other sources there as well.)

We shall refer to the overall structure of Figure 7.2 as a *semi-infinite, $(n+1)$-dimensional, one-dimensionally nonuniform, rectangular grid* or simply as a *semi-infinite grid*. Moreover, we shall call it *grounded* (or *ungrounded*) if all the boxed grids are grounded (respectively, ungrounded). Note that now in the ungrounded case a ground will still exist as a common node for the pure current-source branches. Also, we will call the entire grid *partially grounded* if at least one – but not all – of the boxed grids is grounded. Throughout this and the next two sections, we will impose

Conditions 7.4-1: *The infinite network is a semi-infinite, $(n+1)$-dimensional, one-dimensionally nonuniform, rectangular grid – and in fact a 0-network – of the form shown in Figure 7.2. Moreover, every current source is incident to ground and a floating node of box 0, and the current-source vector $\mathbf{h} = \{h_p : p \in Z^n\}$ is a member of l_{2r}^n.*

The semi-infinite grid can be viewed as a ladder network of "∞-ports" [157, Section 4.2]. These are infinite-dimensional generalizations of the "n-port" concept of finite-network theory [101, Chapter 1]. We define an ∞-*port* as being an infinite 0-network having a countable infinity of designated ordered pairs (n_i^+, n_i^-) of 0-nodes n_i^+, n_i^-, where $i = 1, 2, \ldots$, to which additional circuitry, called *external connections*, may be attached. It is required that the current entering n_i^+ from the external connections be equal to the current leaving n_i^- through the external connections. This requirement is called the *port condition*, and the said current is called the *port current*. The voltage drop from n_i^+ to n_i^- is the *port voltage*. The ∞-port determines a relationship between the vector **i** of all port currents and the vector **v** of all port voltages. When $\mathbf{i} \mapsto \mathbf{v}$ is a function (i.e., a single-valued mapping), that function is called the *resistance operator* of the ∞-port. When $\mathbf{v} \mapsto \mathbf{i}$ is a function, it is called the *conductance operator* of the ∞-port.

We can pair every floating node in any box of Figure 7.2 with ground to obtain an ∞-port. In fact, we may do so even when the boxed grid is ungrounded simply by using the common ground of the source branches. We take each port to be oriented toward ground, that is, the floating node is the first node and ground is the second node of the ordered pair. Moreover, it will always turn out that any external current entering an ∞-port through a floating node can be designated as leaving that ∞-port through ground. Indeed, in the grounded case the ground node is not restraining and therefore does not impose any restriction on the assignment of currents through itself. In the ungrounded case the equations arising from Kirchhoff's current law applied to the floating nodes can be viewed as reappearing in summed form for the ground, and hence the ground does not impose any further restriction beyond these applications of the current law. In both cases, we are free to designate the flow of port currents through ground in this way in order to satisfy the port conditions for each box's ∞-port.

Now, let $\mathbf{v}_k = \{v_{k;p} : p \in Z^n\}$ be the node-voltage vector for the kth boxed grid. This is also the port-voltage vector of the kth box's ∞-port. For $k = 1, 2, \ldots$, let $\mathbf{i}_k = \{i_{k;p} : p \in Z^n\}$ be the vector of currents $i_{k;p}$ flowing from box $k - 1$ to box k through the pth unboxed resistor (i.e., the resistor incident to nodes $n_{k-1;p}$ and $n_{k;p}$). Upon writing Kirchhoff's current law for all the floating nodes in every box using the matrix form, we obtain

$$\mathbf{h} - \mathbf{i}_1 = \mathbf{g}_0 \mathbf{v}_0, \tag{7.18}$$

$$\mathbf{i}_k - \mathbf{i}_{k+1} = \mathbf{g}_k \mathbf{v}_k, \quad k = 1, 2, 3, \ldots . \tag{7.19}$$

Here, as before, $\mathbf{h} = \{h_p : p \in Z^n\}$ is the vector of current sources connected to nodes in box 0, measured with the orientations shown in Figure 7.2. Also, $\mathbf{g}_k$ is the conductance operator for the kth box's ∞-port. The p, q entry in the matrix of $\mathbf{g}_k$ will be denoted by $g_{k;p,q}$. We have $g_{k;p,p} = g_{gk} + 2 \sum_{\nu=1}^{n} g_{k;\nu}$, $g_{k;p,q} = -g_{k;\nu}$ if p and q differ in only their νth components and $\|p - q\|_1 = 1$, and $g_{k;p,q} = 0$ otherwise. As was

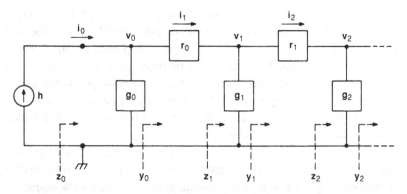

Figure 7.3. The one-ended operator ladder obtained by appropriately connecting ∞-ports. The $\mathbf{i}$'s, $\mathbf{v}$'s, and $\mathbf{h}$ are members of l_{2r}^n, and the $\mathbf{g}$'s and $\mathbf{r}$'s are positive Laurent operators in $[l_{2r}^n; l_{2r}^n]$ for the shunt and series ∞-ports. The $\mathbf{z}$'s and $\mathbf{y}$'s are driving-point resistance and conductance operators measured toward the right.

shown in Section 7.3 just after Equation (7.6), $\mathbf{g}_k$ is a Laurent operator in $[l_{2r}^n; l_{2r}^n]$.

All the resistors between box k and box $k+1$ comprise another ∞-port, where now each port consists of the nodes of one such resistor and the port's orientation is taken to be from box k to box $k+1$. By Ohm's law we have

$$\mathbf{v}_k - \mathbf{v}_{k+1} = \mathbf{r}_k \mathbf{i}_{k+1}, \quad k = 0, 1, 2, \ldots \qquad (7.20)$$

where $\mathbf{r}_k$ is a diagonal operator whose matrix's main diagonal entries are all equal to r_k. Thus, $\mathbf{r}_k = r_k \mathbf{1}$, where $\mathbf{1}$ is the identity operator on l_{2r}^n. $\mathbf{r}_k$ is the resistance operator for this ∞-port.

We can view these ∞-ports as being connected together according to the ladder of ∞-ports shown in Figure 7.3. This conforms with the connections of Figure 7.2 wherein individual nodes are shorted together only if they carry the same index p. The small rectangles in Figure 7.3 represent the said ∞-ports. Since the $\mathbf{g}_k$ and $\mathbf{r}_k$ are all operators in $[l_{2r}^n; l_{2r}^n]$, we will call this network an *operator network* or an *operator ladder*, and in this regard the *shunt* (*series*) ∞-ports are those for $\mathbf{g}_k$ (respectively, $\mathbf{r}_k$). In our subsequent analysis all the $\mathbf{i}_k$ and $\mathbf{v}_k$ and $\mathbf{h}$ as well will be members of l_{2r}^n. Note that the satisfaction of Kirchhoff's laws and Ohm's law by the voltage and current vectors in this operator ladder ensures their satisfaction by the scalar voltages and currents in the semi-infinite grid of Figure 7.2. By virtue of the designated flow of port currents in the shunt ∞-ports and the isolation of the resistors in the series ∞-ports, the port conditions are not violated under the connections of Figure 7.3. (This would not be so were there other internal resistors in any series ∞-port connecting nodes of differing p-indices. Our analysis would then fail.)

7.5 Solving Grounded Semi-infinite Grids: Infinity Imperceptible

As before, we shall say that *infinity is imperceptible* (or *perceptible*) when there is no need (respectively, there is a need) to specify what if anything is connected to 1-nodes of a network in order to determine a unique voltage-current regime. In this section, we will assume that infinity is imperceptible by assuming the following – in addition to Conditions 7.4-1.

Conditions 7.5-1: *The semi-infinite grid is grounded* (so that $g_{gk} > 0$ for all k), *and* $\sum_{k=1}^{\infty}(r_k + g_{gk})$ *diverges.*

Rather than analyzing the operator ladder of Figure 7.3 by manipulating vectors in l_{2r}^n and Laurent operators in $[l_{2r}^n; l_{2r}^n]$, we will invoke the transformation $\mathcal{F}$ discussed in Section 7.1 and will manipulate instead functions in L_2^n and operators that are multiplications by Lebesgue measurable, essentially bounded functions. We have already noted that $\mathcal{F}$ preserves norms; that is, for any $\mathbf{x} \in l_{2r}^n$ and its transform $\mathcal{F}\mathbf{x} \in L_2^n$,

$$\|\mathcal{F}\mathbf{x}\| = \|\mathbf{x}\|. \tag{7.21}$$

Upon applying $\mathcal{F}$ to all the voltage and current vectors in Figure 7.3 and invoking Lemma 7.1-1, we transform the ladder into the one shown in Figure 7.4. Here, $H \in L_2^n$ is given by the Fourier series:

$$H(\theta) = \sum_{p \in Z^n} h_p e^{\iota p \cdot \theta}, \quad \theta \in \Theta^n. \tag{7.22}$$

The I_k and V_k are similarly defined by

$$V_k(\theta) = \sum_{p \in Z^n} v_{k;p} e^{\iota p \cdot \theta}$$

and

$$I_k(\theta) = \sum_{p \in Z^n} i_{k;p} e^{\iota p \cdot \theta}.$$

On the other hand, for each $k = 0, 1, 2, \ldots$, the operator $\tilde{\mathbf{g}}_k = \mathcal{F}\mathbf{g}_k\mathcal{F}^{-1}$ is multiplication by $G_k(\theta)$, which as in Section 7.3 is

$$G_k(\theta) = g_{gk} + 2\sum_{\nu=1}^{n} g_{k;\nu}(1 - \cos\theta_\nu). \tag{7.23}$$

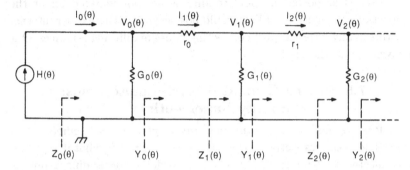

Figure 7.4. The transformed ladder obtained by applying $\mathcal{F}$ to the ladder of Figure 7.3. For each fixed θ the $Z_k(\theta)$ (or $Y_k(\theta)$) denote driving-point resistances (conductances) measured toward the right.

Also, $\tilde{\mathbf{r}}_k = \mathcal{F}\mathbf{r}_k\mathcal{F}^{-1}$ is multiplication by $R_k(\theta)$, but, since $\mathbf{r}_k = r_k\mathbf{1}$, we have $R_k(\theta) = r_k$, that is, the transform $\tilde{\mathbf{r}}_k$ of the operator $\mathbf{r}_k$ is multiplication by the constant r_k.

For each fixed θ, the ladder of Figure 7.4 is linear but in general nonuniform. $Z_k(\theta) = V_k(\theta)/I_k(\theta)$ represents therein the driving-point resistance measured toward the right at the kth port of the ladder. Also, we will have need of the driving-point conductance $Y_k(\theta) = I_{k+1}(\theta)/V_k(\theta)$ measured toward the right just after the shunt element and before the series element of the kth stage.

To solve this transformed ladder, we will follow the procedure indicated in Example 6.8-1 to compute the driving-point immittances using continued fractions. These in turn will yield the $V_k(\theta)$ by means of voltage transfer functions $T_k(\theta)$ as in Section 6.1. Rather than using the cumbersome notation

$$A = \cfrac{1}{a_1 + \cfrac{1}{a_2 + \cfrac{1}{a_3 + \cfrac{1}{\ddots}}}}$$

for a continued fraction, we will employ the simpler – and standard – notation

$$A = \frac{1}{a_1} + \frac{1}{a_2} + \frac{1}{a_3} + \cdots \qquad (7.24)$$

This continued fraction is said to converge if the sequence of its *finite truncations*

$$A^{(m)} = \frac{1}{a_1 +} \frac{1}{a_2 +} \cdots \frac{1}{+ a_m}$$

converges as $m \to \infty$. $A^{(m)}$ is called an *even* (or *odd*) *truncation* if m is even (or odd). We will also employ the following results from the theory of infinite continued fractions [140].

Lemma 7.5-2: *Let all the a_k in (7.24) be positive numbers. If m is an odd positive integer, then*

$$A^{(m+1)} < A^{(m+3)} < A^{(m+2)} < A^{(m)}. \tag{7.25}$$

Thus, the even truncations of (7.24) comprise an increasing sequence and the odd truncations comprise a decreasing sequence. If $\sum a_k = \infty$, then these two sequences converge to the same limit, that is, (7.24) converges as a continued fraction. If $\sum a_k < \infty$, then the two sequences converge to distinct limits, and the sequence of finite truncations of (7.24) diverges by oscillation. Furthermore, $A^{(m)} - A^{(m+1)}$ is the reciprocal of a multinomial in the $a_1, \ldots, a_m$ whose coefficients are all positive numbers.

For the transformed ladder, we have

$$I_0(\theta) = (G_0(\theta) + Y_0(\theta))V_0(\theta),$$

and therefore

$$Z_0(\theta) = \frac{V_0(\theta)}{I_0(\theta)} = \frac{1}{G_0(\theta) + Y_0(\theta)}.$$

Similarly, $Y_0(\theta) = I_1(\theta)/V_0(\theta)$ and $V_0(\theta) = (r_0 + Z_1(\theta))I_1(\theta)$. Consequently,

$$Z_0(\theta) = \frac{1}{G_0(\theta) +} \frac{1}{r_0 + Z_1(\theta)}.$$

Continuing this process of expanding the driving-point resistances and conductances, we obtain $Z_0(\theta)$ as the infinite continued fraction:

$$Z_0(\theta) = \frac{1}{G_0(\theta) +} \frac{1}{r_0 +} \frac{1}{G_1(\theta) +} \frac{1}{r_1 +} \cdots . \tag{7.26}$$

In a similar way, we obtain, for all k

$$Y_k(\theta) = \frac{1}{r_k +} \frac{1}{G_{k+1}(\theta) +} \frac{1}{r_{k+1} +} \frac{1}{G_{k+2}(\theta) +} \cdots . \tag{7.27}$$

Lemma 7.5-3: *Under Conditions 7.5-1, the continued fractions (7.26) and (7.27) converge uniformly for all $\theta \in \Theta^n$; moreover, Z_0 and all the Y_k are continuous positive functions on Θ^n.*

Proof: Consider the difference between any odd truncation of (7.26) or (7.27) and the next larger even truncation. By the last sentence of Lemma 7.5-2, that difference is larger for $\theta = 0$ than it is for any other value of θ because the $G_k(\theta)$ take on their smallest values at $\theta = 0$. Hence, convergence at any $\theta \in \Theta^n$ is no slower than that at $\theta = 0$, whence the uniform convergence. Since every finite truncation of (7.26) and (7.27) is a continuous function of θ, so too are Z_0 and the Y_k. They are also positive because they are bounded below by their even truncations. ♣

The continuity of Z_0 implies that it is bounded on the compact set Θ^n. By Lemma 7.1-1, multiplication by $Z_0(\theta)$ is the transform of a Laurent operator $\mathbf{z}_0 \in [l_2^n; l_2^n]$. Moreover, $\mathbf{z}_0$ is real because (7.26) is real. For any $\mathbf{h} \in l_{2r}^n$, we can determine $\mathbf{v}_0 = \mathbf{z}_0 \mathbf{h} \in l_{2r}^n$ by setting $V_0(\theta) = Z_0(\theta)H(\theta)$ and then applying $\mathcal{F}^{-1}$ to V_0.

To obtain the $V_{k+1}(\theta)$ and thereby the $\mathbf{v}_{k+1}$ for any k, we employ the *voltage-transfer functions* $T_k(\theta) = V_{k+1}(\theta)/V_k(\theta)$. (These are different from the usual open-circuit voltage transfer functions because each stage remains loaded by the rest of the ladder.) By Ohm's law,

$$V_{k+1}(\theta) = V_k(\theta) - r_k I_{k+1}(\theta).$$

By the definition of $Y_k(\theta)$, $I_{k+1}(\theta) = Y_k(\theta)V_k(\theta)$. Hence,

$$V_{k+1}(\theta) = (1 - r_k Y_k(\theta))V_k(\theta).$$

Whence,

$$T_k(\theta) = 1 - r_k Y_k(\theta). \tag{7.28}$$

According to Lemma 7.5-3, Y_k and therefore T_k are continuous functions on Θ^n. They are also real-valued. Since

$$Y_k(\theta) \geq \frac{1}{r_k + \dfrac{1}{G_{k+1}(\theta)}},$$

we have

$$T_k(\theta) \leq \frac{1}{1 + r_k G_{k+1}(\theta)} < 1. \tag{7.29}$$

Now, let $\tilde{t}_k$ be the operator of multiplication by $T_k(\theta)$, and set $\mathbf{t}_k = \mathcal{F}^{-1}\tilde{t}_k\mathcal{F}$. We may call $\mathbf{t}_k$ a *voltage-transfer operator*. By virtue of Lemma 7.1-1 and (7.29), $\|\mathbf{t}_k\| < 1$. Altogether then, we have

Lemma 7.5-4: *For a grounded semi-infinite grid satisfying Conditions 7.4-1 and 7.5-1 and for all $k = 0, 1, 2, \ldots$, the $\mathbf{t}_k$ are Laurent operators in $[l_{2r}^n; l_{2r}^n]$, and $\|\mathbf{t}_k\| < 1$.*

Thus, some formulas, which solve a semi-infinite grounded grid, are

$$V_0(\theta) = Z_0(\theta)H(\theta) \qquad (7.30)$$

and

$$V_k(\theta) = T_{k-1}(\theta) \cdots T_0(\theta)V_0(\theta), \qquad k = 1, 2, \ldots, \qquad (7.31)$$

where H is given by (7.22), Z_0 by (7.26), T_k by (7.28), and Y_k – needed in (7.28) – by (7.27). Upon applying $\mathcal{F}^{-1}$ to the V_k, we get the node-voltage vectors $\mathbf{v}_0$ and

$$\mathbf{v}_k = \mathbf{t}_{k-1} \cdots \mathbf{t}_0 \mathbf{v}_0 \in l_{2r}^n, \qquad k = 1, 2, \ldots$$

for the boxed grids. In particular, the pth node voltage in box k is

$$v_{k;p} = \frac{1}{(2\pi)^n} \int_{\Theta^n} V_k(\theta) e^{-\iota p \cdot \theta} \, d\theta. \qquad (7.32)$$

The node voltages determine in turn the voltage-current regime for the entire semi-infinite grid. That regime satisfies Kirchhoff's laws for a 0-network. Indeed, Kirchhoff's voltage law is automatically satisfied around every 0-loop, and the branch currents, which are generated by Ohm's law from the node voltages, satisfy Kirchhoff's current law at every floating node by virtue of (7.20) and the nodal equations (7.18) and (7.19).

Theorem 7.5-5: *For a grounded semi-infinite grid satisfying Conditions 7.4-1 and 7.5-1, the voltage-current regime determined as just stated from node voltages is identical to the regime given by Theorem 3.5-6.*

Proof: In view of what we have already shown, it needs only to be demonstrated that the total isolated power and the total dissipated power are both finite. Every pure current source h_p feeds a floating node in box 0 and therefore is in parallel with one of the grounding conductances g_{g0}. So, these two elements can be taken as comprising a single branch. Since $\mathbf{h} \in l_{2r}^n$ by Conditions 7.4-1, we have $\sum h_p^2 g_{g0}^{-1} < \infty$; that is, the total isolated power is finite.

As for the total dissipated power, consider the operator ladder of Figure 7.3. The elements $\mathbf{g}_k$ and $\mathbf{r}_k$ are all positive operators in $[l_2^n; l_2^n]$ since G_k and R_k are positive, continuous, bounded functions on Θ^n. In view of Lemma 7.5-3, the same is true for the driving-point resistance operator $\mathbf{z}_0$ and the conductance operators $\mathbf{y}_k$. Now, let $(\cdot, \cdot)$ denote the inner product in l_2^n. An examination of all the terms in $(\mathbf{v}_k, \mathbf{g}_k \mathbf{v}_k)$ shows that it is equal to the sum of all the powers dissipated in the branches of box k, including the grounding branches. On the other hand, $(\mathbf{r}_k \mathbf{i}_{k+1}, \mathbf{i}_{k+1})$ is the sum of all the powers dissipated in the unboxed resistors between box k and box $k + 1$. Furthermore, $(\mathbf{v}_0, \mathbf{h}) = (\mathbf{z}_0\mathbf{h}, \mathbf{h})$ is finite and non-negative, and, by Kirchhoff's laws, it can be expanded as follows.

$$
\begin{aligned}
(\mathbf{v}_0, \mathbf{h}) &= (\mathbf{v}_0, \mathbf{g}_0\mathbf{v}_0) + (\mathbf{v}_0, \mathbf{i}_1) \\
&= (\mathbf{v}_0, \mathbf{g}_0\mathbf{v}_0) + (\mathbf{v}_0 - \mathbf{v}_1, \mathbf{i}_1) + (\mathbf{v}_1, \mathbf{i}_1) \\
&= (\mathbf{v}_0, \mathbf{g}_0\mathbf{v}_0) + (\mathbf{v}_0 - \mathbf{v}_1, \mathbf{i}_1) \\
&\quad + \cdots + (\mathbf{v}_{k-1}, \mathbf{g}_{k-1}\mathbf{v}_{k-1}) + (\mathbf{v}_{k-1}, \mathbf{i}_k) \\
&= (\mathbf{v}_0, \mathbf{g}_0\mathbf{v}_0) + (\mathbf{v}_0 - \mathbf{v}_1, \mathbf{i}_1) + \cdots + (\mathbf{v}_{k-1} - \mathbf{v}_k, \mathbf{i}_k) + (\mathbf{v}_k, \mathbf{i}_k).
\end{aligned}
$$

Moreover,

$$(\mathbf{v}_{k-1} - \mathbf{v}_k, \mathbf{i}_k) = (\mathbf{r}_{k-1}\mathbf{i}_k, \mathbf{i}_k) \geq 0, \quad (\mathbf{v}_{k-1}, \mathbf{i}_k) = (\mathbf{v}_{k-1}, \mathbf{y}_{k-1}\mathbf{v}_{k-1}) \geq 0,$$

and

$$(\mathbf{v}_k, \mathbf{i}_k) = (\mathbf{z}_k\mathbf{i}_k, \mathbf{i}_k) \geq 0.$$

It follows that the partial sums of the infinite series of nonnegative terms

$$(\mathbf{v}_0, \mathbf{g}_0\mathbf{v}_0) + (\mathbf{r}_0\mathbf{i}_1, \mathbf{i}_1) + (\mathbf{v}_1, \mathbf{g}_1\mathbf{v}_1) + (\mathbf{r}_1\mathbf{i}_2, \mathbf{i}_2) + \cdots \qquad (7.33)$$

increase monotonically and are all bounded by $(\mathbf{v}_0, \mathbf{h})$. Hence, (7.33) converges. We have thus shown that the total power dissipated in all the branches of the semi-infinite grid is finite. We can now invoke Theorem 3.5-6 to conclude the proof. ♣

For computational purposes, the continued fractions (7.26) and (7.27) need to be evaluated. Finding closed form expressions for them is an unlikely prospect, save when the semi-infinite grid approaches a uniform one as $k \to \infty$. In the latter case, the characteristic resistance of the uniform structure can be used to terminate truncations of the continued fractions. For example, if the semi-infinite grid is uniform for all $k > k_0$ with the parameters $r_k = r$ and $G_k(\theta) = G(\theta)$, then

$$Z_0(\theta) = \cfrac{1}{G_0(\theta)} + \cfrac{1}{r_0} + \cfrac{1}{\cdots} + \cfrac{1}{G_{k_0}(\theta)} + \cfrac{1}{r_{k_0} + Z(\theta)},$$

where

$$Z(\theta) = -\frac{r}{2} + \left[\left(\frac{r}{2}\right)^2 + \frac{r}{G(\theta)} \right]^{1/2}.$$

$Z_0(\theta)$ can be computed from this. Similar expressions serve for the $Y_k(\theta)$.

If this use of a characteristic resistance is not feasible, an approximate expression for $Z_0(\theta)$ can be obtained by truncating (7.26) either by opening an r_m to get an odd truncation or by shorting a $G_m(\theta)$ to get an even truncation. Let $Z_0^{(m)}(\theta)$ denote an odd (or even) truncation of (7.26) and let $Z_0^{(m)'}(\theta)$ be the very next even (respectively, odd) truncation. Then, since all the terms of (7.26) are positive and have an infinite sum according to Conditions 7.5-1, we can invoke Lemma 7.5-2 to write

$$|Z_0(\theta) - Z_0^{(m)}(\theta)| \leq |Z_0^{(m)}(\theta) - Z_0^{(m)'}(\theta)|, \qquad (7.34)$$

where, as $m \to \infty$, the $Z_0^{(m)}(\theta)$ and $Z_0^{(m)'}(\theta)$ converge monotonically to $Z_0(\theta)$, one increasingly and the other decreasingly. Now, let $v_{0;p}^{(m)}$ be the pth node voltage in box 0 obtained by replacing $Z_0(\theta)$ by $Z_0^{(m)}(\theta)$. By using (7.32), (7.30), (7.34), Schwarz's inequality, and (7.21) – in that order – we find a bound on the error of this approximation to be

$$|v_{0;p} - v_{0;p}^{(m)}| \leq \frac{1}{(2\pi)^n} \int_{\Theta^n} |Z_0^{(m)}(\theta) - Z_0^{(m)'}(\theta)|\,|H(\theta)|\,d\theta \qquad (7.35)$$

$$\leq \|\mathbf{h}\| \left[\frac{1}{(2\pi)^n} \int_{\Theta^n} |Z_0^{(m)}(\theta) - Z_0^{(m)'}(\theta)|^2 \, d\theta \right]^{1/2}. \qquad (7.36)$$

Here is a result that may be of some use for computations. These estimates can be made as sharp as desired by choosing m large enough. In this way, we can compute $v_{0;p}^{(m)}$ as an approximation to $v_{0;p}$ by applying an n-dimensional fast Fourier transform [13, pages 369–379], [104, Chapter 6] to (7.32) with $k = 0$ and $V_0(\theta)$ replaced by $Z_0^{(m)}(\theta)H(\theta)$.

As for the $v_{k;p}$ for other values of k, we can apply the fast Fourier transform to a combination of (7.32), (7.31), and (7.28), with $V_0(\theta)$ replaced $Z_0^{(m)}(\theta)H(\theta)$. This in turn requires estimations of the $Y_k(\theta)$ by truncations of (7.27).

Actually, there is an alternative and exceptionally fast way of computing the $v_{k;p}$ for $k \geq 1$, once the $v_{0;p}^{(m)}$ have been computed for a large enough m and some range of p values. It avoids the estimations of the $Y_k(\theta)$, but unhappily it works for just a few initial values of k. In short, perform an elementary limb analysis. More specifically, apply Ohm's law to the conductances in box 0 and then Kirchhoff's current law to get the currents in the unboxed resistors between box 0 and box 1. This will be possible at all indices p in the said range except for those at the extremities of the range. Then, Ohm's law yields the node voltages $v_{1;p}^{(m)}$ in box 1 for a somewhat smaller range of p values. This procedure can be repeated to get node values in boxes $2, 3, \ldots$. Unfortunately, these computations are unstable, and the exponentially increasing numerical errors overwhelm the true values of the $v_{k;p}$ for k too large.

7.6 Solving Ungrounded Semi-infinite Grids: Infinity Imperceptible

Now for the ungrounded case. (The following arguments are also applicable to partially grounded or grounded grids whose grounding conductances satisfy $\sum g_{gk} < \infty$.) We again consider a semi-infinite grid that fulfills Conditions 7.4-1. Thus, each boxed grid is n-dimensional and uniform, and every current source h_p feeds current from ground to a floating node of box 0. Instead of Conditions 7.5-1, we now impose

Conditions 7.6-1: *The semi-infinite grid is ungrounded (so that $g_{gk} = 0$ for all k). Moreover, only a finite number of the current sources h_p are nonzero, and $\sum h_p = 0$. In addition, the following holds.*

(i) *If $n \geq 3$, then there exists a sequence $\{\alpha_0, \alpha_1, \alpha_2, \ldots\}$ of positive numbers α_k such that $\sum(\alpha_k + r_k) = \infty$ and $g_{k;\nu} \geq \alpha_k$ for every ν.*

(ii) *If $n = 1$ or $n = 2$, then there are two positive numbers α and β such that, for all k and ν, $g_{k;\nu} \geq \alpha$ and $r_k \leq \beta$.*

Note that restriction (ii) implies the conclusion of restriction (i).

Since $g_{gk} = 0$ for all k, it follows from (7.23) and (7.26) that at $\theta = 0$ the odd truncations of (7.26) are infinite in value and the even truncations equal $r_0 + \cdots + r_m$. This singularity in Z_0 is reflected in the fact that the operation $\tilde{z}_0$ of multiplication by $Z_0(\theta)$ is an unbounded operator, and so too is $z_0 = \mathcal{F}^{-1}\tilde{z}_0\mathcal{F}$. In particular, z_0 cannot be applied to every member of l_{2r}^n. However, Conditions 7.6-1 will rescue our analysis in much the same way as this difficulty was surmounted for the ungrounded uniform grids.

Lemma 7.6-2: *Assume Conditions 7.4-1 and 7.6-1. Then, Z_0 is continuous everywhere in Θ^n except at the origin, and, for all k, Y_k is continuous everywhere in Θ^n except possibly at the origin. Moreover, Y_k is bounded everywhere in Θ^n.*

Proof: By (7.23) with $g_{gk} = 0$ and by Conditions 7.6-1,

$$G_k(\theta) \geq 2\alpha_k \sum_{\nu=1}^{n}(1 - \cos\theta_\nu),$$

where $\alpha_k = \alpha$ if $n = 1$ or 2. Given any compact subset Γ^n of Θ^n that does not contain the origin, there is an $\epsilon > 0$ such that $2\sum_{\nu=1}^{n}(1-\cos\theta_\nu) \geq \epsilon$ for all $\theta \in \Gamma^n$. Since $\sum(r_k + \alpha_k) = \infty$, we can now argue as in the proof of Lemma 7.5-3 that, for $\theta \in \Gamma^n$, (7.26) converges no slower than the continued fraction obtained by replacing each $G_k(\theta)$ in (7.26) by $\epsilon\alpha_k$. The latter continued fraction converges because $\sum(r_k + \alpha_k) = \infty$ implies that $\sum(r_k + \epsilon\alpha_k) = \infty$. It follows that (7.26) converges uniformly on Γ^n and that Z_0 is continuous everywhere on Θ^n except at the origin.

The same argument applied to (7.27) shows that Y_k is also continuous everywhere on Θ^n except possibly at the origin. Moreover, $0 \leq Y_k(\theta) < 1/r_k$. ♣

The last result shows that $Y_k(\theta)$ is Lebesgue integrable over Θ^n [145, page 40]. Hence, so too is $T_k(\theta) = 1 - r_k Y_k(\theta)$. Moreover, we again have

$$Y_k(\theta) \geq \frac{1}{r_k} + \frac{1}{G_{k+1}(\theta)}$$

so that

$$T_k(\theta) \leq \frac{1}{1 + r_k G_{k+1}(\theta)} \leq 1. \tag{7.37}$$

Consequently, multiplication by $T_k(\theta)$ is the transform of a Laurent operator $\mathbf{t}_k \in [l_{2r}^n; l_{2r}^n]$. In fact, the conclusion of Lemma 7.5-4 holds once again except that now $\|\mathbf{t}_k\| \leq 1$.

As for multiplication by $Z_0(\theta)$, we have

Lemma 7.6-3: *Under Conditions 7.4-1 and 7.6-1, the function V_0, defined by $V_0(\theta) = Z_0(\theta)H(\theta)$, is a member of L_2^n.*

Proof: By Lemma 7.6-2, V_0 is continuous everywhere except possibly at $\theta = 0$. Consider at first the case where $n \geq 3$. By (7.26), $0 \leq Z_0(\theta) \leq 1/G_0(\theta)$. We can now argue exactly as in Section 7.3 (see the second paragraph after Lemma 7.3-2, and replace X by H) that $|H(\theta)/G_0(\theta)|^2$ is integrable on Θ^n. Therefore, so too is $|V_0(\theta)|^2 = |Z_0(\theta)H(\theta)|^2$.

Now, let $n = 1$ or 2. For each θ, the right-hand side of (7.26) becomes no smaller if, for every k, $G_k(\theta)$ is replaced by $2\alpha \sum_{\nu=1}^{n}(1 - \cos\theta_\nu)$ and r_k is replaced by β. These replacements yield a uniform ladder whose characteristic resistance is

$$C(\theta) = -\frac{\beta}{2} + \left[\left(\frac{\beta}{2}\right)^2 + \frac{\beta}{2\alpha \sum_{\nu=1}^{n}(1 - \cos\theta_\nu)}\right]^{1/2}.$$

Thus, $0 \leq Z(\theta) \leq C(\theta)$ for all θ. As $\|\theta\| \to 0$, $C(\theta) \sim \sqrt{\beta/\alpha}\,\|\theta\|^{-1}$. On the other hand, as in Section 7.3 again, $H(\theta)$ is a continuous function that is asymptotic to $\sum_{\nu=1}^{n} M_\nu \theta_\nu$ as $\|\theta\| \to 0$, where the M_ν are constants. Hence, as $\|\theta\| \to 0$,

$$V_0(\theta) = Z_0(\theta)H(\theta) = O\left(\frac{\sum_{\nu=1}^{n} M_\nu \theta_\nu}{\|\theta\|}\right).$$

For each ν, $\theta_\nu/\|\theta\|$ is bounded on Θ^n. We can conclude again that $V_0 \in L_2^n$. ♣

242   Chapter 7   Grids

We may now solve the ungrounded semi-infinite grid as in the preceding section. Here again, the node-voltage vector $\mathbf{v}_k$ for every boxed grid is found as a member of l_{2r}^n.

Theorem 7.6-4: *For an ungrounded semi-infinite grid satisfying Conditions 7.4-1 and 7.6-1, the formulas (7.30), (7.31), and (7.32) yield node voltages and thereby a voltage-current regime that is the same as the regime dictated by Theorem 3.5-6.*

Proof: The proof of this theorem is much the same as that of Theorem 7.5-5. The only change is that we now transfer the pure current sources into a finite number of floating branches in box 0 in order to obtain conductances across all current sources, as was assumed in Theorem 3.5-6. The transfer can be made by coalescing ground with any node n_0 of box 0 and then proceeding as in Section 3.6. In doing so, the node voltages can be maintained exactly as they were before the transfer; indeed, since only pure current sources were incident to ground before the said coalescence and since $\sum h_p = 0$, we are free to assign to ground any voltage – in our case, n_0's voltage – in order to combine n_0 and ground without altering the other node voltages. The proof now proceeds exactly as before. For example, we again expand $(\mathbf{v}_0, \mathbf{h})$ as before and identify the terms of the expansion as dissipated powers. Moreover, we only need the positivity of $\mathbf{z}_0$ (not its boundedness) in order to identify $(\mathbf{v}_0, \mathbf{h}) = (\mathbf{z}_0\mathbf{h}, \mathbf{h})$ as being finite and nonnegative for $\mathbf{h}$ restricted as stated in Conditions 7.6-1. In this way, we see that the conditions of Theorem 3.5-6 are fulfilled. ♣

As for computational considerations, node voltages can be calculated as in the grounded case using (7.32) and truncations of the continued fractions for Z_0 and the Y_k. Now, however, we may have to be content with (7.35) as an error estimate since the integral in (7.36) may diverge.

This theory for one-dimensionally nonuniform, semi-infinite grids can be used to close a lacuna that was left over in our discussion of uniform grids. We did not show there that the voltage-current regime of a two-dimensional ungrounded grid arising from the node voltages specified in Theorem 7.3-4 agrees with that of Theorem 3.5-6. This we now do. We need only consider the case where there is just one branch current source, for superposition can then be invoked when there are finitely many branch current sources.

Consider a uniform two-dimensional ungrounded grid $\mathbf{N}$ and its grounded version $\mathbf{N}_g$. (Our arguments will work just as well when $\mathbf{N}$

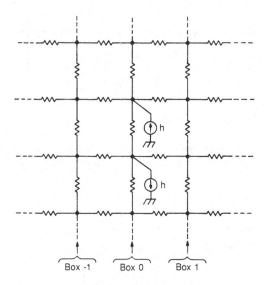

Figure 7.5. The decomposition of a two-dimensional ungrounded grid into a two-way infinite connection of "boxed" one-dimensional subgrids and "unboxed" resistors. The single branch current source, which is confined to box 0, has been expanded into two grounded pure current sources.

is n-dimensional with $n > 2$.) The single branch current source h can be replaced with two pure current sources incident to ground as shown in Figure 7.5. We can also decompose $\mathbf{N}$ into a two-way infinite connection of "boxed" one-dimensional subgrids and "unboxed" resistors; this too is indicated in Figure 7.5. Box 0 contains the branch with the original current source. We take it that the vertical conductances in Figure 7.5 all have the value g and the horizontal resistances all have the value r.

Using $\mathcal{F}$, we transform this network into the ladder of Figure 7.6. We replace $G(\theta)$ for box 0 by two parallel-connected conductances $G(\theta)/2$ and $H(\theta)$ by two parallel-connected current sources $H(\theta)/2$. In effect, we will be analyzing the uniform two-dimensional grid as two identical semi-infinite grids with one-dimensional boxed grids. We need examine only the one on the right. It is driven by the transformed current source

$$\frac{H(\theta)}{2} = \frac{h}{2}(1 - e^{\iota\theta})$$

when the nodes in box 0 having the incident current sources are indexed

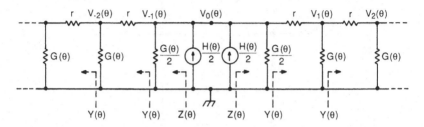

Figure 7.6. The transformed ladder obtained by applying $\mathcal{F}$ to the grid of Figure 7.5. Box 0 has been split in two, and so too have the current sources.

by 0 and 1. Now, $\theta \in R^1$ and $\pi \leq \theta \leq \pi$. Also, the driving-point resistance at the input to the semi-infinite grid is

$$Z(\theta) = \left[\frac{G(\theta)}{2} + Y(\theta) \right]^{-1}$$

where $G(\theta) = 2g(1 - \cos\theta)$ and

$$Y(\theta) = \frac{1}{r} + \frac{1}{G(\theta)} + \frac{1}{r} + \frac{1}{G(\theta)} + \cdots .$$

Hence, $V_0(\theta) = Z(\theta)H(\theta)/2$ and, for $k = 1, 2, \ldots$, $V_k(\theta) = T(\theta)V_{k-1}(\theta)$, where $T(\theta) = 1 - rY(\theta)$. Moreover, these equations continue to hold when $\mathbf{N}$ is replaced by its grounded version $\mathbf{N}_g$, in which case $G(\theta) = g_g + 2g(1 - \cos\theta)$. Since $G(\theta)$ is smaller for $\mathbf{N}$ than it is for $\mathbf{N}_g$, it follows that $Z(\theta)$ for $\mathbf{N}$ bounds the $Z(\theta)$ for $\mathbf{N}_g$, and $T(\theta)$ for $\mathbf{N}$ bounds the $T(\theta)$ for $\mathbf{N}_g$. Furthermore, by our prior order estimate in the ungrounded case, $Z(\theta)H(\theta) = O(1)$ as $\theta \to 0$. This fact coupled with the continuity of $Z(\theta)H(\theta)$ away from the origin shows that in the ungrounded case $V_0(\theta) = Z(\theta)H(\theta)/2$ is Lebesgue integrable over $[-\pi, \pi]$. So too is every $T_k(\theta)$. Therefore, we may apply Lebesgue's theorem dominated convergence to (7.32) to conclude that, as the common value of the grounding conductances in $\mathbf{N}_g$ tends to zero, each node voltage in $\mathbf{N}_g$ converges to the corresponding node voltage in $\mathbf{N}$. This implies that the node voltages for $\mathbf{N}$ obtained from this analysis with two semi-infinite grids are the same as those indicated in Theorem 7.3-4 because the node voltages obtained for $\mathbf{N}_g$ by this analysis coincide with the node voltages for $\mathbf{N}_g$ obtained in Section 7.3; indeed, according to Theorems 7.3-1 and 7.5-5, the latter two sets of node voltages both yield the voltage-current regime of Theorem 3.5-6 for $\mathbf{N}_g$. We can now invoke Theorem 7.6-4 to conclude with

Theorem 7.6-5: *Given a uniform two-dimensional ungrounded grid, the node voltages given by this analysis with two semi-infinite grids coincide with those indicated in Theorem 7.3-4. Moreover, the corresponding voltage-current regime is the one dictated by Theorem 3.5-6.*

Example 7.6-6. Figure 7.6 can be used to compute the driving-point resistance across one branch of a two-dimensional grid of 1 Ω resistors. Apply a 1 A current source across any branch. The resulting voltage across that branch will equal the desired driving-point resistance. As before, $H(\theta) = 1 - e^{\iota\theta}$ and $G(\theta) = 2 - 2\cos\theta$. From the usual technique of determining the characteristic resistance of a uniform ladder, we have

$$Z(\theta) = \left[\frac{G(\theta)}{2} + Y(\theta)\right]^{-1} = [(1 + \cos\theta)^2 + 2 - 2\cos\theta]^{-1/2}.$$

Hence, the sought driving-point resistance is

$$\mathcal{F}^{-1}Z(\theta)\frac{H(\theta)}{2} = \frac{1}{2\pi}\int_{-\pi}^{\pi}\left(\frac{1 - \cos\theta}{3 - \cos\theta}\right)^{1/2} d\theta.$$

A numerical integration for this yields .500, in agreement with the computed value of (7.12) for $n = 2$. (See [171] for other driving-point-resistance expressions for uniform grids obtained from our analysis of semi-infinite grids.) ♣

7.7 Semi-infinite Grids: Infinity Perceptible

The rest of this chapter is devoted to a semi-infinite grid wherein infinity is perceptible in the direction of the grid's one-dimensional nonuniformity. In the preceding two sections, that infinity was rendered imperceptible either by Conditions 7.5-1 or by Conditions 7.6-1, which insured the divergence of

$$\sum_{k=0}^{\infty}\left(r_k + g_{gk} + \sum_{\nu=1}^{n}g_{k;\nu}\right). \tag{7.38}$$

Henceforth, we stipulate that this series converges.

After applying $\mathcal{F}$ to transform the semi-infinite grid of Figure 7.2 into a ladder like that of Figure 7.4, we will employ the techniques of Sections 6.8 through 6.11 to examine trajectories in the voltage-current plane for that ladder. Those trajectories will now converge, but there is a complication in the present case: The voltages $V_k(\theta)$ and $I_k(\theta)$ are

complex-valued. In order to use monotonicity arguments as before, we
have to examine the real and imaginary parts of those trajectories separately. All this requires more notation.

   $[R^2; R^2]$ is the space of linear (and therefore continuous) mappings of
R^2 into R^2. Such a mapping is represented by a 2×2 matrix of real
numbers; we use the same symbol for both the mapping and its matrix
representation. Similarly, $[C^2; C^2]$ denotes the space of linear mappings
of C^2 into C^2. Any $F \in [R^2; R^2]$ is also a *real* mapping in $[C^2; C^2]$. We
assign to F the uniform operator norm:

$$\|F\| = \sup\{\|Fx\| : x \in C^2, \|x\| = 1\}.$$

   $l_2^{n,2}$ will denote the complex Hilbert of vectors $\mathbf{x} = \{x_p : p \in Z^n\}$,
where $x_p = (x_{p,1}, x_{p,2}) \in C^2$, such that the inner product for any $\mathbf{x}, \mathbf{y} \in$
$l_2^{n,2}$ is

$$(\mathbf{x}, \mathbf{y}) = \sum_{p \in Z^n} (x_{p,1} \bar{y}_{p,1} + x_{p,2} \bar{y}_{p,2}),$$

where the overbar denotes the complex conjugate. [This use of $(\cdot, \cdot)$ for
both a point in C^2 and for an inner product will be clarified whenever
the need arises.] The real Hilbert space $l_{2r}^{n,2}$ is defined in the same way
except that now $x_p \in R^2$. On the other hand, $L_2^{n,2}$ is the complex Hilbert
space of (equivalence classes of) functions on Θ^n with values in C^2 such
that the inner product for any $X, Y \in L_2^{n,2}$ is

$$(X, Y) = \frac{1}{(2\pi)^n} \int_{\Theta^n} (X_1(\theta) \bar{Y}_1(\theta) + X_2(\theta) \bar{Y}_2(\theta))\, d\theta,$$

where again the subscripts 1 and 2 denote components with respect to C^2.

   The Fourier-series transformation $\mathcal{F} \colon \mathbf{x} \mapsto X$ extends directly to $l_2^{n,2}$
through the definition (7.2), where now $x_p \in C^2$ is the pth component
of $\mathbf{x} \in l_2^{n,2}$ and X is a member of $L_2^{n,2}$. The inverse transformation $\mathcal{F}^{-1}$
is defined by (7.3). $\mathcal{F}$ is now a Hilbert-space isomorphism from $l_2^{n,2}$ onto
$L_2^{n,2}$, with again preservation of the norm. If $\mathbf{x}$ is real (i.e., $\mathbf{x} \in l_{2r}^{n,2}$),
then X has an even real part X^r and an odd imaginary part X^i with
respect to θ, where

$$X^r(\theta) = \sum_{p \in Z^n} x_p \cos(p \cdot \theta) \qquad (7.39)$$

and

$$X^i(\theta) = \sum_{p \in Z^n} x_p \sin(p \cdot \theta). \qquad (7.40)$$

In fact, the evenness of X^r and the oddness of X^i as functions of θ also constitute sufficient conditions for **x** to be real. Another result we shall need is

Lemma 7.7-1: *If* $\mathbf{c} = \{c_p : p \in Z^n, c_p \in C^1\}$ *and if* $\sum_{p \in Z^n} \|p\| \, |c_p| < \infty$, *where* $\|p\|^2 = p_1^2 + \cdots + p_n^2$, *then* $\mathbf{c}$ *is a member of* l_2^n, *and, for* $C = \mathcal{F}\mathbf{c}$, *the mapping* $\theta \mapsto C(\theta)$ *is a continuous function for all* θ.

Proof: There is an integer $P > 1$ such that, whenever $\|p\| > P$, we have $\|p\| \, |c_p| < 1$. Hence,

$$\sum_{\|p\|>P} |c_p|^2 = \sum_{\|p\|>P} \|p\|^{-2} \|p\|^2 |c_p|^2 \leq \sum_{\|p\|>P} \|p\| \, |c_p| < \infty.$$

Thus, $\mathbf{c} \in l_2^n$.

From (7.2) we can obtain, for any θ and $\theta + \Delta\theta$ in Θ^n, the estimate:

$$|C(\theta) - C(\theta + \Delta\theta)| \leq \sum_{p \in Z^n} |c_p| \, |p \cdot \Delta\theta| \left| \frac{1 - e^{\iota p \cdot \Delta\theta}}{p \cdot \Delta\theta} \right|. \qquad (7.41)$$

But $|1 - e^{\iota p \cdot \Delta\theta}| \, / \, |p \cdot \Delta\theta|$ is bounded for all $\Delta\theta$ by a constant M. Moreover, $|p \cdot \Delta\theta| \leq \|p\| \, \|\Delta\theta\|$. Hence, the right-hand side of (7.41) is bounded by $M \, \|\Delta\theta\| \sum \|p\| \, |c_p|$, which tends to zero as $\|\Delta\theta\| \to 0$. ♣

We now consider another matter: the kinds of sources that will be allowed for the semi-infinite grid. Rather than having only pure source branches feeding the nodes of box 0 from ground, we allow resistances in those branches, but, when this happens, those resistances are all required to appear for all nodes of box 0 and to have the same value. Moreover, we take it that those source branches are either all in the Thevenin form or all in the Norton form. In the former case, we choose polarities in such a fashion that the voltage $v_{0,p}$ at the pth node of box 0 and current $i_{0,p}$ entering that node from the source branch are related by

$$v_{0,p} = e_{a,p} - r_a i_{0,p}, \qquad (7.42)$$

where $e_{a,p}$ is the voltage source in that branch and r_a is its positive resistance not depending upon p. When that source branch is in the Norton form, we have

$$i_{0,p} = h_{a,p} - v_{0,p}/r_a, \qquad (7.43)$$

where $h_{a,p} = e_{a,p}/r_a$. The resistors r_a will enter our analysis in the same way as do the unboxed resistors. The values $r_a = 0$ for (7.42) and $r_a = \infty$ for (7.43) are allowed.

In contrast to a "pure" source, we say that a source is *mixed* if there is a resistance in its branch. Moreover, to signify that the resistances in a set of source branches all have the same value, we say that the sources are *uniformly mixed* (it is not required that the sources have the same value).

This terminology relates to that for various boundary value problems. A pure voltage source corresponds to a Dirichlet condition, a pure current source to a Neumann condition, and a source with a resistance (in either the Thevenin or Norton form) to a mixed boundary condition. In the last case, we are requiring that the boundary conditions be uniformly mixed with respect to p. We will refer to these sources as the *boundary conditions at $k = 0$*.

Because infinity is perceptible in the direction of the grid's nonuniformity, we have to specify what is connected at the grid's infinite extremities in that direction if a unique voltage-current regime is to ensue. Specifically, each one-ended path passing only through nodes of the same index p is a representative of a 0-tip for the semi-infinite grid of Figure 7.2. We set up a 1-node as the set consisting of that 0-tip and one elementary tip of a source branch, whose other elementary tip is contained in the ground. (Ground remains a 0-node.) That 1-node is also assigned the index p. We will see later on that its node voltage $v_{\omega,p}$ is the limit of the node voltages along any representative of its 0-tip. The current leaving that 1-node through the source branch is $i_{\omega,p}$. We now choose the source polarity (see the right-hand side of Figure 7.7) to get

$$v_{\omega,p} = e_{b,p} + r_b i_{\omega,p} \qquad (7.44)$$

for the Thevenin form and

$$i_{\omega,p} = -h_{b,p} + v_{\omega,p}/r_b \qquad (7.45)$$

for the Norton form. Here, r_b is the source-branch's resistance, and $e_{b,p}$ and $h_{b,p} = e_{b,p}/r_b$ are the voltage-source and current-source values. Again we take it that such 1-nodes and incident source branches have been designated for all $p \in Z^n$; the source branches have generally differing source values, but they all have the same resistance r_b, with $r_b = 0$ for (7.44) and $r_b = \infty$ for (7.45) being allowed. Moreover, they are all in the Thevenin form or all in the Norton form. We have hereby set up additional boundary conditions, which are now imposed at infinity in the direction of the grid's nonuniformity.

Let us now gather together all the assumptions that will be imposed on the semi-infinite grid. We have already explained most of them, but some will be discussed later on. These conditions are assumed tacitly in all our forthcoming lemmas regarding semi-infinite grids with infinity perceptible.

Conditions 7.7-2:   *The infinite network is a semi-infinite, $(n + 1)$-dimensional, one-dimensionally nonuniform, rectangular grid – of the form shown in Figure 7.2. (Hence, the floating resistors r_k and floating conductances $g_{k;\nu}$ are all positive, and the grounding conductances g_{gk} are all nonnegative.) However, it is now a 1-network with a 1-node for each index p; that 1-node shorts a 0-tip to an elementary tip of a grounding source branch as designated above. All other 1-nodes are singletons. Also, there are grounding source branches incident to the nodes of box 0. (When the source branches are pure, any one of them may be an open circuit or a short circuit.) There are no other source branches. Moreover, the following restrictions hold.*

(1)   *The series (7.38) converges.*

(2)   *The sources incident to the nodes of box 0 satisfy exactly one of the following four restrictions:*

    a.   *They are all in the Thevenin form such that (7.42) holds, and all have the same resistance $r_a > 0$. Also,*

$$\sum_{p \in Z^n} \|p\| \, |e_{a,p}| < \infty.$$

    b.   *They are all in the Norton form such that (7.43) holds, and all have the same conductance $r_a^{-1} > 0$. Also,*

$$\sum_{p \in Z^n} \|p\| \, |h_{a,p}| < \infty.$$

    c.   *They are all pure voltage sources so that (7.42) holds with $r_a = 0$. Also,*

$$\sum_{p \in Z^n} \|p\| \, |e_{a,p}| < \infty.$$

    d.   *They are all pure current sources so that (7.43) holds with $r_a = \infty$. Also,*

$$\sum_{p \in Z^n} \|p\| \, |h_{a,p}| < \infty.$$

(3)   *The sources incident to the p-indexed 1-nodes satisfy restriction 2 with index a replaced by index b, (7.42) replaced by (7.44), and (7.43) replaced by (7.45).*

(4)   *If only pure current sources exist at the nodes of box 0 and also at the p-indexed 1-nodes, then $g_{gk} > 0$ for at least one k.*

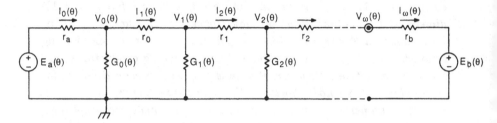

Figure 7.7. The transformed ladder when infinity is perceptible. This is an ω-cascade. Now, in addition to a source branch at the input, a source branch at infinity is needed to determine a unique voltage-current regime.

7.8 Forward and Backward Mappings

An application of $\mathcal{F}$ to the semi-infinite grid of Figure 7.2 yields the ladder of Figure 7.7, which by virtue of the 1-nodes specified above, is in fact an ω-cascade. Its kth stage consists of a shunting conductance $G_k(\theta)$ followed by a series resistance r_k and defines for each θ a forward mapping $F_k(\theta)$ from the input voltage-current pair $X_k(\theta) = (V_k(\theta), I_k(\theta))$ to the output pair $X_{k+1}(\theta) = (V_{k+1}(\theta), I_{k+1}(\theta))$. Upon applying $\mathcal{F}$ to (7.18), (7.19), and (7.20), we obtain again

$$I_k(\theta) - I_{k+1}(\theta) = G_k(\theta)V_k(\theta) \tag{7.46}$$

and

$$V_k(\theta) - V_{k+1}(\theta) = r_k I_{k+1}(\theta), \tag{7.47}$$

where $k = 0, 1, 2, \ldots$ and $G_k(\theta)$ is given by (7.23). Hence, for fixed θ, $F_k(\theta)$ is a real member of $[C^2; C^2]$ having the matrix

$$F_k(\theta) = \begin{bmatrix} 1 + r_k G_k(\theta) & -r_k \\ -G_k(\theta) & 1 \end{bmatrix}. \tag{7.48}$$

Its determinant equals 1, and so $F_k(\theta)$ is nonsingular. The backward mapping $B_k(\theta)$ of the kth stage is the inverse of $F_k(\theta)$:

$$B_k(\theta) = \begin{bmatrix} 1 & r_k \\ G_k(\theta) & 1 + r_k G_k(\theta) \end{bmatrix}. \tag{7.49}$$

Suppressing θ, we can view F_k and B_k as mappings from C^2-valued functions on Θ^n to other such functions. Since each G_k is a continuous function of θ, F_k is a member of $[L_2^{n,2}; L_2^{n,2}]$, and so too is B_k.

Now, let θ be fixed again. Given any $X_0(\theta) = (V_0(\theta), I_0(\theta)) \in C^2$, the forward mappings define a trajectory $\{X_0(\theta), X_1(\theta), X_2(\theta), \ldots\}$, where $X_{k+1}(\theta) = F_k(\theta)X_k(\theta)$. To show that the trajectory converges, we will use the monotonicity arguments of Section 6.8 (see especially the proof of Theorem 6.8-4) on its real part $\{X_0^r(\theta), X_1^r(\theta), X_2^r(\theta), \ldots\}$ and on its imaginary part $\{X_0^i(\theta), X_1^i(\theta), X_2^i(\theta), \ldots\}$ separately. Here, $X_k^r(\theta) = (V_k^r(\theta), I_k^r(\theta))$ and $X_k^i(\theta) = (V_k^i(\theta), I_k^i(\theta))$ are defined in accordance with (7.39) and (7.40). Since $F_k(\theta)$ is real, the real and imaginary parts of $F_k(\theta)X_k(\theta)$ are respectively $F_k(\theta)X_k^r(\theta)$ and $F_k(\theta)X_k^i(\theta)$. Hence, we need merely show that the real sequence

$$\{X_0(\theta), F_0(\theta)X_0(\theta), F_1(\theta)F_0(\theta)X_0(\theta), \ldots\}$$

converges whenever $X_0(\theta) \in R^2$.

As before, Q_j is the open jth quadrant in R^2, and $\bar{Q}_j$ is the closure of Q_j. Also, $V = \{(v, 0) \in R^2 : v > 0\}$ and $I = \{(0, i) \in R^2 : i > 0\}$. The next lemma follows readily from (7.46) and (7.47). Its first three conclusions are almost the same as Conditions 6.8-2.

Lemma 7.8-1: *Let the mapping $F_k(\theta)$ be applied only to points in R^2. If either $g_{gk} \neq 0$ or $\theta \neq 0$ or both, then the following three conditions hold.*

(1) $F_k(\theta)$ *has no fixed points other than the origin.*
(2) $F_k(\theta)$ *has the following directions throughout R^2.*
   a. *If $X_{k+1}(\theta) \in Q_1$, then $X_k(\theta) - X_{k+1}(\theta) \in Q_1$; if $X_{k+1}(\theta) \in V$, then $X_k(\theta) - X_{k+1}(\theta) \in Q_1 \cup I$; if $X_{k+1}(\theta) \in I$, then $X_k(\theta) - X_{k+1}(\theta) \in Q_1 \cup V$.*
   b. *If $X_k(\theta) \in Q_2$, then $X_{k+1}(\theta) - X_k(\theta) \in Q_2$; if $X_k(\theta) \in (-V)$, then $X_{k+1}(\theta) - X_k(\theta) \in Q_2 \cup I$; if $X_k(\theta) \in I$, then $X_{k+1}(\theta) - X_k(\theta) \in Q_2 \cup (-V)$.*
   c. *If $X_{k+1}(\theta) \in Q_3$, then $X_k(\theta) - X_{k+1}(\theta) \in Q_3$; if $X_{k+1}(\theta) \in (-V)$, then $X_k(\theta) - X_{k+1}(\theta) \in Q_3 \cup (-I)$; if $X_{k+1}(\theta) \in (-I)$, then $X_k(\theta) - X_{k+1}(\theta) \in Q_3 \cup (-V)$.*
   d. *If $X_k(\theta) \in Q_4$, then $X_{k+1}(\theta) - X_k(\theta) \in Q_4$; if $X_k(\theta) \in V$, then $X_{k+1}(\theta) - X_k(\theta) \in Q_4 \cup (-I)$; if $X_k(\theta) \in (-I)$, then $X_{k+1}(\theta) - X_k(\theta) \in Q_4 \cup V$.*
(3) *If $X_k(\theta) \neq Y_k(\theta)$ and if $Y_k(\theta) - X_k(\theta) \in \bar{Q}_4$, then $Y_{k+1}(\theta) - X_{k+1}(\theta) \in Q_4$ and $\|Y_k(\theta) - X_k(\theta)\| < \|Y_{k+1}(\theta) - X_{k+1}(\theta)\|$.*

When $\theta = 0$ and $g_{gk} = 0$, these three conditions have to be modified as follows.

(1′) *The fixed points of $F_k(\theta)$ are all the points on the real axis $V \cup \{0\} \cup (-V)$.*
(2′) *If $X_k(\theta) \in Q_1 \cup I \cup Q_2$, then $X_k(\theta) - X_{k+1}(\theta) \in V$. If $X_k(\theta) \in Q_3 \cup (-I) \cup Q_4$, then $X_{k+1}(\theta) - X_k(\theta) \in V$.*

(3') *If* $Y_k(\theta) - X_k(\theta) \in Q_4 \cup (-I)$, *then* $Y_{k+1}(\theta) - X_{k+1} \in Q_4$ *and*
$\|Y_k(\theta) - X_k(\theta)\| < \|Y_{k+1}(\theta) - X_{k+1}(\theta)\|$. *If* $Y_k(\theta) - X_k(\theta) \in V$,
then $Y_{k+1}(\theta) - X_{k+1}(\theta) = Y_k(\theta) - X_k(\theta)$.

[This lemma can be strengthened in several ways; e.g., in (2)a, $Q_1 \cup I$ can be replaced by I, and $Q_1 \cup V$ by Q_1. But this weaker version is needed for Lemma 7.10-2.]

The next lemma holds for complex as well as real voltage-current pairs. Its proof is much like that of Lemma 6.8-8 and is therefore omitted. It need only be noted that $0 \le G_k(\theta) \le g_{gk} + 4\sum_{\nu=1}^{n} g_{k;\nu}$ before invoking Conditions 7.7-2 and then following the prior proof.

Lemma 7.8-2: *Under Conditions 7.7-2 there exists a sequence* $\{N_0, N_1, N_2, \ldots\}$ *of real positive numbers* N_k, *which are independent of* θ *and satisfy* $\sum_{k=0}^{\infty} N_k < \infty$, *such that, for each* $\theta \in \Theta^n$, $\|F_k(\theta)X_k(\theta) - X_k(\theta)\| \le N_k \|X_k(\theta)\|$ *whenever* $X_k(\theta) \in C^2$.

The next step is to examine the trajectory

$$\{X_0(\theta), X_1(\theta), X_2(\theta), \ldots\} \tag{7.50}$$

resulting from an arbitrary choice of $X_0(\theta) \in R^2$. We may use the proof of Theorem 6.8-4, making appropriate changes in notation again and using restrictions 1' and 2' of Lemma 7.8-1 when $\theta = g_{gk} = 0$, to conclude that (7.50) converges. This conclusion continues to hold for any $X_0(\theta) \in C^2$ because it holds for the real and imaginary parts separately. In fact, we have

Lemma 7.8-3: *For fixed* θ *and any choice of* $X_0(\theta) \in C^2$, *the trajectory* *(7.50) converges in* C^2. *Moreover, for all* k,

$$\|X_k(\theta)\| \le \|X_0(\theta)\| \exp\left(\sum_{j=0}^{\infty} N_j\right), \tag{7.51}$$

where the N_k *are the constants of Lemma 7.8-2.*

By virtue of the last lemma, we can let $X_\omega(\theta)$ be the limit in C^2 of (7.50). Thus, for each θ, $F^\omega(\theta)$ can be defined as the mapping $X_0(\theta) \mapsto X_\omega(\theta)$, that is, as the strong operator limit of the composite mappings $F_{0,k}(\theta) = F_{k-1}(\theta)F_{k-2}(\theta)\cdots F_0(\theta)$. It follows that each element in the 2×2 matrix for $F_{0,k}(\theta)$ converges to the corresponding element in the 2×2 matrix for $F^\omega(\theta)$. Moreover, if X_0 is taken as any member of $L_2^{n,2}$, then F^ω becomes defined as the mapping $X_0 \mapsto X_\omega$, that is, as a mapping of $L_2^{n,2}$ into the space of complex-valued functions on Θ^n. We may send $k \to \infty$ in (7.51) to obtain

$$\|X_\omega(\theta)\| \leq \|X_0(\theta)\| \exp\left(\sum_{j=0}^{\infty} N_j\right). \tag{7.52}$$

By the theorem of dominated convergence, this shows that $\|X_\omega(\cdot)\|^2$ is Lebesgue integrable and that F^ω maps $L_2^{n,2}$ into $L_2^{n,2}$. Since all of the F_k are real mappings, so too is F^ω, that is, F^ω maps $L_{2r}^{n,2}$ into $L_{2r}^{n,2}$. But (7.52) implies even more; namely, F^ω is a bounded linear mapping of $L_2^{n,2}$ into $L_2^{n,2}$.

Furthermore, we have seen that the determinant $\det F_k(\theta)$ of the matrix representation of $F_k(\theta)$ is equal to 1. Therefore, for the composite mapping $F_{0,k}(\theta)$, we have $\det F_{0,k}(\theta) = 1$ too. Since a determinant is a continuous function of its elements, $\det F^\omega(\theta) = 1$ as well. Hence, for each θ, $F^\omega(\theta)$ has an inverse $B^\omega(\theta)$.

We summarize these results on F^ω with

Lemma 7.8-4: *For each $\theta \in \Theta^n$, the 2×2 matrix $F^\omega(\theta)$ exists as the entrywise limit of the composite mappings $F_{0,k}(\theta)$ as $k \to \infty$. Also, $\det F^\omega(\theta) = 1$. In addition,*

$$F^\omega: X_0 \mapsto (\theta \mapsto F^\omega(\theta)X_0(\theta))$$

is a real member of $[L_2^{n,2}; L_2^{n,2}]$ satisfying

$$\|F^\omega\| \leq \exp\left(\sum_{j=0}^{\infty} N_j\right).$$

This inequality for $\|F^\omega\|$ is a consequence of (7.52).

We shall now show that all four entries of the matrix $F^\omega(\theta)$ are continuous functions of θ, and similarly for $B^\omega(\theta)$. First note that this is true for the composite mappings $F_{0,k}(\theta)$ because it is true for every $F_k(\theta)$.

Lemma 7.8-5: $\|F_{0,k+1}(\theta) - F_{0,m}(\theta)\|$ *tends to zero uniformly for all $\theta \in \Theta^n$ as k and m tend to infinity independently.*

Proof: We are free to choose $k > m$ so that

$$F_{0,k+1}(\theta) - F_{0,m}(\theta) = [F_k(\theta) \cdots F_m(\theta) - 1]F_{0,m}(\theta),$$

where 1 denotes the identity matrix. According to (7.51), we have

$$\|F_{0,m}(\theta)\| \leq \exp\left(\sum_{j=0}^{\infty} N_j\right) < \infty,$$

and therefore $\|F_{0,\,m}(\theta)\|$ is bounded uniformly for all m and θ. Thus, we need merely show that

$$\|F_k(\theta)\cdots F_m(\theta) - 1\| \tag{7.53}$$

tends to zero uniformly for all θ.

We may set $F_j(\theta) = D_j(\theta) + 1$, where

$$D_j(\theta) = \begin{bmatrix} r_j G_j(\theta) & -r_j \\ -G_j(\theta) & 0 \end{bmatrix}.$$

Furthermore, the expression within the norm symbol of (7.53) may be written as

$$
\begin{aligned}
& (D_k(\theta) + 1)\cdots(D_m(\theta) + 1) - 1 \\
&= (D_k(\theta) + 1)\cdots(D_{m+1}(\theta) + 1)D_m(\theta) \\
&\quad + (D_k(\theta) + 1)\cdots(D_{m+1}(\theta) + 1) - 1 \\
&= (D_k(\theta) + 1)\cdots(D_{m+1}(\theta) + 1)D_m(\theta) + \cdots \\
&\quad + (D_k(\theta) + 1)D_{k-1}(\theta) + (D_k(\theta) + 1) - 1 \\
&= F_k(\theta)\cdots F_{m+1}(\theta)D_m(\theta) + \cdots + F_k(\theta)D_{k-1}(\theta) + D_k(\theta).
\end{aligned}
$$

Since $\|F_j(\theta)\| \le N_j + 1$, (7.53) is bounded by

$$(N_k + 1)\cdots(N_{m+1} + 1)\|D_m(\theta)\| + \cdots + (N_k + 1)\|D_{k-1}(\theta)\| + \|D_k(\theta)\|$$

$$\le [\|D_m(\theta)\| + \cdots + \|D_k(\theta)\|] \prod_{j=m+1}^{\infty} (N_j + 1)$$

$$\le \left(\sum_{j=m}^{\infty} \|D_j(\theta)\| \right) \exp\left(\sum_{j=m+1}^{\infty} N_j \right)$$

By Lemma 7.8-2, $\|D_j(\theta)\| \le N_j$ and $\sum_{j=m}^{\infty} N_j \to 0$ as $m \to \infty$. This proves what we wished to show about (7.53) and establishes Lemma 7.8-5. ♣

Lemma 7.8-6: *The entries in the matrices $F^\omega(\theta)$ and $B^\omega(\theta)$ are continuous functions of $\theta \in \Theta^n$.*

Proof: For any $\theta_1, \theta_2 \in \Theta^n$ and any fixed $x \in C^2$,

$$\|F^\omega(\theta_1)x - F^\omega(\theta_2)x\| = \|F^\omega(\theta_1)x - F_{0,\,k}(\theta_1)x\|$$
$$+ \|F_{0,\,k}(\theta_1)x - F_{0,\,k}(\theta_2)x\| + \|F_{0,\,k}(\theta_2)x - F^\omega(\theta_2)x\|.$$

It follows from Lemma 7.8-5 that given any $\epsilon > 0$ there is a k such that the first and third differences on the right-hand side are each less than $\epsilon/3$ for all choices of θ_1 and θ_2 in Θ^n. Fix k this way. Since the four entries of the matrix $F_{0,\,k}(\theta)$ are continuous – and therefore uniformly continuous – with regard to all θ in the compact set Θ^n, there is a

$\delta > 0$ such that the second difference is also less than $\epsilon/3$ whenever $\|\theta_1 - \theta_2\| < \delta$. Hence, $F^\omega(\theta)x$ is a continuous C^2-valued function on Θ^n. By choosing x equal to $(1,0)$ and then to $(0,1)$, it can be seen that the four entries of $F^\omega(\theta)$ are continuous with respect to $\theta \in \Theta^n$.

That the four entries of the 2×2 matrix $B^\omega(\theta)$ are also continuous with respect to $\theta \in \Theta^n$ now follows from the facts that $B^\omega(\theta) = [F^\omega(\theta)]^{-1}$ and $\det F^\omega(\theta) = 1$. ♣

7.9 Solving Semi-infinite Grids: Infinity Perceptible

In order to find the desired solution, we shall first verify that the transformed boundary condition due to the sources at the input to the semi-infinite grid is mapped by F^ω into a locus in the voltage-current plane having a unique intersection with the transformed boundary condition due to the sources at the 1-nodes. (Remember that we are presently dealing with just the real parts – or just the imaginary parts – of transformed voltages and currents.) Because of the linearity of the grid and the uniformity of the boundary conditions, that locus and the transformed boundary conditions are now straight lines, in contrast to the situation discussed in Section 6.11. Thus, we need merely examine the slopes of those lines.

Let $\theta \in \Theta^n$ be arbitrary but fixed. Assume that there is a transformed resistive voltage source in Thevenin form at $k = 0$, and set

$$S_a(\theta) = \{(V_0(\theta), I_0(\theta)) \colon V_0(\theta) = E_a(\theta) - r_a I_0(\theta)\}.$$

Assume a similar source at $k = \omega$ and set

$$S_b^*(\theta) = \{(V_\omega(\theta), I_\omega(\theta)) \colon V_\omega(\theta) = E_b(\theta) + r_b I_\omega(\theta)\}.$$

These are straight lines in the (v, i)-plane. (See Figure 7.8.) Upon applying the mapping $F^\omega(\theta)$ to $S_a(\theta)$, we obtain another straight line $F^\omega(\theta)S_a(\theta)$. In order to prove that a solution for the ladder exists, we have to show that $F^\omega(\theta)S_a(\theta)$ intersects $S_b^*(\theta)$. Note that, with respect to the v-axis, $S_a(\theta)$ has the negative slope $-1/r_a$, whereas $S_b^*(\theta)$ has the positive slope $1/r_b$.

Consider any negatively sloped line $S \colon v = e - ri$, where $r > 0$. Any point x on that line can be written as $x = (e - ri, i)$. Upon applying $F_k(\theta)$ to S, we obtain the point

$$([1 + r_k G_k(\theta)]e - [r + r_k + rr_k G_k(\theta)]i \ , \ -G_k(\theta)e + [1 + rG_k(\theta)]i).$$

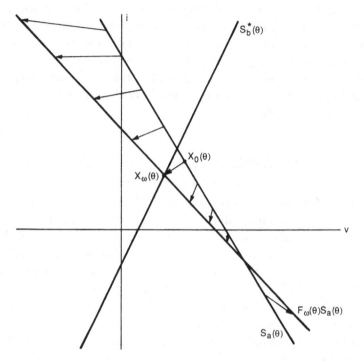

Figure 7.8. The construction for the input $X_0(\theta) = (V_0(\theta), I_0(\theta))$ and output $X_\omega(\theta) = (V_\omega(\theta), I_\omega(\theta))$ for the transformed ladder of Figure 7.7.

As i varies, the latter point traces another straight line whose slope is given by the continued fraction

$$-\frac{1}{r_k} + \frac{1}{G_k(\theta)} + \frac{1}{r}. \qquad (7.54)$$

Thus, $F_k(\theta)$ maps a line with the negative slope $-1/r$ into a line with the negative slope (7.54). By induction, we can conclude that the composite mapping $F_{0,k}(\theta) = F_{k-1}(\theta)\cdots F_0(\theta)$ maps $S_a(\theta)$ into a line with the negative slope

$$-\sigma_k = -\frac{1}{r_{k-1}} + \frac{1}{G_{k-1}(\theta)} + \cdots + \frac{1}{r_0} + \frac{1}{G_0(\theta)} + \frac{1}{r_a}. \qquad (7.55)$$

Since $G_j(\theta) \le g_{gj} + 4\sum_{\nu=1}^{n} g_{j;\nu}$ and since r_j and the $g_{j;\nu}$ are positive whereas g_{gj} is nonnegative, it follows that, for all k, σ_k is uniformly bounded below and above by positive numbers according to

$$\left(r_a + \sum_{j=0}^{\infty} r_j\right)^{-1} \leq \sigma_k \leq \frac{1}{r_a} + \sum_{j=0}^{\infty}\left(g_{gj} + 4\sum_{\nu=1}^{n} g_{j;\nu}\right). \qquad (7.56)$$

Because $F^{\omega}(\theta)$ is the strong operator limit of the $F_{0,k}(\theta)$ and because the two sides of (7.56) are independent of k, we can also conclude that $F^{\omega}(\theta)S_a(\theta)$ is a negatively sloped straight line with the slope $-\sigma_w$, where σ_w is also bounded according to (7.56). Consequently, $F^{\omega}(\theta)S_a(\theta)$ intersects $S_b^*(\theta)$. The unique intersection point $X_w(\theta)$ specifies the solution at $k = \omega$ for the ladder:

$$X_\omega(\theta) = [F^{\omega}(\theta)S_a(\theta)] \cap S_b^*(\theta). \qquad (7.57)$$

See again Figure 7.8. The equations

$$X_0(\theta) = B^{\omega}(\theta)X_\omega(\theta) \qquad (7.58)$$

and

$$X_k(\theta) = F_{0,k}(\theta)X_0(\theta), \quad k = 1, 2, 3, \ldots \qquad (7.59)$$

then specify the solution at all the other nodes of the ladder. Upon combining the separate analyses for the real and imaginary parts, we obtain the complex transforms of the boxed node voltages and the unboxed currents. However, in order to transfer this solution to the grid of Figure 7.2, we will establish later on some continuity properties as θ varies throughout Θ^n.

Note that all this continues to hold when $S_b^*(\theta)$ is a vertical line $v = E_b(\theta)$ (a pure voltage source at $k = \omega$) or a horizontal line $i = -H_b(\theta)$ (a pure current source at $k = \omega$) if $S_a(\theta)$ is a negatively sloped line.

Consider now the case where $S_a(\theta)$ is a vertical line. This means that $r_a = 0$, i.e., there is a pure voltage source at $k = 0$. $G_0(\theta)$ can now be deleted because it no longer affects the voltages and currents within the ladder. The result is the ladder with r_a replaced by r_0 and with the node indices starting at $k = 1$. Thus, a solution for the ladder exists as before. Upon restoring $G_0(\theta)$, we can say that $F^{\omega}(\theta)S_a(\theta)$ is still a negatively sloped line and that Equations (7.57), (7.58), and (7.59) continue to hold even though $S_a(\theta)$ is vertical.

Finally, consider the case where $S_a(\theta)$ is a horizontal line (a pure current source at $k = 0$). Now, $r_a = \infty$, and the term $1/r_a$ on the right-hand side of (7.56) disappears; nonetheless, we have a uniform upper bound on σ_k for all k. However, the lower bound is no longer positive. A positive lower bound will exist for all θ so long as at least one of the

g_{gk} is positive, say, $g_{gm} > 0$. In this case, for every $k = m + 1, m + 2, \ldots$

$$\left(\frac{1}{g_{gm}} + \sum_{j=m}^{\infty} r_j \right)^{-1} \leq \sigma_k \leq \sum_{j=0}^{n} \left(g_{gj} + 4 \sum_{\nu=1}^{n} g_{j;\,\nu} \right). \tag{7.60}$$

We can again conclude that $F^{\omega}(\theta)S_a(\theta)$ is negatively sloped and that the above solution holds.

An exceptional case arises when $g_{gk} = 0$ for all k and there are pure current sources at the input and output, say, $H_a(\theta)$ and $H_b(\theta)$, both measured as feeding current from ground into the nodes at $k = 0$ and $k = \omega$ respectively. If $H_a(0) \neq -H_b(0)$, then the ladder network has no solution at $\theta = 0$. Indeed, $G_k(0) = 0$ for every k, and Figure 7.7 reduces to a series connection through $r_0, r_1, r_2, \ldots$ and the two current sources. So, we must have $I_0(\theta) = I_{\omega}(\theta)$. However, $I_0(0) = H_a(0)$, whereas $I_{\omega}(0) = -H_b(0) \neq H_a(0)$. A difficulty has arisen. We shall simply avoid this situation by requiring that, whenever we have two pure current sources at $k = 0$ and $k = \omega$, at least one of the g_{gk} is positive. (This avoided situation is a singular case requiring an approach like that of Section 7.6.)

As the last step, we shall convert the transformed solution, which exists for every θ, into the solution for the semi-infinite grid. The conditions on the source values imposed in restriction 2 of Conditions 7.7-2 in conjunction with Lemma 7.7-1 ensure that the transforms of the source values [that is, $E_a(\theta)$, $H_a(\theta)$, $E_b(\theta)$, and $H_b(\theta)$] are continuous functions of θ. Once again we analyze separately the real and imaginary parts of the transformed variables.

The source line $S_a(\theta)$ shifts parallel to itself continuously as θ varies, and, by the continuity of $F^{\omega}(\theta)$ (Lemma 7.8-6), $F^{\omega}(\theta)S_a(\theta)$ also changes its position in a continuous fashion as θ varies, where now the slope of $F^{\omega}(\theta)S_a(\theta)$ will change continuously too but will remain negative. Similarly, $S_b^*(\theta)$ shifts parallel to itself continuously as θ varies. As a result, the unique intersection point $X_{\omega}(\theta) \in R^2$ between $F^{\omega}(\theta)S_a(\theta)$ and $S_b^*(\theta)$ depends continuously upon θ. By Lemma 7.8-6 again, $B^{\omega}(\theta)$ is also continuous with respect to θ, and so $X_0(\theta) = B^{\omega}(\theta)X_{\omega}(\theta)$ is an R^2-valued continuous function of θ too. Finally, the continuity of each $F_{0,\,k}(\theta)$ implies that $X_k(\theta)$ is an R^2-valued continuous function of θ as well.

The next step is to take Fourier coefficients of the first component $V_k(\theta)$ of $X_k(\theta)$ to get the node voltages in box k, but we need to digress for a moment. In general, the boundary conditions lead to complex-valued $E_a(\theta)$ or $H_a(\theta)$ and complex-valued $E_b(\theta)$ or $H_b(\theta)$. It was the reality of the mappings $F_k(\theta)$ that allowed us to examine the real and imaginary parts of these boundary functions separately in order to use the monotonicity properties of the trajectories. We now combine these two analyses.

When $\mathbf{e}_a = \{e_{a,p} : p \in Z^n\}$ is given as any member of l_{2r}^n, each $E_a(\theta)$ has an even real part and an odd imaginary part, and the same is true for the other possible boundary sources $H_a(\theta)$, $E_b(\theta)$, and $H_b(\theta)$. Moreover, $F_k(\theta)$ is not only real but also an even function of θ. It can therefore be shown that $X_\omega(\theta)$ and every $X_k(\theta)$ has an even real part and an odd imaginary part. [For instance, when dealing with the odd imaginary parts, if $X_\omega^i(\theta)$ is in the first quadrant, $X_\omega^i(-\theta)$ will be in the third quadrant.] Thus, $V_k(\theta)$ has these properties as well so that its Fourier coefficients (7.32) are all real. We have hereby found a solution for the semi-infinite grid with a perceptible infinity.

There is one further matter we should verify. For each p, is the pth 1-node's voltage the limit of the node voltages along any representative of the pth 0-tip? Yes. For instance, consider the case of resistive Thevenin's sources at $k = \omega$. By virtue of Lemma 7.8-5, $X_k(\theta)$ tends to $X_\omega(\theta)$ uniformly for all θ. This ensures the desired convergence of the Fourier coefficients, namely, for each fixed p and as $k \to \infty$, we have $v_{k,p} \to v_{\omega,p}$ and $i_{k,p} \to i_{\omega,p}$, where $v_{\omega,p} - r_b i_{\omega,p} = e_{b,p}$ as in (7.44).

We summarize everything with the following conclusion.

Theorem 7.9-1: *Assume Conditions 7.7-2. Then, there is a unique set of node voltages $v_{k,p}$ for which the following is true.*

a. *Kirchhoff's voltage law for 0-loops, Kirchhoff's current law at floating ordinary nodes, and Ohm's law all hold for the corresponding voltage-current regime.*
b. *For every k, $\mathbf{v}_k = \{v_{k,p} : p \in Z^n\} \in l_{2r}^n$.*
c. *The boundary conditions at $k = 0$ hold; that is, for every $p \in Z^n$ we have $e_{a,p} = v_{0,p} + r_a i_{0,p}$ if $r_a \geq 0$, or $i_{0,p} = h_{a,p}$ if $r_a = \infty$.*
d. *The boundary conditions at $k = \omega$ are approached as $k \to \infty$; that is, for every $p \in Z^n$ we have $v_{k,p} \to v_{\omega,p}$ and $i_{k,p} \to i_{\omega,p}$, where $e_{b,p} = v_{\omega,p} - r_b i_{\omega,p}$ if $r_b \geq 0$, or $i_{\omega,p} = -h_{b,p}$ if $r_b = \infty$.*

Those node voltages are the Fourier coefficients of the functions given by (7.57), (7.58), and (7.59).

The uniqueness of the $v_{k,p}$ can also be argued as follows. Assume there are two sets $\{v_{k,p}\}$ and $\{v'_{k,p}\}$ of floating-node voltages that satisfy the conditions of the theorem. Set $u_{k,p} = v_{k,p} - v'_{k,p}$. For the $u_{k,p}$, the sources are all equal to zero. It follows that the straight lines $S_a(\theta)$,

$F^\omega(\theta)S_a(\theta)$, and $S_b^*(\theta)$ all pass through the origin. Thus, in the transform domain the trajectory corresponding to the $u_{k,p}$ starts at the origin and converges to the origin. However, there is only one such trajectory, the identically zero trajectory. Therefore, $v_{k,p} = v'_{k,p}$.

Corollary 7.9-2: *For the voltage-current regime corresponding to the node voltages of Theorem 7.9-1, the power generated by the sources is equal to the power dissipated in the resistors.*

Proof: For the sake of definiteness, let us assume that all sources are resistive and in Thevenin's form. The other cases are argued similarly.

Using an expansion like that in the proof of Theorem 7.5-5, we can write the following, where now $(\cdot,\cdot)$ denotes the inner product in l_{2r}^n.

$$(e_a, i_0) = (r_a i_0, i_0) + (v_0, g_0 v_0) + (r_0 i_1, i_1) + \cdots$$
$$+ (v_{k-1}, g_{k-1} v_{k-1}) + (r_{k-1} i_k, i_k) + (v_k, i_k)$$

As before, $(v_j, g_j v_j) + (r_j i_{j+1}, i_{j+1})$ is the power dissipated in the jth stage, a nonnegative quantity. By virtue of Lemma 7.8-5 and the fact that $\mathcal{F}$ is an isomorphism from l_2^n onto L_2^n, we have that $v_k \to v_\omega = e_b + r_b i_\omega$ and $i_k \to i_\omega$ in l_{2r}^n as $k \to \infty$. Consequently,

$$(e_a, i_0) + (e_b, -i_\omega) =$$
$$(r_a i_0, i_0) + (r_b i_\omega, i_\omega) + \sum_{j=0}^{\infty} [(v_j, g_j v_j) + (r_j i_{j+1}, i_{j+1})].$$

The left-hand side is the generated power, and the right-hand side is the dissipated power. ♣

7.10 Transfinite Grids

In addition to the boundary conditions at box 0 of a semi-infinite grid, we have so far placed boundary conditions at infinity as well, that is, at $k = \omega$. As we shall see, the latter boundary conditions can be moved beyond infinity by extending the grid transfinitely in the direction of the one-dimensional nonuniformity and using the techniques introduced in Sections 6.10 and 6.11.

For instance, if k_0 is any transfinite ordinal less than ω^2, we can set up a transfinite semi-infinite grid whose box indices extend through $k = 0, 1, \ldots, \omega, \omega + 1, \ldots, k_0$. Instead of a box at $k = k_0$ we may have sources, in addition to the sources at $k = 0$. This structure is in fact a transfinite cascade, each stage of which is a boxed grid followed by unboxed resistors; let us call such a stage an *el-section grid*. At $k = \omega$, the 0-tip of index $p \in Z^n$ of the first $\vec{\omega}$-cascade is shorted by a 1-node

to the pth floating 0-node of box ω, and similarly for all the limit ordinals less than k_0. At $k = k_0$ boundary conditions like those indicated
in restriction 3 of Conditions 7.7-2 are imposed. Now, if this transfinite
cascade satisfies a suitably modified version of those conditions – with
restriction 1 holding for every $\bar{\omega}$-cascade within the transfinite cascade
– then the transformed trajectory

$$\{X_0(\theta), X_1(\theta), \ldots, X_\omega(\theta), X_{\omega+1}(\theta), \ldots, X_{k_0}(\theta)\} \qquad (7.61)$$

will again converge for each $\theta \in \Theta^n$, and the analysis can be carried along
to obtain a result like that of Theorem 7.9-1. This works because there
are only a finite number of limit-ordinal indices in (7.61) so that the
overall cascade can be decomposed into a finite number of $\bar{\omega}$-cascades,
possibly followed by a finite number of el-section grids. Consequently,
(7.61) can be decomposed into a finite number of convergent infinite
sequences, possibly followed by a finite sequence.

A convergent infinity of convergent infinite sequences is the next possibility toward which this gambit points. Let $k_0 = \omega^2$. Then, the transfinite semi-infinite grid is an ω^2-cascade of el-section grids. We shall argue
that the analyses of Sections 7.8 and 7.9 extend to this case. Before doing
so, however, let us assert still more: Those analyses can be extended to
ω^p-cascades of el-section grids, where p is any natural number, or ω, or a
still larger ordinal. Rather than repeating everything in this generality,
let us simply present the analysis for an ω^2-cascade. How to proceed for
an ω^p-cascade, where $p > 2$, should then be evident.

Actually, the analysis of an ω^2-cascade is similar to the analysis of an
ω-cascade; at various points we need merely cite some prior arguments.
Let us define an ω^2-*cascade of el-section grids* to be the following transfinite extension of the structure whose initial part is illustrated in Figure
7.2. The indices k number the boxes consecutively starting at $k = 0$
and continuing on through all the ordinals less than ω^2. Between $k = l$
and $k = l + \omega$, where l is any limit ordinal less than ω^2, the structure
is a semi-infinite, $(n + 1)$-dimensional, one-dimensionally nonuniform,
rectangular grid as defined in Section 7.7. Thus, for every k, $r_k > 0$,
$g_{k;\nu} > 0$, and $g_{gk} \geq 0$. For each limit ordinal l with $\omega \leq l < \omega^2$ and for
each $p \in Z^n$, a 1-node shorts the pth node in box l to the pth 0-tip of
the preceding semi-infinite grid. Moreover, for each $p \in Z^n$, we define
a pth 1-*tip* as the 1-tip having a representative 1-path whose embraced
0-nodes are the pth 0-nodes in the boxes (and therefore whose embraced
1-nodes are the pth 1-nodes for each l). For each p, we designate a pth
2-node containing two elements, one being the pth 1-tip and the other

being the elementary tip of a source branch whose other elementary tip is contained in the ground. These are the only connections that will be allowed at $k = \omega^2$. Finally, there are sources connected from ground to the 0-nodes of box 0. We use the notation (7.42) and (7.43) for the sources at $k = 0$. The sources at $k = \omega^2$ are denoted by (7.44) and (7.45) except that now ω is replaced by ω^2. There are no other sources. The degenerate case of a short circuit or an open circuit is allowed for any source. This ends the definition of an ω^2-cascade of el-section grids.

Let A be the set of all finite and transfinite ordinals from $k = 0$ up to but not including ω^2. Also, let $\{x_k : k \in A\}$ be a subset of R^m (or of C^m). As in Section 6.10, we say that the *transfinite series* $\sum_{k \in A} x_k$ *is summable* and *has the sum* $S \in R^m$ (respectively, $S \in C^m$) if, given any $\epsilon > 0$, there is a finite subset M of A such that $\|S - \sum_{k \in M} x_k\| < \epsilon$.

The following is assumed throughout the rest of this section.

Conditions 7.10-1:   *The infinite network is an ω^2-cascade of el-section grids such that the following is satisfied.*

(1)   *The transfinite series*

$$\sum_{k \in A} \left(r_k + g_{gk} + \sum_{\nu=1}^{n} g_{k;\nu} \right) \qquad (7.62)$$

   is summable.

(2)   *Restrictions 2,3, and 4 of Conditions 7.7-2 hold except that the sources beyond infinity are now incident to the p-indexed 2-nodes at $k = \omega^2$; also, ω is replaced by ω^2 in (7.44) and (7.45).*

An application of the Fourier-series transformation $\mathcal{F}$ converts this structure into an ω^2-cascade each stage of which consists of a shunt element $G_k(\theta)$ followed by a series element r_k. Once again, we may examine separately the real and imaginary parts of any transformed voltage-current trajectory because the forward mapping of every stage is real. So, let $X_0(\theta) \in R^2$. Then, $X_k(\theta) \in R^2$ too – if $X_k(\theta)$ exists.

For each natural number j, consider the ω-cascade between $k = \omega \cdot j$ and $k = \omega \cdot (j + 1)$. By virtue of restriction 1 of Conditions 7.10-1, the series (7.38) for that ω-cascade converges. All that is needed for the discussion of Section 7.8 is now in force. Therefore, all the lemmas and results in that section hold for the jth ω-cascade. For instance, it is characterized by a forward mapping $F_j^\omega(\theta) : X_{\omega \cdot j}(\theta) \mapsto X_{\omega \cdot (j+1)}(\theta)$, where $F_j^\omega(\theta) \in [R^2; R^2]$, and the four entries in the 2×2 matrix for $F_j^\omega(\theta)$ are continuous functions of θ. Moreover, $\det F_j^\omega(\theta) = 1$, and therefore $F_j^\omega(\theta)$

has an inverse, namely, the backward mapping $B_j^\omega(\theta): X_{\omega\cdot(j+1)}(\theta) \mapsto X_{\omega\cdot j}(\theta)$. The four entries of $B_j^\omega(\theta)$ are also continuous functions of θ. Furthermore, there is a constant N_j^ω, not depending on θ, such that, for all $x \in C^2$,

$$\|F_j^\omega(\theta)x - x\| \le N_j^\omega \|x\|. \tag{7.63}$$

Finally, the properties stated in Lemma 7.8-1 extend directly to $F_j^\omega(\theta)$. In particular, we have

Lemma 7.10-2: *Let $F_j^\omega(\theta)$ be applied only to points in R^2. If either $\theta \ne 0$, or $g_{gk} \ne 0$ for at least one k with $\omega \cdot j \le k < \omega \cdot (j+1)$, or both, then $F_j^\omega(\theta)$ satisfies parts 1, 2, and 3 of Lemma 7.8-1. If $\theta = 0$ and $g_{gk} = 0$ for all such k, then $F_j^\omega(\theta)$ satisfies parts 1', 2', and 3' of Lemma 7.8-1.*

In this way, the original ω^2-cascade can be viewed as an ω-cascade of ω-cascades, where the latter ω-cascades play the roles of the individual stages in the analysis of a single ω-cascade. However, to extend the prior analysis to the present ω^2-cascade, we need the next lemma. In its proof, A is the set of all ordinals from 0 up to but not including ω^2 as before, $\sum_{j=0}^\infty$ denotes the usual summation over all natural numbers j, and $\sum_{k=\omega\cdot j}^{\omega\cdot(j+1)}$ denotes a summation over all ordinals from $\omega \cdot j$ up to but not including $\omega \cdot (j+1)$. (Thus, $\sum_{j=0}^\infty = \sum_{j=0}^\omega$.)

Lemma 7.10-3: *Under Conditions 7.10-1, the constants N_j^ω of (7.63) can be so chosen that $\sum_{j=0}^\infty N_j^\omega < \infty$.*

Proof: For the kth stage in the jth $\bar\omega$-cascade, that is, for $\omega \cdot j \le k < \omega \cdot (j+1)$, we have – as in the proof of Lemma 6.8-8 – that, for any $x \in C^2$ and all $\theta \in \Theta^n$,

$$\|F_k(\theta)x - x\| \le (\gamma_k + r_k + \gamma_k r_k)\|x\|, \tag{7.64}$$

where $\gamma_k = g_{gk} + 4\sum_{\nu=1}^n g_{k;\nu}$. As for any $F_j^\omega(\theta)$, we may write

$$\|F_j^\omega(\theta)x - x\| \le \|F_{\omega\cdot j}(\theta)x - x\| + \|F_{\omega\cdot j+1}(\theta)F_{\omega\cdot j}(\theta)x - F_{\omega\cdot j}(\theta)x\|$$
$$+ \|F_{\omega\cdot j+2}(\theta)F_{\omega\cdot j+1}(\theta)F_{\omega\cdot j}(\theta) - F_{\omega\cdot j+1}(\theta)F_{\omega\cdot j}(\theta)x\| + \cdots.$$

As in (7.51), we get from (7.64) that, for any natural number m,

$$\|F_{\omega\cdot j+m}(\theta)\cdots F_{\omega\cdot j}(\theta)x\| \le \|x\|e^{Q_j},$$

where

$$Q_j = \sum_{k=\omega\cdot j}^{\omega\cdot(j+1)} (\gamma_k + r_k + \gamma_k r_k). \tag{7.65}$$

Upon combining the last three inequalities and passing to a limit, we get

$$\|F_j^\omega(\theta)x - x\| \le \|x\| Q_j e^{Q_j}. \tag{7.66}$$

So, we may choose $N_j^\omega = Q_j e^{Q_j}$. Set $M = \sum_{j=0}^\infty Q_j = \sum_{k \in A}(\gamma_k + r_k + \gamma_k r_k)$. M is a finite quantity because of restriction 1 of Conditions 7.10-1 and the nonnegativity of the r_k, g_{gk}, and $g_{k;\nu}$. Whence, $\sum_{j=0}^\infty N_j^\omega = \sum Q_j e^{Q_j} \le e^M \sum_{j=0}^\infty Q_j = M e^M < \infty.$ ♣

Lemmas 7.10-2 and 7.10-3 may be used as in the proof of Theorem 6.8-4 to get

Lemma 7.10-4: *Under Conditions 7.10-1, for every $X_0(\theta) \in R^2$, the trajectory (7.61), where $k_0 = \omega^2$, converges in R^2 in the sense of the summability of its increments.*

This result yields a mapping $F^{\omega^2}(\theta): X_0(\theta) \mapsto X_{\omega^2}(\theta)$, where $X_{\omega^2}(\theta) \in C^2$ is the unique limit of the trajectory that starts at $X_0(\theta) \in C^2$. If the N_j^ω are chosen according to Lemma 7.10-3, the proof of Lemma 7.10-3 can be repeated to replace (7.66) by

$$\|F^{\omega^2}(\theta)x - x\| \le N^{\omega^2}\|x\|, \tag{7.67}$$

where $N^{\omega^2} = Q'e^{Q'}$ and $Q' = \sum_{j=0}^\infty N_j^\omega$. All the arguments of Section 7.8 can also be repeated with the ω-cascades in the present ω^2-cascade taking the roles of the el-section grids in the ω-cascade of Section 7.8. The following is obtained.

Lemma 7.10-5: *Assume Conditions 7.10-1 and choose $N_j^\omega = Q_j e^{Q_j}$, where Q_j is given by (7.65). Then, the mapping $F^{\omega^2}(\theta) \in [R^2; R^2]$ satisfies (7.67). Moreover, the four entries of the 2×2 matrix for $F^{\omega^2}(\theta)$ are continuous functions of θ, and $\det F^{\omega^2}(\theta) = 1$. Finally, $F^{\omega^2}(\theta)$ satisfies the conclusions of Lemma 7.8-1 when it is applied to points in R^2. Specifically, if either $\theta \ne 0$, or $g_{gk} \ne 0$ for at least one $k \in A$, or both, then $F^{\omega^2}(\theta)$ satisfies parts 1, 2, and 3 of that lemma. If $\theta = 0$ and $g_{gk} = 0$ for all $k \in A$, then $F^{\omega^2}(\theta)$ satisfies parts 1', 2', and 3' of that lemma.*

In view of Lemma 7.10-5, $F^{\omega^2}(\theta)$ has an inverse $B^{\omega^2}(\theta)$ with $\det B^{\omega^2} = 1$ and with the four entries in the matrix for $B^{\omega^2}(\theta)$ being continuous functions of θ. $B^{\omega^2}(\theta)$ is the backward mapping of the transformed ω^2-cascade.

The next step is to impose the boundary conditions at $k = 0$ and at $k = \omega^2$. For each fixed θ, the transformed boundary conditions are again straight lines in the voltage-current plane, and so too are their images under the forward and backward mappings. Thus, we need merely examine the slopes of those straight lines. The analysis of Section 7.9 extends to the ω^2-cascade. The only change is that we now consider each ω-cascade consecutively in order to extend the bounds (7.56) to the slopes of $S_a(\theta)$, $F_0^\omega(\theta)S_a(\theta)$, $F_1^\omega(\theta)F_0^\omega(\theta)S_a(\theta)\cdots$ to obtain finally the following bound on the slope $-\sigma_{\omega^2}$ of $F^{\omega^2}(\theta)S_a(\theta)$.

$$\left(r_a + \sum_{k \in A} r_k\right)^{-1} \le \sigma_{\omega^2} \le \frac{1}{r_a} + \sum_{k \in A}\left(g_{gk} + 4\sum_{\nu=1}^{n} g_{k;\nu}\right) \qquad (7.68)$$

The cases where $r_a = 0$ and $r_a = \infty$ are handled as before. When $r_a = r_b = \infty$, the singular case arises if $g_{gk} = 0$ for every $k \in A$. This can again be avoided by requiring that $g_{gk} > 0$ for at least one k. Thus, the solution in the transform domain is given by the following:

$$X_{\omega^2}(\theta) = \left(F^{\omega^2}(\theta)S_a(\theta)\right) \cap S_b^*(\theta) \qquad (7.69)$$

$$X_0(\theta) = B^{\omega^2}(\theta)X_{\omega^2}(\theta) \qquad (7.70)$$

$$X_k(\theta) = F_{0,k}(\theta)X_0(\theta), \quad k \in A \qquad (7.71)$$

The symbol $F_{0,k}(\theta)$ in (7.71) denotes the composite mapping

$$F_0^\omega(\theta)\cdots F_{j-1}^\omega(\theta)F_{\omega\cdot j}(\theta)F_{\omega\cdot j+1}(\theta)\cdots F_{k-1}(\theta),$$

where j is a natural number and $\omega \cdot j$ is the largest limit ordinal no larger than k. ($F_{0,k}(\theta) = F_0(\theta)\cdots F_{k-1}(\theta)$ when $k < \omega$, and $F_{0,k}(\theta) = F_0^\omega(\theta)\cdots F_{j-1}^\omega(\theta)$ when $k = \omega \cdot j$.)

By Lemma 7.7-1 and the restraints on the source values given in restriction 2 of Conditions 7.7-2, the dependence of $S_a(\theta)$ and $S_b^*(\theta)$ on θ is again continuous. The same continuity also holds for all the forward and backward mappings. Hence, $X_k(\theta)$ is continuous with respect to θ for every $k = 0, \ldots, \omega^2$. This allows us to take Fourier coefficients $v_{k,p}$ of the transformed solution $X_k(\theta)$ to conclude with the next theorem. Here, $v_{k,p}$ is the node voltage at the pth node $n_{k,p}$ in the kth box for $0 \le k < \omega^2$ and is the output voltage at the pth 2-node for $k = \omega^2$. Also, $i_{k,p}$ is the current entering $n_{k,p}$ from the preceding part of the cascade; when k is a limit ordinal, that current is taken to be the sum of all the currents leaving $n_{k,p}$ through all the branches incident to $n_{k,p}$.

Theorem 7.10-6: *Assume Conditions 7.10-1. Then, there is a unique set of node voltages $v_{k,p}$, where $k = 0, \ldots, \omega^2$ and $p \in Z^n$, for which the following is true.*

   a. *Kirchhoff's voltage law for 0-loops, Kirchhoff's current law at floating ordinary nodes, and Ohm's law all hold for the corresponding voltage-current regime.*

b. *For every $k = 0, \ldots, \omega^2$, we have that $\mathbf{v}_k = \{v_{k,p} \colon p \in Z^n\} \in l_{2r}^n$.*

c. *The boundary conditions at $k = 0$ hold; that is, for every $p \in Z^n$, we have $e_{a,p} = v_{0,p} + r_a i_{0,p}$ if $r_a \geq 0$, or $i_{0,p} = h_{a,p}$ if $r_a = \infty$.*

d. *For any limit ordinal $\omega \cdot j$ less than ω^2 and for $k = \omega \cdot (j-1) + m$, we have $v_{k,p} \to v_{\omega \cdot j, p}$ as m increases through all the natural numbers.*

e. *The boundary conditions at $k = \omega^2$ are approached in the following manner. For every $p \in Z^n$ and as j increases through all the natural numbers, $v_{\omega \cdot j, p} \to v_{\omega^2, p}$ and $i_{\omega \cdot j, p} \to i_{\omega^2, p}$, where $e_{b,p} = v_{\omega^2, p} - r_b i_{\omega^2, p}$ if $r_b \geq 0$, or $i_{\omega^2, p} = -h_{b,p}$ if $r_b = \infty$.*

Those node voltages are the Fourier coefficients of the functions given by (7.69), (7.70), and (7.71).

Corollary 7.9-2 extends readily to this transfinite cascade.

7.11 Grids with Two-dimensionally Transfinite Nonuniformities

?

Chapter 8

Applications

In this last chapter we shall survey a number of instances where infinite electrical networks are useful models of physical phenomena, or serve as analogs in some other mathematical disciplines, or are realizations of certain abstract entities. We shall simply describe those applications without presenting a detailed exposition. To do the latter would carry us too far afield into quite a variety of subjects. However, we do provide references to the literature wherein the described applications can be examined more closely.

Several examples are presented in Sections 8.1 and 8.2 that demonstrate how the theory of infinite electrical networks is helpful for finding numerical solutions of some partial differential equations when the phenomenon being studied extends over an infinite region. The basic analytical tool is an operator version of Norton's representation, which is appropriate for an infinite grid that is being observed along a boundary. In effect, the infinite grid is replaced by a set of terminating resistors and possibly equivalent sources connected to the boundary nodes. In this way, the infinite domain of the original problem can be reduced to a finite one – at least so far as one of the spatial dimensions is concerned. This can save computer time and memory-storage requirements.

In Section 8.3 we describe two classical problems in the theory of random walks on infinite graphs and state how infinite-electrical-network theory solves those problems. Indeed, resistive networks are analogs for random walks. Furthermore, our extension of infinite graphs into transfinite graphs, given in Section 3.2 and Chapter 5, leads to a generalization of random walks. This matter is briefly explored for random walks on transfinite ladders. Also, some literature is cited regarding the connection between infinite electrical networks and the classification of Riemann surfaces.

In the final section of this book, Section 8.4, we call attention to an abstraction, the operator network, that has been attracting some interest in recent years. Finite or infinite networks, whose parameters are linear, continuous, positive operators on a Hilbert space, are easy enough to handle with regard to existence and uniqueness theories when those operators are all invertible, but networks of noninvertible operators present substantial theoretical problems. In any case, this book may at times be pertinent in this regard because some operators are realizable by infinite networks of scalar elements. Thus, an operator network might be viewed as an interconnection of infinite subnetworks. Section 7.4 provided one example of this. We briefly discuss this matter and cite a number of references as an entree into that subject.

8.1 Surface Operators

Perhaps the most natural way by which infinite electrical networks arise as models of physical phenomena is through finite-difference discretizations of various partial differential equations. This was pointed out in Section 1.7, wherein two examples of such discretizations were given. Whenever the domain of analysis is infinite in extent, the so-called *exterior problem* is at hand, and an infinity of sample points for the dependent variables might well be chosen. As a result, an infinite electrical network emerges as an approximating discrete analog for the original continuous phenomenon.

To be sure, such infinite networks might be avoided in several ways. First of all, only a finite number of sample points may be chosen simply to force a finite analog to an inherently infinite system. Similarly, the domain itself might be rendered finite by truncating it with a surface; this can work if sufficiently accurate boundary conditions on the truncating surface can be ascertained. As an alternative to a finite-difference approach, one might employ a finite-element technique whereby finite elements are used over a finite domain and a finite number of infinite elements are used to approximate an exterior infinite region. Still another approach is to use a Green's function to obtain a solution over the infinite domain for a delta-function source located at an arbitrary point and then to employ an integral representation for the solution induced by a distribution of sources.

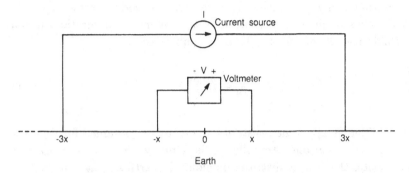

Figure 8.1. The Wenner configuration of voltage and current probes. The four probes are aligned and equally spaced with the voltage probes inside the current probes.

A possible alternative to these standard techniques emerges from our present ability to solve infinite electrical networks. Whether solutions based upon infinite-electrical-network models will ever comprise another standard method remains to be seen. The purpose of this and the next section is to present several examples showing that infinite electrical networks are at least pertinent to problems of this nature.

Example 8.1-1. *Surface resistivity measurements*: For our first practical application of an infinite network, let us consider a common problem in geophysical exploration. In order to judge the composition of the earth from measurements taken at the earth's surface, several different methods can be used. One of them, called the *resistivity method*, estimates the resistivity of the earth at various depths and locations by injecting and extracting a current I through a pair of current probes and measuring the resulting voltage V between a pair of voltage probes. This yields an estimate of the earth's average resistivity for one particular configuration of probes. Many such measurements for different probe configurations and placements provide some evidence about the earth's structure and composition.

One arrangement of voltage and current probes that is used is called the *Wenner configuration* and is illustrated in Figure 8.1. The probes are arranged in a straight line and are equally spaced with the voltage probes outside of the current probes. This setup is expanded and contracted by increasing and decreasing x in Figure 8.1 and is shifted over the surface to obtain a variety of voltage-current pairs V, I. Each pair is used to compute an *apparent resistivity* ρ_a, which is defined as the value a uniform earth would have to have in order to yield the same ratio V/I for the chosen spread and location of the Wenner configuration. It is a fact [26, page 9] that for the Wenner configuration

$$\rho_a = \frac{4\pi x V}{I}. \tag{8.1}$$

Thus, different apparent resistivity values ρ_a are obtained for different spreads and locations. Actually, ρ_a is a proxy for a *surface operator* that relates the voltage distribution along the earth's surface to currents injected and extracted at that surface.

The next step is to interpret these data, and this is usually done by a matching technique. A library of different variations in ρ_a for different conjectured formations of the earth are computed, and a formation yielding a variation in ρ_a close to the measured variation is viewed as a possible answer. This means that the so-called *direct problem* has to be solved many times; that is, given a configuration of probes and a conjectured variation in conductivity within the earth, compute the value of ρ_a. What is needed is a solution to Laplace's equation, $\nabla \cdot (\sigma \nabla \phi) = 0$, in three rectangular coordinates x, y, z where z measures depth into the earth. Also, σ is the earth's conductivity and ϕ is electric potential.

This is where the theory of infinite electrical networks is helpful. After all, to an observer at the earth's surface, whose measuring instruments may be spread apart by perhaps a kilometer or so, the earth truly appears to be a semi-infinite medium. Thus, its discretized version can well be a semi-infinite grid. Our theories yield an especially concise solution to the direct problem if the conjectured structure of the earth is a horizontal layering. In this case, we have a semi-infinite three-dimensional grid whose resistances vary vertically but not horizontally. Moreover, it is an excellent assumption that infinity is imperceptible, for otherwise the earth's resistivity would have to exhibit a severe anisotropy between the horizontal and vertical directions, with horizontal resistivity increasing and vertical resistivity decreasing with depth; this is most unlikely. Thus, the Laurent matrix $\mathbf{z}_0$ of Section 7.6 is precisely the surface operator we need.

Let us take uniform horizontal increments $\Delta x = \Delta y$ and uniform vertical increments Δz to obtain a three-dimensional grid as a discretization of the horizontally layered earth. (See Section 1.7 for this derivation in the two-dimensional case.) With regard to Figure 8.1, set $x = \mu \Delta x$, where μ is a positive integer. Then, by (7.26) the transform Z_0 of z_0 is the continued fraction

$$Z_0(\theta_1, \theta_2) = \cfrac{(\Delta x)^2}{2\sigma_1(2 - \cos\theta_1 - \cos\theta_2) + \cfrac{1}{\rho_2\zeta}}$$
$$+ \cfrac{1}{2\sigma_3(2 - \cos\theta_1 - \cos\theta_2) + \cfrac{1}{\rho_4\zeta + }} \cdots$$

where $\zeta = (\Delta z/\Delta x)^2$, $\sigma_k = \rho_k^{-1}$, and ρ_k is the earth's resistivity at the depth $k\Delta z/2$ with $k = 1, 2, 3, \ldots$. (The lowered plus signs denote a continued fraction; see Section 7.5.) Then, (7.32) and (8.1) yield a concise solution to our direct problem:

$$\rho_a = \frac{16\mu}{\pi\Delta x\Delta z} \int_0^\pi \int_0^\pi Z_0(\theta_1, \theta_2) \sin\mu\theta_1 \sin 3\mu\theta_1 \, d\theta_1 d\theta_2$$

[See [185] for a more detailed derivation of these formulas. Here is a correction for that reference: Replace Δz by $\Delta z/2$ in the two equations just before (2.1).] ♣

Example 8.1-2. *Borehole resistivity measurements*: Another form of the resistivity method for geophysical exploration occurs when the voltage and current probes are located down in a borehole. The current probes are rings pressed against the sides of the cylindrical borehole, and the voltage probes measure the voltages along those sides. Because of the incursion of the drilling mud into the earth from the borehole, the earth's resistivity varies radially from the borehole but usually not in the polar direction – nor possibly in the vertical direction. For a given configuration of voltage and current probes, the measured apparent resistivity ρ_a of the earth is again that resistivity a uniform earth would have to have in order to yield the same measured voltages for the given injected and extracted currents. The problem that now arises is the computation of ρ_a under a given probe configuration and a conjectured radial variation of resistivity.

Once again, our infinite-electrical-network theory yields a concise solution. Now, however, a discretization in cylindrical coordinates is appropriate. Because the currents are uniformly injected and extracted with respect to the polar coordinate, the discretized model reduces to a two-dimensional, semi-infinite, rectangular grid that is uniform in the vertical direction but varies in the radial direction. Furthermore, the discretization error can be substantially reduced by normalizing with respect to a grid derived from a uniform earth rather than with respect to the earth itself. The result is a discrete apparent resistivity ρ_{da} in place of ρ_a.

As for the discretization, we choose uniform vertical increments and possibly varying radial increments. Let (j, k) be the indices for the sample points, with $j = 1, 2, \ldots$ for the radial direction and $k = \ldots,$ $-1, 0, 1, \ldots$ for the vertical direction. Given a conjectured radial variation in resistivity, we obtain a grid whose vertical conductance between nodes (j, k) and $(j, k + 1)$ will be denoted by G_j for all k and whose radial resistance between nodes (j, k) and $(j + 1, k)$ will be denoted by $R_{j+1/2}$ again for all k. A typical result is as follows. Let a pair of current probes inject a current at node $(1, k)$ and extract it at node $(1, -k)$. Let a pair of voltage probes measure the voltage drop from node $(1, m)$ to node $(1, n)$, where $0 < m < n$. Then, it follows from the theory of Section 7.6 that

$$\rho_{da}(k, m, n) = \frac{\int_{-\pi}^{\pi} Z_D(\theta) \sin k\theta \, (\sin m\theta - \sin n\theta) \, d\theta}{\int_{-\pi}^{\pi} U_D(\theta) \sin k\theta \, (\sin m\theta - \sin n\theta) \, d\theta}$$

where

$$Z_D(\theta) = \frac{1}{(2 - 2\cos\theta)G_1} + \frac{1}{R_{3/2}} + \frac{1}{(2 - 2\cos\theta)G_2} + \frac{1}{R_{5/2}} + \cdots$$

and $U_D(\theta)$ is what $Z_D(\theta)$ becomes when resistivity is set equal to 1 everywhere [184]. ♣

Example 8.1-3. *Surface electromagnetic measurements*: Another way of investigating the structure of the earth through equipment located at its surface is to measure electric and magnetic fields produced by a transient electromagnetic source. One arrangement of measuring equipment for this purpose is the *Turam configuration* [15, pages 141–142]. It consists of a long straight cable at the earth's surface excited by a generator connected to the ends of the cable through a wide loop. The generator produces a transient voltage. As a result, the electric and magnetic fields in the vicinity of the midpoint of the cable form a transient polarized electromagnetic wave with the electric field **E** directed parallel to the cable and the magnetic field **H** circulating around the cable perpendicularly to **E**. These fields are essentially constant along the cable's direction at least in the vicinity of the cable's midpoint. Because of this, we can use the two-dimensional notation introduced in Section 1.7.

We choose an x, y, z rectangular coordinate system with the origin at the cable's midpoint. The y-axis lies along the cable and the z-axis is directed positively downward into the earth. Thus, the x-axis is perpendicular to the cable and along the earth's surface. Let us use the

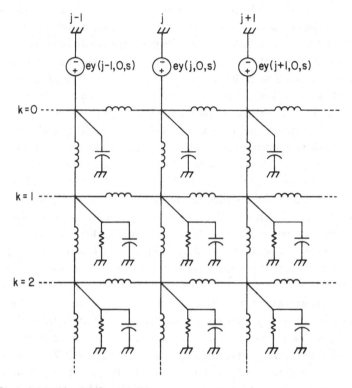

Figure 8.2. The RLC grid that represents the earth under a polarized electromagnetic excitation.

discretization described in Section 1.7. Maxwell's equations reduce to (1.13), (1.14), and (1.15), and the forward and backward finite differences yield (1.16), (1.17), and (1.18), which are difference equations with respect to the spatial coordinates and differential equations with respect to time t. This is realized by the rectangular RLC grid indicated in Figure 1.12. For our application, the grid can be redrawn as in Figure 8.2. The cable is taken to lie one increment Δz above the earth's surface, and thus the grounding conductances are absent for $k = 0$. Note that, unlike the preceding two examples, we now have a grounded grid. As a result, the singularity inherent in the analysis of an ungrounded grid does not now appear.

The single component E_y of the electric field and the two components H_x and H_z of the magnetic field are measured at various points along the x axis as functions of time for a given transient excitation of the cable. This data can be used to check a conjectured formation of the earth by

solving the "direct problem" for the grid of Figure 8.2. That grid is derived from the assumed spatial variations in the conductivity σ, dielectric
permittivity ϵ, and magnetic permeability μ of the earth. In particular, the
grid determines a surface impedance operator, which relates the electric
and magnetic fields to each other. A good match with the measured data
is taken to be evidence in support of the conjectured formation.

Once again, our theory of infinite resistive grids readily yields a solution to this direct problem when the earth is horizontally layered.
The Laplace transformation is used to convert the sampled values of
$E_y(x, z, t)$ along the line at $z = 0$ into spatially discretized, transformed
fields $ey(j, 0, s)$, where $x = j\Delta x$ and $j = \ldots, -1, 0, 1, \ldots$; s is the independent variable for Laplace transforms. For s real and positive, the
grid of Figure 8.2 consists of real positive parameters – effectively resistors. So, our theory of one-dimensionally nonuniform, cubic grids can
be used to relate the voltage sources $ey(j, 0, s)$ at $k = 0$ to the currents
at $k = 0$ and thereby to the spatially discretized, transformed, x and
z components of the magnetic field at the surface. This relationship is
given by a transformed surface impedance operator, which can be written down immediately as a continued fraction obtained from the conjectured parameter values in the earth. Finally, use can be made of any
one of several standard algorithms to convert numerically Laplace transforms into transients. It is hoped all this yields a good match between
the computed, transient, electric, and magnetic fields and the measured
ones. Otherwise, more direct problems for other conjectured formations
need to be solved if a proper match is to be found.

See [175] and [176] for a detailed exposition of these ideas. ♣

8.2 Domain Contractions

Our theory of one-dimensionally nonuniform, infinite grids is useful even
when the grid contains a two-dimensional or higher dimensional nonuniformity so long as the latter nonuniformity is restricted to a finite region.
The surface operators, discussed in the preceding section, can be used
to replace the infinite grid by a finite electrical network that contains
the higher-dimensional nonuniformity. Finite network theory can then
be used to solve the latter network, after which the solution can be extended to the original infinite grid. In short, this technique contracts
the infinite domain of an exterior problem down to a finite one, and
yields certain advantages for numerical computations. Let us present
them through some applications again.

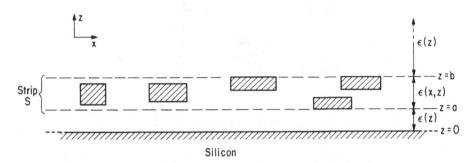

Figure 8.3. A possible two-dimensional configuration of interconnection wires above a silicon chip. The conducting bodies are cross-hatched. The wires are all contained in the strip S between $z = a$ and $z = b$. The permittivity ϵ of the dielectric surrounding the wires is allowed to vary only in the vertical z direction above ($b < z < \infty$) and below ($0 < z < a$) the strip S.

Example 8.2-1. *Electrical capacitance computations*: A common problem in the design of integrated circuits on a semiconductor chip is the determination of the capacitance coefficients of the interconnection wires residing just above the chip. The problem is really a three-dimensional one, but a two-dimensional analysis is simpler and displays the basic ideas equally well. Figure 8.3 shows a cross-sectional view of several interconnection wires above a silicon chip, which in this analysis is a grounded perfectly conducting plane. Rectangular coordinates (x, z) are used with the origin on that grounded plane. The wires are also perfectly conducting and embedded in a dielectric medium whose permittivity ϵ above and below the wires varies with z in general, but not with x. The problem reduces to solving Laplace's equation $\nabla \cdot (\epsilon \nabla \phi) = 0$ to obtain the electric potential $\phi(x, z)$ in the vicinity of the wires when $\phi(x, z)$ is a given constant value on each wire and $\phi(x, 0) = 0$ on the ground plane. We are confronted with an infinite domain above the silicon and between the wires.

Here, too, a finite-difference procedure can be used to change Laplace's equation into an infinite capacitive grid between the wires and above the silicon – as was indicated in Section 1.7. (The grid is capacitive because conductivity α is now replaced by dielectric permittivity ϵ. However, thinking in terms of capacitors instead of resistors involves simply an unessential change in terminology.) The conventional procedure for finding a solution is to render the domain – and thereby the grid – finite by bounding it with an artificial truncating surface at zero potential, as

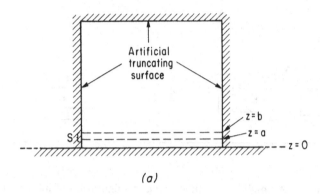

(a)

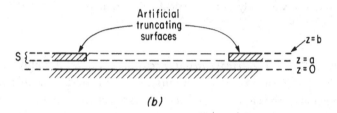

(b)

Figure 8.4. (a) In the customary truncation for a finite-difference analysis, the truncating surface renders the domain finite, perhaps with a rectangular boundary as shown. (b) With the use of infinite-network theory, the domain of analysis remains infinite. Only the strip S is rendered finite by means of the indicated truncating surfaces.

indicated in Figure 8.4(a). This introduces a domain-truncation error. Alternatively, we can and will leave almost all of the domain infinite and truncate it only within the strip S, as shown in Figure 8.4(b). This too introduces a truncation error, but it is far less because the thin truncating surfaces of Figure 8.4(b) distort the field in the vicinity of the wires far less than does the truncating surface of Figure 8.4(a).

To make computation feasible, we have to replace the infinite grid occupying the domain of Figure 8.4(b) by an equivalent finite one. This can be done by using a surface operator to remove the infinite grid above $z = b$ and replace it by a set of terminating capacitors connected to the nodes along $z = b$. These terminating capacitors exactly duplicate the effect of the infinite grid above $z = b$. They can be obtained as follows.

As before, the surface operator is a continued fraction:

$$Y_b(\theta) = \cfrac{1}{\delta_1} + \cfrac{1}{(2 - 2\cos\theta)\gamma_2} + \cfrac{1}{\delta_3} + \cfrac{1}{(2 - 2\cos\theta)\gamma_4} + \cdots$$

where $-\pi \leq \theta \leq \pi$. The coefficients δ_k and γ_{k+1} are readily computed from the discretization step. Moreover, the continued fraction converges because the dielectric permittivity $\epsilon(z)$ becomes the constant value for air above the insulating layer on the chip. As for the terminating capacitors c_m, we obtain

$$c_m = -\frac{1}{2\pi} \int_{-\pi}^{\pi} Y_b(\theta) \cos m\theta \, d\theta, \quad m = 1, 2, 3, \ldots$$

The capacitance c_m has a positive value for every m; a capacitor of value c_m is connected between every two nodes on $z = b$ that are m nodal increments apart. Moreover, all the terminating capacitors that are connected between nodes on $z = b$ and the truncating surfaces in Figure 8.4(b) are summed to get a grounding capacitor at that node.

A similar analysis can be performed to replace the sideways infinite grid between $z = 0$ and $z = a$ by terminating capacitors along $z = a$. In this way, we get a finite network of capacitors connected to the nodes between $z = a$, $z = b$, and the truncating surfaces. That finite network can be solved numerically. The determined values of the node voltages adjacent to each wire yield the charges on the wires and thereby the desired capacitance coefficients [178]. A three-dimensional version of this is given in [187].

The computational benefit furnished by this procedure is that it can save substantial amounts of memory storage and computer time by reducing – perhaps by an order of magnitude or more – the vertical dimension of the truncated nodal array. Indeed, let d be the ratio of number of nodes in the truncated strip S in Figure 8.4(b) divided by the number of nodes within the rectangular surface of Figure 8.4(a). If the horizontal widths of both truncations are the same and the same uniform increments Δx and Δz are chosen for both cases, d becomes the ratio of the height of S to the height of the rectangular surface. As compared to the conventional approach, our procedure, based upon infinite-network theory, reduces storage requirements and the number of multiplications and additions by about a factor of d [178, page 989].

There is another computational advantage. As was mentioned before, the strip truncation of Figure 8.4(b) disturbs the electric field in the

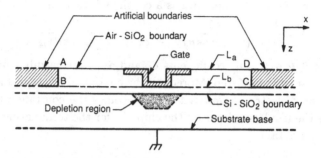

Figure 8.5. A cross-sectional view of the gate and depletion region
in a MOSFET. The cross-hatched areas with corners A,B and C,D
are artificial surfaces used for truncating the strip in which a two-
dimensional nonuniformity, namely, the gate appears. The line BC
simply denotes the bottom level of the gate.

vicinity of the wire less than does the rectangular truncation of Figure
8.4(a). Hence, for the same accuracy the x-direction length of the truncated strip can be made less than that of the rectangular surface. This
decreases d.

Still another benefit accrues to our method when it is used for design
purposes, that is, when one wishes to compute the capacitance coefficients for many different configurations of the interconnection wires. The
advantage arises from the fact that the terminating capacitors do not
depend upon the shape of the wires or upon the dielectric permittivity ϵ
within S. Thus, if one wishes to repeat the computational procedure for
different geometries within S, it is only necessary to compute the terminating capacitors once and for all; the repetitions of the computation
need only be made for the analysis within S. In this way the aforementioned computational advantages are magnified many times. ♣

Example 8.2-2. *MOSFET threshold voltages*: Another design problem for integrated circuits requires the prediction of the threshold voltage for a proposed design of an electronic device, called a MOSFET.
That voltage is the voltage on a contact, called the "gate," at which
the device turns on. A typical configuration for the gate and depletion
region in a cross-sectional view is shown in Figure 8.5. The gate (the
U-shaped cross-hatched region) is a metal contact embedded in insulating material SiO_2. It lies above and close to – but not in contact with –
the silicon chip. The substrate base is a grounded conducting coating in
contact with the lower surface of the silicon chip Si. When the gate carries a sufficiently large voltage V_g, a depletion region (shown dotted) is

induced in the silicon below the gate contact, and a conducting channel is formed along the upper surface of the depletion region. The problem is to predict at what voltage V_g the conducting channel first appears.

The presence of the channel can be ascertained once the electric potential $\phi(x, z)$ is determined within the silicon. The latter is governed by a nonlinear form of Yukawa's potential equation (1.8) with a zero source term:

$$\nabla \cdot (\epsilon \nabla \phi) = \beta(\phi). \qquad (8.2)$$

Here, ϵ is again dielectric permittivity. Also, $\beta(\phi) = ae^{\kappa\phi} - be^{-\kappa\phi} + c$, where a, b, c, and κ denote constants. Moreover, $\beta(\phi)$ is effectively zero everywhere in the silicon except in the depletion region. Also, (8.2) continues to hold throughout the insulating material SiO_2 and in the air above the chip when $\beta(\phi)$ is set equal to zero in those regions. This boundary-value problem is actually an exterior problem because the fringing electric field throughout the air and toward the sides within the silicon should be taken into account. Thus, (8.2) needs to be solved throughout the infinite region above the substrate base, around the gate, and upward through the air.

To handle the nonlinearity in $\beta(\phi)$, an iterative procedure can be employed, each step of which requires a solution of (8.2) with $\beta(\phi)$ replaced by an assumed spatial distribution of electric charges throughout the depletion region. This replacement converts the nonlinear Yukawa's equation into a linear Poisson's equation. At each iteration, the solution of Poisson's equation determines the next replacement for $\beta(\phi)$. The iterations are continued until convergence is achieved. This is a computationally intensive procedure, especially if the fringing field is taken into account.

Here, too, infinite electrical network theory is helpful. For each step of the iteration, the domain-contraction procedure described in the preceding example can be used. A strip between the lines AD and BC is set up to contain the region where a two-dimensional nonuniformity (namely, the gate) appears. This strip is truncated by the (cross-hatched) grounded surfaces with the corners A, B and C, D. Then through the technique described in the preceding example, the medium above the line AD in Figure 8.5 is replaced by a set of terminating capacitors connected to sampling nodes along line AD.

As for the medium below line BC, a complication, which is the main point of this example, arises now. The charges within the depletion region are sources for the field. That is, at each iteration Poisson's equation with an assumed charge distribution holds within the depletion region, and Laplace's equation holds outside of it. This means that the equivalent transformed ladder, like the one shown in Figure 7.4, has to be augmented with transformed charge sources connected between ground and the upper nodes up to a node corresponding to the depth of the depletion region. The result is that the medium below line BC is replaced by terminating capacitors, and in addition equivalent charge sources are connected between ground and the nodes along BC to account for the charges in the depletion region. These arise from a Norton's equivalent representation at the input to the transformed ladder.

As a result of all this, the infinite domain above the substrate base is contracted to the finite domain within the rectangle $ABCD$ of Figure 8.5. In fact, the problem is reduced to solving a finite capacitive network having nodes within that rectangle and with charge sources at the nodes on the finite line between B and C. The two parts of the network on opposite sides of the gate are coupled by the terminating capacitors that extend across the gate. Once the network is solved, the node voltages within the silicon can be obtained very rapidly through an elementary limb analysis. This in turn determines the depletion region for the next iteration. In this way, fringing in the infinite regions above the chip and laterally into the silicon can be taken into account. The procedure has other advantages as well [3].   ♣

Example 8.2-3.   *Polarized electromagnetic waves on buried two-dimensional anomalies*: The surface operator discussed in Example 8.1-3 can be used to solve the direct problem of a buried anomaly in a horizontally layered earth under polarized electromagnetic Turam excitation at the earth's surface. The anomaly is assumed to vary only two-dimensionally, being uniform in the third dimension. The polarization of the electromagnetic wave is as before with the E_y component directed along the axis of the anomaly. With regard to geophysical exploration, this is an unrealistic model – except when the searcher knows beforehand that the anomaly extends uniformly in one direction and also knows what that direction is. In most cases a buried anomaly is three-dimensional. Nonetheless, an analysis of this two-dimensional model can provide some insight into the behavior of polarized electromagnetic waves within the earth.

Once again, a discretization of the polarized Maxwell equation leads to a grounded grid like that of Figure 8.2. This is redrawn schematically in Figure 8.6, where $k = 1$ is the index for the surface of the earth and the two-dimensional anomaly lies between the horizontal levels $k = m$

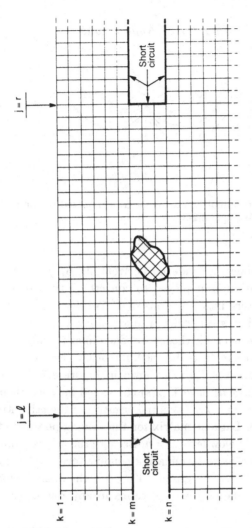

Figure 8.6. The grid for Example 8.2-3 obtained from a discretization for a polarized electromagnetic wave impinging upon a buried anomaly in a horizontally layered earth. The columns of the grid are indexed by j, and the rows by k.

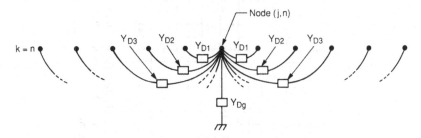

Figure 8.7. The terminating conductances connected to one node on level $k = n$ for Example 8.2-3.

and $k = n$. The nodes are indexed horizontally by j. As before, the Laplace transformation is employed to convert all time functions and differential parameters into functions of s. We again have in effect a resistive network when s is restricted to a real positive value.

Next, our infinite network theory for one-dimensionally nonuniform grids can be used to replace the medium below level $k = n$ by the terminating conductances $Y_{Dp}(s)$, shown in Figure 8.7. In particular, the nodes (i, n) and (j, n) have $Y_{Dp}(s)$ connected between them, where $p = |i - j| \neq 0$. Moreover, $Y_{Dp}(s)$ is the negative of the pth entry on the principal row of the Laurent matrix for the driving-point conductance operator for the grid below level $k = n$. Because the grid is grounded, we have in addition one more terminating conductance $Y_{Dg}(s)$ connected between each node (j, n) and ground. $Y_{Dg}(s)$ is the main-diagonal term plus the sum of all the off-diagonal terms in any row of the Laurent matrix.

As for the medium above level $k = m$, an operator version of Norton's theorem might be used as in the preceding example to account for the source at the earth's surface. In particular, if the electric field is measured along a wide stretch of the earth's surface, that field may be represented by voltage sources at the nodes along $k = 1$, and the operator version of Norton's theorem may be applied along $k = m$ to the grid from $k = m$ to $k = 1$ terminated with those sources along $k = 1$. This yields terminating conductances and equivalent sources connected to the nodes along level $k = m$. Those terminations at a single node (j, m) are shown in Figure 8.8. On the other hand, if the electric field is measured only in the vicinity of the exciting cable, the strip for analysis must be expanded to include all levels between $k = 1$ and $k = n$; in addition, the air above $k = 1$ must be taken into account with another infinite grid. That in

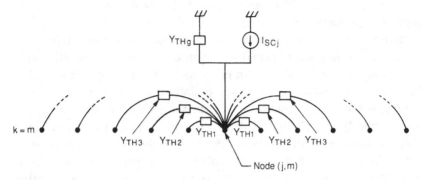

Figure 8.8. The terminating conductances and equivalent current source connected to one node on level $k = m$ for Example 8.2-3.

turn can be represented just by terminating conductances like those of Figure 8.7.

In both cases the domain for analysis is reduced to a single strip containing the anomaly. For definiteness consider the former case where the strip lies between $k = m$ and $k = n$. Upon truncating the strip sideways at $j = l$ and $j = r$, as indicated by the short circuits of Figure 8.6, we finally obtain a finite network, whose solution can be obtained routinely. An elementary limb analysis or alternatively the use of transfer ratios for the ladder above $k = m$ yields the Laplace-transformed, electric and magnetic fields at the surface as functions of real positive s. Finally, an appropriate inversion of the Laplace transformation yields the desired transient fields. The details of this analysis can be found in [176]. ♣

Example 8.2-4. *Cylindrical and spherical coordinates*: In the prior examples our technique for contracting domains has been applied only to rectangular coordinate systems. Nonetheless, it works for cylindrical and spherical coordinates, too, so long as the operational calculus used to decompose the grid into a transformed ladder is appropriately modified.

Consider a physical situation wherein a three-dimensional anomaly exists within a cylindrically layered medium, namely, the flow of petroleum into a borehole through a perforated completion zone [183]. For a steady Newtonian flow, the governing equation is Poisson's: $\nabla \cdot (\sigma \nabla P) = -H$. Here, $\sigma = \kappa/\mu$, where κ is medium permeability and μ is fluid viscosity; σ can be taken to be constant except within the completion zone, which is a bounded three-dimensionally varying medium adjacent to the perforated hole in the side of the oil well's pipe. Also, P is pressure and H

is a source function, which can be taken to be the flow at the hole in the pipe. The problem is to compute P.

Here again, we have an exterior problem. Upon using finite differences for Poisson's equation in cylindrical coordinates, we get a resistive grid of concentric cylindrical uniform grids, all connected together by radial branches. The resistances vary only radially from the borehole's center except in the completion zone, wherein the resistances vary vertically and circularly as well. Let C denote a cylindrical grid centered along the borehole's axis and large enough to contain the completion zone. The grid outside C can be represented by a ladder of ∞-ports. Each series ∞-port consists of the radial resistors between two consecutive cylindrical grids. Each shunt ∞-port consists of the resistors on any one cylinder and a hypothetical ground node as well.

The next step is to construct an operational calculus to convert the conductance operators of the ∞-ports into multiplications by functions. Actually, the polar coordinate measuring circular displacements around the borehole leads to a finite analog of a Laurent matrix, namely, the circulant matrix [103], [132]. The discrete-Fourier-series transformation [105, page 297] transforms circulant matrices in the same way as the Fourier-series transformation transforms Laurent matrices. By combining these transformations [183, Section 4], one obtains an operational calculus that converts the grid outside the cylinder C into a ladder, whose parameters are multiplications by functions. This allows us to proceed exactly as before to represent the medium outside C by terminating conductances connected to the nodes on C. A finite network is then achieved when the infinite cylinder within C is truncated vertically.

The analysis of an infinite spherical grid with a bounded three-dimensional anomaly is more complicated [188]. In addition to the radial coordinate r, we now have a coordinate θ measuring displacements in latitude and a coordinate ϕ measuring displacements in longitude. Finite differences for Poisson's equation in spherical coordinates lead to concentric spherical grids connected together by radial resistors. We can take a finite number of uniform increments in ϕ, and similarly for θ. This leads to circulant matrices with respect to the displacements in ϕ, but no such matrices arise for θ. Nevertheless, we can proceed as before for the grid outside a sphere S containing the three-dimensional anomaly. That grid is now a ladder of n-ports, where $n = k \times l$ is the number of nodes on each sphere, k being the number of sample points along the θ-coordinate and l being that number along the ϕ-coordinate. The discrete-Fourier-series transformation applied for the ϕ-coordinate converts the grid into a ladder whose parameters are multiplications by functions that map the transform variable into the space of $k \times k$ matrices. Despite this complication, terminating conductances connected to the nodes on S can still be derived as representatives of the infinite medium outside S [188]. For this case, no further medium truncation is needed.

In summary, infinite-electrical-network theory also provides a domain-contraction technique for analyzing certain exterior problems in cylindrical and spherical coordinates. ♣

8.3 Random Walks

Infinite electrical networks also arise surprisingly enough in probability theory, in particular, as analogs of random walks on infinite graphs. Nash-Williams pointed this out in an early paper [100, pages 183–184]. Doyle and Snell [43] give a tutorial exposition of this relationship.

The random walks of concern to us have been defined up to now on an ordinary, countably infinite network – or, in our terminology, on a 0-connected 0-network $\mathbf{N}^0$. Infinite nodes are allowed in $\mathbf{N}^0$ so long as they are restraining. Self-loops and parallel branches are allowed too. $\mathbf{N}^0$ is purely resistive, having no sources. The conductance g_j of each branch, rather than relating voltages and currents, now determine "transition probabilities" as follows. An object W, called a *random walker*, wanders through $\mathbf{N}^0$ proceeding from node to adjacent node in a random way. Upon reaching a node n_k, W chooses the next branch b_j to traverse and thereby the next node n_l according to the *transition probability*

$$p_{kl} = \frac{g_j}{\gamma_k}, \tag{8.3}$$

where γ_k is the sum of the conductances of all branches incident to n_k, g_j is the conductance of b_j, and n_k and n_l are the two nodes of b_j. Since n_k is restraining, p_{kl} is a positive number no larger than 1. The alternating sequence of nodes and branches traversed in this way is called a *random walk*. It need not be a path because a node or branch may reappear in the sequence. When there are no parallel branches – or better still when parallel branches are combined by summing their conductances, a random walk is a particular case of a Markov chain [72] with a countably infinite state space consisting of the nodes of $\mathbf{N}^0$. Random walks arise in many scientific disciplines.

A question commonly asked about a random walk may be stated in the following way. Partition the node set of $\mathbf{N}^0$ into three sets: $\mathcal{N}_e, \mathcal{N}_f, \mathcal{N}_g$. Assume a random walker starts at a node $n_0 \in \mathcal{N}_f$. What is the probability that W reaches a node of $\mathcal{N}_e$ before W reaches a node of $\mathcal{N}_g$?

The answer can be found electrically. Short together all the nodes of $\mathcal{N}_g$ and refer to the resulting node n_g as ground. It is taken to be at 0 V. Connect a star network of pure 1-V voltage sources between n_g and the nodes of $\mathcal{N}_e$ to provide 1-V node voltages at the latter nodes. Let $\mathbf{M}^0$ denote the resulting network. Since all the nodes of $\mathbf{N}^0$ are restraining, we can transfer all the pure voltage sources through the nodes of $\mathcal{N}_e$ in accordance with Section 3.6 and then invoke Theorem 3.7-1 to conclude that $\mathbf{M}^0$ has a unique voltage-current regime whenever Conditions 3.6-5 are satisfied. This will certainly be so when $\mathcal{N}_e$ is a finite set. Since $\mathbf{N}^0$ is 0-connected, it also follows that every node of $\mathbf{M}^0$ has a unique node voltage with respect to ground.

Moreover, every node voltage in $\mathbf{M}^0$ lies between 0 V and 1 V. Indeed, suppose there is a node with the voltage $u_1 > 1$. We can partition the node set into a set $\mathcal{N}_1$ whose nodes have voltages no less than u_1 and another node set $\mathcal{N}_2$ whose nodes have voltages less than u_1. Since $\mathbf{M}^0$ is 0-connected, there is at least one branch with one node in $\mathcal{N}_1$ and the other in $\mathcal{N}_2$. For every such branch, a positive current flows through it from $\mathcal{N}_1$ to $\mathcal{N}_2$. Since all nodes other than n_g are restraining, Kirchhoff's current law holds at every node of $\mathcal{N}_1$. We can now infer that there must be at least one source connected to a node of $\mathcal{N}^1$ or to an infinite extremity of the subnetwork induced by $\mathcal{N}_1$. But there are no such sources. Our supposition is false. By a similar argument, no node voltage is less than 0 V.

It happens to be a fact that the probability p_0 of the walker W in $\mathbf{N}^0$ starting at a node $n_0 \in \mathcal{N}_f$ and reaching a node of $\mathcal{N}_e$ before reaching a node of $\mathcal{N}_g$ is equal to the node voltage u_0 at n_0 in the electrical analog $\mathbf{M}^0$ [43]. This is one way by which infinite electrical networks relate to random walks.

There is a related problem. Consider a random walker W in $\mathbf{N}^0$ that starts at a node n_0 and proceeds without stopping. There are two possibilities: Either it is a certainty that W will return to n_0 at least once or there is a positive probability that W will never return to n_0. In the first case the network $\mathbf{N}^0$ is called *recurrent* and in the second case *transient*. (It turns out that recurrence or transience is independent of the choice of n_0.) Finding conditions under which $\mathbf{N}^0$ is either recurrent or transient has been an active area of research. Polya [110] first proposed it in somewhat different form [43, page 88], and an advanced treatment is given in [125]. (This question is related to other research areas in mathematics such as the classification of Riemann surfaces as parabolic or hyperbolic or the existence of harmonic functions on such surfaces. For a sampling of this literature, see [24], [25], [32], [39], [42], [56], [57], [58], [61], [68], [69], [71], [90], [91], [108], [109], [117], [119], [122], [123], [129], [130], [137], [148], [150]–[153].)

Here, too, the theory of infinite electrical networks can be used to solve the problem [100]. We now convert $\mathbf{N}^0$ into a 1-network by designating all of its 0-tips as members of a single 1-node n_g, which in this case is taken to be ground. In other words, we short together all the infinite extremities of $\mathbf{N}^0$. Next, a 1-V pure voltage source is connected between n_0 and n_g with the positive terminal at n_0. The result of all this is a 1-network $\mathbf{N}^1$. The driving-point conductance G_{0g} as measured between n_0 and n_g in $\mathbf{N}^1$ is equal to the current i_0 passing through the source and into n_0. It is another fact [100] that $\mathbf{N}^0$ is recurrent if $G_{0g} = 0$ and transient if $G_{0g} > 0$. In fact, the probability that W will never return to its starting node n_0 is G_{0g}/γ_0, where γ_0 is the sum of the conductances of the branches incident to n_0.

Example 8.3-1. The last result sheds some more light on the behavior of ungrounded, n-dimensional, uniform grids. It is well known that such a grid is recurrent if $n = 1$ or 2, and transient if $n \geq 3$. (See [43, Chapter 6]; it is of no consequence in this regard that we are allowing branches parallel to different coordinates to have different conductances – in contrast to [43].) With n_0 chosen as any 0-node, we have $G_{0g} = 0$ for $n = 1$ or 2, that is, there is no current in the 1-V source nor in any branch of $\mathbf{N}^1$. (This is a consequence of the singularities discussed in Section 7.3.) What happens is that every 0-node voltage becomes 1 V. On the other hand, for $n \geq 3$ there are currents throughout $\mathbf{N}^1$, and node voltages differ.

Another way to view this situation is to replace the 1-V voltage source between n_0 and n_g by a 1-A current source. In order to invoke Theorem 3.7-1 for the existence of a unique voltage-current regime, we need to transfer this current source into resistive branches in such a way that the total isolated power of the transferred sources is of finite power. If it exists, that distribution of current sources will by itself satisfy Kirchhoff's current law at all 0-nodes except at n_0 (see Section 3.6). The result is called a "flow of finite power from n_0 to infinity" [43, page 142]. We can now conclude that such a transference of the current sources can be made for $n \geq 3$ but not for $n = 1$ or 2. ♣

It appears that this analogy extends readily enough to k-networks and that a new problem concerning the recurrence or transience of transfinite networks might be posed and solved by using our theory of infinite electrical networks. To do so, we must first define a "transfinite random walk." But how can a walker starting at a 0-node reach a k-node where $k > 0$? Let us speculate with an example.

Example 8.3-2.   Consider a transfinite linear ladder, in particular, an ω^2-cascade of el-sections as in Figure 1.7(a) with node indices as in Figure 6.19. Let n_p denote the node of index p, where $0 \leq p \leq \omega^2$. Thus, n_0 is the input 0-node, n_ω is the first 1-node, and n_{ω^2} is the output 2-node. Also, let n_g be the ground node (i.e., the bottom line in Figure 6.19). Assign to the 0th ω-cascade between n_0 and n_ω the resistance values indicated in Figure 1.7(a). This makes n_ω perceptible from n_0. Also, assign to the mth ω-cascade between nodes $n_{\omega \cdot m}$ and $n_{\omega \cdot (m+1)}$, where $1 \leq m < \omega$, the resistance values obtained by dividing the resistance values of Figure 1.7(a) by 10^m. This makes n_{ω^2} perceptible from n_0. We shall now argue that a walker W starting at n_0 has a positive probability of reaching n_{ω^2} before reaching n_g through a certain kind of transfinite random walk.

First of all, isolate the mth ω-cascade by disconnecting the transfinite ladder at the nodes $n_{\omega \cdot m}$ and $n_{\omega \cdot (m+1)}$ (i.e., split each node in two so that the output of the preceding ω-cascade is isolated from the input to the succeeding ω-cascade). Let $n'_{\omega \cdot m}$ and $n'_{\omega \cdot (m+1)}$ be the input and output nodes of the isolated ω-cascade. In accordance with the electrical analog of a random walk, append a 1-V voltage source between $n'_{\omega \cdot (m+1)}$ and ground n_g with the positive terminal at the former node. Let the input at $n'_{\omega \cdot m}$ be open-circuited. This produces a voltage-current regime whose input voltage-current pair $x = (v_{\omega \cdot m}, 0)$ and output voltage-current pair $y = (1, i_{\omega \cdot (m+1)})$ can be constructed as shown in Figure 8.9. In contrast to Figure 6.23, the loci S_a, S_b^*, and $T = f_m^\omega(S_a)$ are now straight lines by virtue of the linearity of the ladder. Here, f_m^ω denotes the forward mapping of the mth ω-cascade. Because of the open-circuit at the input and the pure voltage source at the output, S_a is the abscissa and S_b^* is the vertical line at $v = 1$. Also, T has a negative slope and passes through the origin. It follows that $0 < v_{\omega \cdot m} < 1$. Invoking the analogy between electrical networks and random walks (even though the 1-V source is impressed at a 1-node rather than at a 0-node), we are led to believe that there is a positive probability $v_{\omega \cdot m}$ that a random walker W starting at $n'_{\omega \cdot m}$ will reach $n'_{\omega \cdot (m+1)}$ before reaching n_g.

This argument might be extended to the ω^2-cascade if we specify as follows how W chooses the next branch once a limit node $n_{\omega \cdot m}$ has been reached. W simply examines the two conductive branches incident to

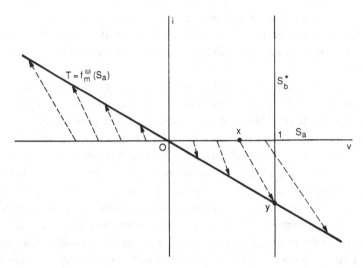

Figure 8.9. The construction of the input voltage-current pair x and output voltage-current pair y for the mth ω-cascade of Example 8.3-2 – in accordance with Section 6.11 and Figure 6.23.

$n_{\omega \cdot m}$ and uses the proportionality rule (8.3), thereby ignoring the 0-tip embraced by $n_{\omega \cdot m}$. This means that, once W reaches $n_{\omega \cdot m}$, W is certain to embark on a random walk in the mth ω-cascade without returning to the preceding $(m + 1)$th ω-cascade. As a result, the probability of reaching n_{ω^2} before reaching n_g through this new kind of random walk starting at n_0 is the product $\prod_{m=0}^{\infty} p_m$, where $p_m = v_{\omega \cdot m}$. This product is *not* the same as the node voltage v_0 induced at n_0 by a 1-V source connected to the output of the ω^2-cascade because p_m is determined electrically by first isolating the mth ω-cascade.

An electrical analog can be devised for this kind of transfinite walk by inserting unity-gain isolating amplifiers [34, page 179] between the ω-cascades. Each amplifier's output is connected to the output 1-node of the preceding ω-cascade and the amplifier's input is connected to the input 0-node of the succeeding ω-cascade. Then, the above product of probabilities $\prod_{m=0}^{\infty} p_m$ is the node voltage at n_0 when a 1-V voltage source is impressed from ground to n_{ω^2} as a limiting case. ♣

It appears from this example that the idea of a "transfinite random walk" might be a viable concept, based by analogy on the theory of transfinite electrical networks in conjunction with the use of isolating amplifiers. According to our rule for proceeding from a node of higher rank, every k-node with $k > 0$ would have to embrace a 0-node. On the other hand, other kinds of transfinite random walks might be feasible if appropriate rules for proceeding from nodes of higher ranks could be devised.

Finally, let us point to another way that the literature on random walks might relate to our transfinite networks. We have just seen in Example 8.3-1 that a "flow of finite power from a given 0-node n_0 to infinity" is needed if we are to invoke Theorem 3.7-1 and thereby assert the transience of a uniform grid. Such a flow can be interpreted as a 1-basic current in $\mathcal{I}$ if we short n_0 to the network's extremities (i.e., if we append a 1-node containing n_0 and all the 0-tips). On the other hand, we achieved an existence and uniqueness theorem for the voltage-current regime of a 1-network by expanding the space $\mathcal{K}^0$ to include all finite-powered 1-basic currents in addition to all 0-loop currents. This suggests that what the above-cited literature is touching on is the question of how rich can the space $\mathcal{K}^0$ be for various 1-networks. But the same question can be posed for k-networks. Is there an analogous body of results on both k-networks and transfinite random walks waiting to be discovered?

8.4 Operator Networks

The decomposition of a one-dimensionally nonuniform grid into a ladder of ∞-ports, a procedure discussed in Section 7.4, leads to another kind of electrical network, one whose parameters are operators on a Hilbert space. To be sure, the network and the operators are very specialized: a ladder of Laurent operators on l_{2r}^n. Such operators commute, but they are not required to be invertible.

This idea can be generalized. An ∞-port, which internally is an infinite resistive network without sources, may define a resistance operator $r : i \mapsto v$ that maps any $i \in l_{2r}$ into $v = ri \in l_{2r}$ [161, Section III]. This can be one element of a finite or infinite network, whose voltages and currents are vectors in l_{2r} and whose resistances are operators in $[l_{2r}; l_{2r}]$. That network can be constructed by interconnecting ∞-ports at their port terminals. However, in order for the resistance-operator representation of each ∞-port to remain valid, the port conditions must not be violated by the interconnections, that is, the current entering one terminal of a port must equal the current leaving its other terminal. This can always be assured by isolating every port with an ideal transformer [143, pages 21–22]. On the other hand, certain kinds of ∞-ports, such as the ladder of Laurent operators discussed in Section 7.4, may not need ideal transformers. In any case, the abstract idea of a finite or infinite electrical network, whose elements are operators on a Hilbert space, can be realized in this way. In this construct the resistance operators need not commute. In fact, a general theory for infinite operator networks, much like that of Chapters 3 and 5, can be established if the resistance operators are all invertible and if the ∞-port interconnections simulate connections among ordinary branches. Let us sketch it.

Let $\mathcal{H}$ be a real Hilbert space. Starting with a k-graph, we assign to every branch the resistive structure described in Section 1.4. Now, however, every branch voltage v_j, branch current i_j, voltage source e_j, and current source h_j is a member of $\mathcal{H}$. Moreover, every branch resistance r_j is a positive invertible continuous linear operator mapping $\mathcal{H}$ onto $\mathcal{H}$. Thus, a conductance $g_j = r_j^{-1}$ for each branch exists as an invertible continuous linear operator too [94, page 416]. With $(\cdot, \cdot)_{\mathcal{H}}$ denoting the inner product of $\mathcal{H}$, the positivity of r_j means that $(r_j x, x)_{\mathcal{H}} \geq 0$ for all $x \in \mathcal{H}$. Moreover, with $w = r_j x$, we have $(r_j x, x) = (w, g_j w) = (g_j w, w)$. Hence, g_j is positive too. Let us take every branch in the Thevenin form, namely, a series connection of the operator r_j and a (possibly zero) voltage source $e_j \in \mathcal{H}$. Power relations are now given in terms of the inner product of $\mathcal{H}$ rather than the product in R^1. Thus, the total power dissipated in all the resistances is $\sum (r_j i_j, i_j)_{\mathcal{H}}$, and the total power supplied by all the voltage sources is $\sum (e_j, i_j)_{\mathcal{H}}$. Moreover, the total isolated power is $\sum (e_j, g_j e_j)_{\mathcal{H}}$. Such a network will be called an *operator network* or an *operator k-network*, in conformity with the terminology of Section 7.4.

The development, which led to Theorem 5.5-1, can now be followed exactly as before (see Section 3.3 for 1-networks) except for some changes in notation. Thus, the Hilbert space $\mathcal{I}$ now has the inner product $(\mathbf{i}, \mathbf{s})_{\mathcal{I}} = \sum (r_j i_j, s_j)_{\mathcal{H}}$. Similarly, the pairing of a voltage vector $\mathbf{u}$ with a current vector $\mathbf{i}$, which may define $\mathbf{u}$ as a functional, reads as follows: $\langle \mathbf{u}, \mathbf{i} \rangle = \sum (u_j, i_j)_{\mathcal{H}}$. Note also that, since r_j and g_j are positive operators, they are both self-adjoint and have unique positive square roots [89, page 157]. This allows us to extend the proof of Lemma 3.3-4 in the context of k-networks. For example, with two applications of Schwarz's inequality – one for members of $\mathcal{H}$ and the other for infinite series – (3.8) will now read as follows, wherein $\| \cdot \|_{\mathcal{H}}$ and $\| \cdot \|_{\mathcal{I}}$ are the norms for $\mathcal{H}$ and $\mathcal{I}$ respectively.

$$\sum |(e_j, i_j)_{\mathcal{H}}| = \sum |(g_j^{1/2} e_j, r_j^{1/2} i_j)_{\mathcal{H}}| \leq \sum \|g_j^{1/2} e_j\|_{\mathcal{H}} \|r_j^{1/2} i_j\|_{\mathcal{H}}$$
$$\leq \left[\sum \|g_j^{1/2} e_j\|_{\mathcal{H}}^2 \sum \|r_j^{1/2} i_j\|_{\mathcal{H}}^2\right]^{1/2} = \left[\sum (e_j, g_j e_j)_{\mathcal{H}}\right]^{1/2} \|\mathbf{i}\|_{\mathcal{I}}.$$

On the other hand, no change in notation is needed for the operator R, which is again defined by $R\mathbf{i} = (r_1 i_1, r_2 i_2, \ldots)$. An almost word-for-word repetition of the arguments in Sections 3.3 and 5.5 involving the subspaces $\mathcal{K}^0$ and $\mathcal{K}$ of $\mathcal{I}$ leads to

Theorem 8.4-1: *Let $\mathbf{N}$ be an operator k-network, every branch of which is in the Thevenin form, let every r_j be a positive invertible continuous linear mapping of $\mathcal{H}$ onto $\mathcal{H}$, let every e_j be a member of $\mathcal{H}$, and let $\sum(e_j, g_j e_j)_{\mathcal{H}} < \infty$. Then, there exists a unique $\mathbf{i} \in \mathcal{K}$ such that*

$$\langle \mathbf{e} - R\mathbf{i}, \mathbf{s} \rangle = \sum (e_j - r_j i_j, s_j)_{\mathcal{H}} = 0$$

for all $\mathbf{s} \in \mathcal{K}^0$. Furthermore, the conclusions of Corollaries 3.3-6 and 3.3-7 also hold when their notations are suitably altered.

An entirely equivalent dual theory can be devised as well. $\mathbf{N}$ is again an operator k-network but with all branches in the Norton form. The total isolated power is now $\sum(r_j h_j, h_j)_{\mathcal{H}}$ where $r_j = g_j^{-1}$. We define $\mathcal{V}$ to be the space of all voltage vectors $\mathbf{v} = (v_1, v_2, \ldots)$ such that $\sum(v_j, g_j v_j)_{\mathcal{H}} < \infty$ and $\langle \mathbf{v}, \mathbf{s} \rangle = \sum(v_j, s_j)_{\mathcal{H}} = 0$ for all $\mathbf{s} \in \mathcal{K}^0$. Also, we set $G\mathbf{v} = (g_1 v_1, g_2 v_2, \ldots)$. These expansions of our definitions along with essentially the same proofs lead to the following generalization of Theorems 3.5-2, 3.5-3, and 3.5-4.

Theorem 8.4-2: *Let $\mathbf{N}$ be an operator k-network, every branch of which is in the Norton form, let every g_j be a positive invertible continuous linear mapping of $\mathcal{H}$ onto $\mathcal{H}$, let every h_j be a member of $\mathcal{H}$, and let $\sum(r_j h_j, h_j)_{\mathcal{H}} < \infty$. Then, there exists a unique $\mathbf{v} \in \mathcal{V}$ such that*

$$\langle \mathbf{w}, \mathbf{h} - G\mathbf{v} \rangle = \sum (w_j, h_j - g_j v_j)_{\mathcal{H}} = 0$$

for all $\mathbf{w} \in \mathcal{V}$. Moreover, under the Thevenin–Norton conversion of all branches, the voltage-current regime dictated by this conclusion is exactly the same as that dictated by Theorem 8.4-1.

Most of the other results of Chapters 3 and 5 can also be lifted to this Hilbert-space setting without much alteration so long as the invertibility of all the r_j is maintained. Moreover, fundamental theories for infinite electrical networks consisting entirely of inductors or entirely of capacitors can be established by copying the theory for infinite resistive networks; this requires no more than a change of terminology (see the Appendix in [161]). An ∞-port consisting solely of inductors (or capacitors) may yield an operator element l (or c) that relates voltage v and current i according to $v = l\, di/dt$ (or $i = c\, dv/dt$), where l (or c) is an operator on l_{2r}. Upon interconnecting such ∞-ports while maintaining the port conditions and consistency with branch connections, we obtain still another abstraction derived from infinite electrical networks: the RLC operator network, whose resistors, inductors, and capacitors are operators in l_{2r} [161]. Our point is that infinite networks of ordinary electrical parameters provide one approach to realizing the operator-network abstraction.

However, if the r_j are not all invertible, severe complications can arise. One way of obtaining a noninvertible r_j with a resistive ∞-port is to connect shorts internally across some of the ports; this precludes the surjectivity of r_j. How to determine the resistance operators for combinations of ∞-ports connected according to the simplest electrical circuits becomes problematic. The series connection for two noninvertible resistance operators r_1 and r_2 defined on all of l_{2r} is apparent – just add them to get the overall resistance. However, the resistance for the parallel connection is formally $(r_1^{-1} + r_2^{-1})^{-1}$, a meaningless formula now. Assigning a significance to the parallel connection is a difficult chore, even when r_1 and r_2 are finite-dimensional operators. Duffin and his students have examined this and other such problems for a variety of network connections; see [9] and the references therein.

During the past two decades a fairly substantial literature on operator networks has arisen. As our final note, let us list a sampling of that literature. [But first two caveats: (i) We only list papers that deal with operators on infinite-dimensional spaces; the literature dealing with finite-dimensional operators – the so-called "n-port theory" – is older and massive. (ii) This is merely a sampling; there are undoubtedly other pertinent works.] See [4]–[8], [10]–[12], [30], [40], [41], [49], [50], [60], [82], [83], [97], [98], [102], [159], [161], [162], [164], [185].

Bibliography

[1] A. Abian, *The Theory of Sets and Transfinite Arithmetic*, W.B. Saunders Co., Philadelphia, 1965.

[2] P. Amstutz, Sur la solution élémentaire d'une équation de Laplace discrétisé, *Ann. Telecommun.*, **22** (1967), 149–152.

[3] H.K. An and A.H. Zemanian, Computationally efficient determination of threshold voltages in narrow-channel MOSFET's including fringing and inversion effects, *IEEE Trans. Electron Devices*, **36** (1989), 534–541.

[4] W.N. Anderson, Jr., T.D. Morley, and G.E. Trapp, Ladder networks, fixpoints, and the geometric mean, *Circuits, Systems, and Signal Processing*, **2** (1983), 259–268.

[5] W.N. Anderson, Jr., T.D. Morley, and G.E. Trapp, Symmetric function means of positive operators, *Linear Alg. and Its Appl.*, **60** (1984), 129–143.

[6] W.N. Anderson, Jr., T.D. Morley, and G.E. Trapp, Infinite networks and quadratic optimal control, *Circuits, Systems, and Signal Processing*, **9** (1990), 229–238.

[7] W.N. Anderson, Jr. and M. Schreiber, The infimum of two projections, *Acta Sci. Math. (Szeged)*, **33** (1972), 165–168.

[8] W.N. Anderson, Jr. and G.E. Trapp, Shorted operators. II, *SIAM J. Appl. Math.*, **28** (1975), 60–71.

[9] W.N. Anderson, Jr. and G.E. Trapp, Matrix operations induced by electrical network connections – a survey, *Constructive Approaches to Mathematical Models*, Editors: C.V. Coffman and G.J. Fix, Academic Press, New York, 1979.

[10] T. Ando, *Topics in Operator Inequalities*, Lecture Notes, Hokkaido University, Sapporo, Japan, 1978.

[11] T. Ando, An inequality between symmetric function means of positive operators, *Acta Sci. Math.*, **45** (1983), 19–22.

[12] T. Ando and J.W. Bunce, The geometric mean, operator inequalities, and the wheatstone bridge, *Linear Alg. and Its Appl.*, **97** (1987), 77–91.

[13] A. Antoniou, *Digital Filters: Analysis and Design*, McGraw-Hill Book Co., New York, 1979.

[14] K. Barbey, W. Hackenbroch, and H. Willie, Partially translation invariant linear systems, *Integral Equations and Operator Theory*, **3** (1980), 311–322.

[15] A.E. Beck, *Physical Principles of Exploration Methods*, John Wiley, New York, 1981.

[16] V. Belevitch, On the realizability of non-rational positive real functions, *Circuit Theory and Applications*, **1** (1973), 17–30.

[17] V. Belevitch, On the asymptotic behavior of meromorphic RL-impedances, *Networks and Signal Theory*, Editors: J.K. Skwirzynski and J.O. Scanlon, P. Peregrimes, London, 1973, 240–247.

[18] V. Belevitch, On the convergence of infinite resistive ladders, *Proc. European Conference on Circuit Theory*, Lausanne, 1978, 29–45; published by Georgi, St. Saphorin, Switzerland.

[19] V. Belevitch, A class of nonrational impedances, *Circuit Theory and Applications*, **6** (1978), 315–319.

[20] V. Belevitch, The Gauss hypergeometric ratio as a positive real function, *SIAM J. Math. Anal.*, **13** (1982), 1024–1040.

[21] V. Belevitch and J. Boersma, The Bessel ratio $K_{\nu+1}(z)/K_\nu(z)$ as a passive impedance, *Philips J. Res.* **34** (1979), 163–173.

[22] V. Belevitch and J. Boersma, On Stieltjes transforms involving Γ-functions, *Mathematics of Computation*, **38** (1982), 223–226.

[23] S.K. Berberian, *Introduction to Hilbert Space*, Oxford University Press, New York, 1961.

[24] K.A. Berman and M. Konsowa, A counterexample to the conjecture of Woess on simple random walks on trees, *Proc. Amer. Math. Soc.*, **105** (1989), 443–449.

[25] K.A. Berman and M.H. Konsowa, Random paths and cuts, electrical networks, and reversible Markov chains, *SIAM J. Disc. Math.*, **3** (1990), 311–319.

[26] P.K. Bhattacharya and H.P. Patra, *Direct Current Geoelectric Sounding*, Elsevier Publishing Co., Amsterdam, 1968.

[27] M.T. Boudjelkha and J.B. Diaz, Half space and quarter space Dirichlet problems for the partial differential equation $\Delta u - \lambda_2 u = 0$ - Part I, *J. Appl. Anal.*, **1** (1972), 297–324.

[28] F.H. Branin, The algebraic-topological basis for network analogies and the vector calculus, *Proc. Symp. Generalized Networks*, Polytechnic Institute of Brooklyn, New York, 1966.

[29] A. Brown and P.R. Halmos, Algebraic properties of Toeplitz operators, *J. für Mathematik*, **213** (1963), 89–102.

[30] C.A. Butler, *Approximation of Generalized Schur Complements*, Ph.D. dissertation, University of Illinois, Urbana, Illinois, 1987.

[31] G.A. Campbell, Physical theory of electric wave-filter, *Bell Syst. Tech. J.*, **1** (1922), 1–32.

[32] D.I. Cartwright and W. Woess, Infinite graphs with nonconstant Dirichlet finite harmonic functions, preprint, 1989.

[33] L.O. Chua, *Nonlinear Network Theory*, McGraw-Hill Book Co., New York, 1969.

[34] L.O. Chua, C.A. Desoer, and E.S. Kuh, *Linear and Nonlinear Circuits*, McGraw-Hill Book Co., New York, 1987.

[35] R. Courant, K. Friedrichs, and H. Lewy, Über die partiellen Differenzgleichungen der mathematischen Physik, *Mathematische Annalen*, **100** (1928), 32–74; translated: On the partial difference equations of mathematical physics, *IBM J. Research and Development*, **11** (1967), 215–234.

[36] B. Davies and B. Martin, Numerical inversion of the Laplace transform: a survey and comparison of methods, *J. Comp. Phys.*, **33** (1979), 1–32.

[37] L. DeMichele and P.M. Soardi, A Thomson's principle for infinite, nonlinear, resistive networks, *Proc. Amer. Math. Soc.*, in press.

[38] P. Dienes, *The Taylor Series*, Oxford University Press, Oxford, England, 1931.

[39] J. Dodziuk, Difference equations, isoperimetric inequality and transience of certain random walks, *Trans. Amer. Math. Soc.*, **284** (1984), 787–794.

[40] V. Dolezal, *Nonlinear Networks*, Elsevier, New York, 1977.

[41] V. Dolezal, *Monotone Operators and Applications in Control and Network Theory*, Elsevier, New York, 1979.

[42] P.G. Doyle, Electric currents in infinite networks, preprint, 1988.

[43] P.G. Doyle and J.L. Snell, *Random Walks and Electric Networks*, The Carus Mathematical Monographs, The Mathematical Association of America, 1984.

[44] R.J. Duffin, Discrete potential theory, *Duke Math. J.*, **20**, (1953), 233–251.

[45] R.J. Duffin, Basic properties of discrete analytic functions, *Duke Math. J.*, **23** (1956), 335–364.

[46] R.J. Duffin, Yukawan potential theory, *J. Math. Anal. Appl.*, **35** (1971), 105–130.

[47] R.J. Duffin and E.P. Shelly, Difference equations of polyharmonic type, *Duke Math. J.*, **25** (1958), 209–238.

[48] N. Dunford and J.T. Schwartz, *Linear Operators, Part I*, Interscience, New York, 1966.

[49] W. Fair, Noncommutative continued fractions, *SIAM J. Math. Anal.*, **2** (1971), 226–232.

[50] P.A. Fillmore and J.P. Williams, On operator ranges, *Advances in Mathematics*, **7** (1971), 254–281.

[51] H. Flanders, Infinite networks: I – Resistive networks, *IEEE Trans. Circuit Theory*, **CT-18** (1971), 326–331.

[52] H. Flanders, Infinite networks: II – Resistance in an infinite grid, *J. Math. Anal. Appl.*, **40** (1972), 30–35.

[53] H. Flanders, A new proof of R. Foster's averaging formula in networks, *Linear Algebra and Its Applications*, **8** (1974), 35–37.

[54] R.M. Foster, The average impedance of an electrical network, *Contributions to Applied Mechanics*, Reissner Anniv. Vol., Edwards Bros. Ann Arbor, Michigan, 1949, 333–340.

[55] R.M. Foster, An extension of a network theorem, *IRE Trans. Circuit Theory*, **CT-8** (1961), 75–76.

[56] P. Gerl, Random walks on graphs, in *Probability Measures on Groups VIII*, H. Heyer (Editor), Lecture Notes in Mathematics, **1210** (1986), Springer-Verlag, New York, 285–303.

[57] P. Gerl, Random walks on graphs with a strong isoperimetric inequality, *J. Theoretical Prob.*, **1** (1988), 171–187.

[58] P. Gerl and W. Woess, Simple random walks on trees, *Eur. J. Combinatorics*, **7** (1986), 321–331.

[59] M.L. Glasser and I.J. Zucker, Extended Watson integrals for the cubic lattices, *Proc. Natl. Acad. Sci.*, USA, **74** (1977), 1800–1801.

[60] W.L. Green and T.D. Morley, Parallel sums and norm convergence, *Circuits, Systems, and Signal Processing*, **9** (1990), 213–222.

[61] D. Griffeath and T.M. Liggett, Critical phenomena for Spitzer's reversible nearest particle systems, *Annals of Probability*, **10** (1982), 881–895.

[62] W. Hackenbroch, Integraldarstellung einer Klasse dissipativer linearer Operatoren, *Math. Z.*, **109** (1969), 273–287.

[63] R. Halin, Some path problems in graph theory, *Abh. Math. Sem. Univ. Hamburg*, **44** (1975), 175–186.

[64] P.R. Halmos, *Naive Set Theory*, Springer-Verlag, New York, 1974.

[65] H.A. Heilbron, On discrete harmonic functions, *Proc. Cambridge Phil. Soc.*, **45** (1949), 194–206.

[66] C.S. Hsu, H.C. Yee, and W.H. Cheng, Determination of global regions of asymptotic stability for difference dynamical systems, *J. Appl. Mech.*, **44** (1977), 147–153.

[67] W.H. Ingram and C.M. Cramlet, On the foundations of electrical network theory, *J. Math. and Phys.*, **23** (1944), 134–155.

[68] M. Kanai, Rough isometries, and combinatorial approximations of geometries of non-compact riemannian manifolds, *J. Math. Soc. Japan*, **37** (1985), 391–413.

[69] M. Kanai, Rough isometries and the parabolicity of riemannian manifolds, *J. Math. Soc. Japan*, **38** (1986), 227–238.

[70] L.V. Kantorovich and G.P. Akilov, *Functional Analysis in Normed Spaces*, Macmillan Co., New York, 1964.

[71] T. Kayano and M. Yamasaki, Dirichlet finite solutions of Poisson equations on an infinite network, *Hiroshima Math. J.* **12** (1982), 569–579.

[72] J.G. Kemeny, J.L. Snell, and A.W. Knapp, *Denumerable Markov Chains*, Second Edition, Springer-Verlag, New York, 1976.

[73] A.E. Kennelly, *Artificial Lines and Nets*, Wiley, New York, 1912.

[74] G. Kirchhoff, Ueber die Auflösung der Gleichungen, auf welche man bei der Untersuchung der linearen Vertheilung galvanischer Ströme geführt wird, *Annalen der Physik und Chemie*, **72** (1847), 497–508.

[75] M.A. Krasnoselskii and Ya.B. Rutickii, *Convex Functions and Orlicz Spaces*, P. Noordhoff, Groningen, The Netherlands, 1961.

[76] G. Kron, Equivalent circuit of the field equations of Maxwell – I, *Proc. IRE*, **32** (1944), 289–299.

[77] G. Kron, Equivalent circuits of the electric field, *J. Appl. Mech.*, **11** (1944), A149–A161.

[78] G. Kron, Electric circuit models of Schrodinger equation, *Phys. Rev.*, **67** (1945), 39–43.

[79] G. Kron, Numerical solution of ordinary and partial differential equations by means of equivalent circuits, *J. Appl. Phys.*, **16** (1945), 172–186.

[80] G. Kron, Equivalent circuits of compressible and incompressible fluid flow fields, *J. Aeronaut. Sci.*, **12** (1945), 221–231.

[81] G. Kron, Electric circuit models of partial differential equations, *Electrical Engineering*, **67** (1948), 672–684.

[82] F. Kubo, Conditional expectations and operations derived from network connections, *J. Math. Anal. Appl.*, **80** (1981), 477–489.

[83] F. Kubo and T. Ando, Means of positive linear operators, *Math. Ann.*, **246** (1980), 205–224.

[84] L. Lavatelli, The resistive net and finite difference equations, *Amer. J. Phys.*, **40** (1972), 1246–1257.

[85] K.J. Lindberg, *Contractive Projections in Orlicz Sequence Spaces*, Ph.D. Thesis, University of California, Berkeley, 1971.

[86] K.J. Lindberg, On subspaces of Orlicz sequence spaces, *Studia Mathematica*, **45** (1973), 119–146.

[87] J. Lindenstrauss and L. Tzafriri, On Orlicz sequence spaces, *Israel J. Math.*, **10** (1971), 379–390.

[88] J. Lindenstrauss and L. Tzafriri, *Classical Banach Spaces I*, Springer-Verlag, New York, 1977.

[89] L.A. Liusternik and V.J. Sobolev, *Elements of Functional Analysis*, Ungar, New York, 1961.

[90] T. Lyons, A simple criterion for transience of a reversible Markov chain, *Annals of Probability*, **11** (1983), 393–402.

[91] F.Y. Maeda, A remark on parabolic index of infinite networks, *Hiroshima Math. J.*, **7** (1977), 147–152.

[92] J.F. McAllister, Jr., Equivalent circuits of the electromagnetic field, *Gen. Elec. Rev.*, **47** (1944), 9–14.

[93] H.M. Melvin, On the concavity of resistance functions, *J. Appl. Phys.*, **27** (1956), 658–659.

[94] A.N. Michel and C.J. Herget, *Mathematical Foundations in Engineering and Science*, Prentice-Hall, Englewood Cliffs, N.J., 1981.

[95] G.J. Minty, Monotone networks, *Proc. Royal Soc. London, Series A*, **257** (1960), 194–212.

[96] L. Mirsky, *An Introduction to Linear Algebra*, Oxford University Press, London, 1955.

[97] T.D. Morley, Parallel summation, Maxwell's principle, and the infimum of projections, *J. Math. Anal. Appl.*, **70** (1979), 33–41.

[98] T.D. Morley, Shorts of block operators and infinite networks – A note on the shorted operator: II, *Circuits, Systems, and Signal Processing*, **9** (1990), 161–170.

[99] H. Nakano, Modulared sequence spaces, *Proc. Japan Acad.*, **27** (1951), 508–512.

[100] C.St.J.A. Nash-Williams, Random walk and electric currents in networks, *Proc. Cambridge Phil. Soc.*, **55** (1959), 181–194.

[101] R.W. Newcomb, *Linear Multiport Synthesis*, McGraw-Hill Book Co., New York, 1966.

[102] K. Nishio and T. Ando, Characterizations of operations derived from network connections, *J. Math. Anal. Appl.*, **53** (1976), 539–549.

[103] B. Noble, *Applied Linear Algebra*, Prentice-Hall, Englewood Cliffs, N.J., 1969.

[104] A.V. Oppenheim and R.W. Schafer, *Digital Signal Processing*, Prentice-Hall, Englewood Cliffs, N.J., 1975.

[105] A.V. Oppenheim, A.S. Willsky, with I.T. Young, *Signals and Systems*, Prentice-Hall, Englewood Cliffs, N.J., 1983.

[106] A.M. Panov, Behavior of the trajectories of a system of finite-difference equations in a neighborhood of a singular point, *Uchenyi Zapiski Sverdlovsk Uralski Gosudarstvenii Universitet*, **19** (1956), 89–99 (in Russian).

[107] B. Peikari, *Fundamentals of Network Analysis and Synthesis*, Prentice-Hall, Englewood Cliffs, New Jersey, 1974.

[108] M.A. Picardello and W. Woess, Martin boundaries of random walks: ends of trees and groups, *Trans. Amer. Math. Soc.*, **302** (1987), 185–205.

[109] M.A. Picardello and W. Woess, Harmonic functions and ends of graphs, *Proc. Edinburgh Math. Soc.*, **31** (1988), 457–461.

[110] G. Polya, Über eine Aufgabe der Wahrscheinlichkeitsrechnung betreffend die Irrfahrt im Strassennetz, *Mathematische Annalen*, **84** (1921), 149–160.

[111] J.W.S. Rayleigh (J.W. Strutt), On the theory of resonance, *Philosophical Transactions of the Royal Society of London*, **161** (1871), 77–118.

[112] P.I. Richards, A special class of functions with positive real part in a half plane, *Duke Math J.*, **14** (1947), 777–786.

[113] P.I. Richards, General impedance-function theory, *Quart. Appl. Math.*, **6** (1948), 21–29.

[114] A. Robinson, *Nonstandard Analysis*, North Holland Publishing Co., Amsterdam, 1974.

[115] J.P. Roth, An application of algebraic topology to numerical analysis: On the existence of a solution to the network problem, *Proc. Nat. Acad. Sciences*, **41** (1955), 518–521.

[116] B. Rotman and G.T. Kneebone, *The Theory of Sets and Transfinite Numbers*, Oldbourne Book Co., London, 1966.

[117] H.L. Royden, Harmonic functions on open Riemann surfaces, *Trans. Amer. Math. Soc.*, **73** (1952), 40–94.

[118] J.E. Rubin, *Set Theory*, Holden Day, San Francisco, 1967.

[119] E. Schlesinger, Infinite networks and Markov chains, preprint, 1989.

[120] R. Schwindt, *Lineare Transformationen vektorwertige Funktionen*, Ph.D. Thesis, University of Cologne, Germany, 1965.

[121] C.E. Shannon and D.W. Hagelbarger, Concavity of resistance functions, *J. Appl. Phys.*, **27** (1956), 42–43.

[122] P.M. Soardi, Parabolic networks and polynomial growth, *Colloquium Mathematicum*, **LX/LXI** (1990), 65–70.

[123] P.M. Soardi and W. Woess, Uniqueness of currents in infinite resistive networks, *Discrete Appl. Math.*, **31** (1991), 37–49.

[124] W.W. Soroka, *Analog Methods in Computation and Simulation*, McGraw-Hill, New York, 1955.

[125] F. Spitzer, *Principles of Random Walk*, second edition, Springer-Verlag, New York, 1976.

[126] A. Stöhr, Über einige lineare partiellen Differenzgleichungen mit konstanten Koeffizienten, *Mathematische Nachrichten*, **3** (1949/1950), 208–242, 295–315, 330–357.

[127] J.L. Synge, The fundamental theorem of electrical networks, *Quart. Appl. Math.*, **9** (1951), 113–127; **11** (1953), 215.

[128] W.M. Telford, L.P. Geldart, R.E. Sheriff, and D.A. Keys, *Applied Geophysics*, Cambridge University Press, New York, 1976.

[129] C. Thomassen, Resistances and currents in infinite electrical networks, *J. Combinatorial Theory, Series B*, in press, 1989.

[130] C. Thomassen, Transient random walks, harmonic functions, and electric currents in infinite resistive networks, preprint, 1989.

[131] W. Thomson and P.G. Tait, *Principles of Mechanics and Dynamics: Part I*, Dover Publications, New York, 1962.

[132] G.E. Trapp, Inverse of circulant matrices and block circulant matrices, *Kyungpook Math. J.*, **13** (1973), 11–20.

[133] A.C. Tucker and A.H. Zemanian, A matroid related to finitely chainlike, countably infinite networks, *Networks*, **12** (1982), 453–457.

[134] M.M. Vainberg, *Variational Method and Method of Monotone Operators in the Theory of Nonlinear Equations*, John Wiley, New York, 1973.

[135] B. Van der Pol, The finite-difference analogy of the periodic wave

equation and the potential equation, Appendix IV in *Probability and Related Topics in Physical Sciences*, M. Kac, Editor, Interscience Publishers, New York, 1959, 237–257.

[136] B. Van der Pol and H. Bremmer, *Operational Calculus*, Cambridge University Press, Cambridge, England, 1959.

[137] N.Th. Varopoulos, Isoperimetric inequalities and Markov chains, *J. Functional Analysis*, **63** (1985), 215–239.

[138] A. Vazsonyi, Numerical method in the theory of vibrating bodies, *J. Appl. Phys.*, **15** (1944), 598–606.

[139] K.W. Wagner, Die Theorie der Kettenleiten nebst Anwendungen, *Arch. Elektrotech.*, **3** (1915), 315–332.

[140] H.S. Wall, *Analytic Theory of Continued Fractions*, Van Nostrand, New York, 1948.

[141] G.N. Watson, Three triple integrals, *Quarterly J. Math. (Oxford Series)*, **10** (1939), 266–276.

[142] E. Weber, *Linear Transient Analysis, Volume II*, John Wiley, New York, 1956.

[143] L. Weinberg, *Network Analysis and Synthesis*, McGraw-Hill, New York, 1962, 175–176.

[144] D.V. Widder, *The Laplace Transform*, Princeton University Press, Princeton, New Jersey, 1946.

[145] J.H. Williamson, *Lebesgue Integration*, Holt, Rinehart, and Winston, New York, 1962.

[146] H. Willie, *Periodisch Invariante Lineare Übertragungssysteme*, Ph.D. Thesis, University of Regensburg, 1978.

[147] R.J. Wilson, *Introduction to Graph Theory*, Oliver and Boyd, Edinburgh, 1972.

[148] W. Woess, Transience and volumes of trees, *Archiv der Mathematik*, **86** (1986), 184–192.

[149] J.Y.T. Woo, On modular sequence spaces, *Studia Mathematica*, **48** (1973), 271–289.

[150] M. Yamasaki, Extremum problems on an infinite network, *Hiroshima Math. J.*, **5** (1975), 223–250.

[151] M. Yamasaki, Parabolic and hyperbolic infinite networks, *Hiroshima Math. J.*, **7** (1977), 135–146.

[152] M. Yamasaki, Quasiharmonic classification of infinite networks, *Discrete Appl. Math.*, **2** (1980), 339–344.

[153] M. Yamasaki, Ideal boundary limit of discrete Dirichlet functions, *Hiroshima Math. J.*, **16** (1986), 353–360.

[154] A.H. Zemanian, *Distribution Theory and Transform Analysis*, McGraw-Hill Book Co., New York, 1965; republished by Dover Publications, New York, 1987.

[155] A.H. Zemanian, Inversion formulas for the distributional Laplace transformation, *SIAM J. Appl. Math.*, **14** (1966), 159–166.

[156] A.H. Zemanian, *Generalized Integral Transformations*, Wiley Interscience Publishers, New York, 1968; republished by Dover Publications, New York, 1987.

[157] A.H. Zemanian, *Realizability Theory for Continuous Linear Systems*, Academic Press, New York, 1972.

[158] A.H. Zemanian, Countably infinite networks that need not be locally finite, *IEEE Trans. Circuits and Systems*, **CAS-21** (1974), 274–277.

[159] A.H. Zemanian, Infinite networks of positive operators, *Circuit Theory and Applications*, **2** (1974), 69–74.

[160] A.H. Zemanian, Relaxive Hilbert ports, *SIAM J. Control*, **12** (1974), 106–123.

[161] A.H. Zemanian, Passive operator networks, *IEEE Trans. Circuits and Systems*, **CAS-21** (1974), 184–193.

[162] A.H. Zemanian, Continued fractions of operator-valued analytic functions, *J. Approximation Theory*, **11** (1974), 319–326.

[163] A.H. Zemanian, Connections at infinity of a countable resistive network, *Circuit Theory and Applications*, **3** (1975), 333–337.

[164] A.H. Zemanian, The voltage and current ratios of RLC operator networks, *J. Math. Anal. Appl.*, **55** (1976), 394–406.

[165] A.H. Zemanian, The complete behavior of certain·infinite networks under Kirchhoff's node and loop laws, *SIAM J. Appl. Math.*, **30** (1976), 278–295.

[166] A.H. Zemanian, The limb analysis of countably infinite electrical networks, *J. Combinatorial Theory, Series B*, **24** (1978), 76–93.

[167] A.H. Zemanian, Countably infinite, nonlinear, time-varying, active, electrical networks, *SIAM J. Math. Anal.*, **10** (1979), 944–960.

[168] A.H. Zemanian, Countably infinite, time-varying, electrical networks, *SIAM J. Math. Anal.*, **10** (1979), 1193–1198.

[169] A.H. Zemanian, A justification of the characteristic-impedance method for lumped periodic cascades, *IEEE Trans. Circuits and Systems*, vol. CAS-29 (1982), 323–327.

[170] A.H. Zemanian, Nonuniform semi-infinite grounded grids, *SIAM J. Math. Anal.*, **13** (1982), 770–788.

[171] A.H. Zemanian, A classical puzzle: The driving-point resistances of infinite grids, *IEEE Circuits and Systems Magazine*, **6** (1984), 7–9.

[172] A.H. Zemanian, The driving-point immittances of infinite cascades of nonlinear three-terminal networks: Part I – Uniform cascades and characteristic immittances, *IEEE Trans. Circuits and Systems*, **CAS-31** (1984), 326–336.

[173] A.H. Zemanian, The driving-point immittances of infinite cascades of nonlinear three-terminal networks: Part II – Nonuniform cascades, *IEEE Trans. Circuits and Systems*, **CAS-31** (1984), 336–341.

[174] A.H. Zemanian, The driving-point immittances of infinite cascades of nonlinear resistive two-ports, *IEEE Trans. Circuits and Systems*, **CAS-31** (1984), 960–968.

[175] A.H. Zemanian, Operator-valued transmission lines as models of a horizontally layered earth under transient two-dimensional electromagnetic excitation, *IEEE Trans. Antennas and Propagation*, **AP-33** (1985), 346–350.

[176] A.H. Zemanian, Operator-valued transmission lines in the analysis of two-dimensional anomalies imbedded in a horizontally layered earth under transient polarized electromagnetic excitation, *SIAM J. Appl. Math.*, **45** (1985), 591–620.

[177] A.H. Zemanian, Infinite electrical networks with finite sources at infinity, *IEEE Trans. Circuits and Systems*, **CAS-34** (1987), 1518–1534.

[178] A.H. Zemanian, A finite-difference procedure for the exterior problem

inherent in capacitance computations for VLSI interconnections, *IEEE Trans. Electron Devices*, **35** (1988), 985–992.

[179] A.H. Zemanian, Transfinite cascades, *IEEE Trans. Circuits and Systems*, in press, 1990.

[180] A.H. Zemanian, An electrical gridlike structure excited at infinity, *Mathematics of Control, Signals, and Systems*, in press.

[181] A.H. Zemanian, *Boundary conditions at infinity for a discrete form of* $\nabla \cdot (\sigma \nabla \Phi) = \alpha \Phi, \alpha \geq 0$, State University of New York at Stony Brook, CEAS Technical Report 523, August, 1988.

[182] A.H. Zemanian, Transfinite graphs and electrical networks, *Trans. Amer. Math. Soc.*, in press.

[183] A.H. Zemanian and H.K. An, Finite-difference analysis of borehole flows involving domain contractions around three-dimensional anomalies, *Appl. Math. Comput.*, **26** (1988), 45–75.

[184] A.H. Zemanian and B. Anderson, Modeling of borehole resistivity measurements using infinite electrical grids, *Geophysics*, **52** (1987), 1525–1534.

[185] A.H. Zemanian and P. Subramaniam, A theory for ungrounded electrical grids and its application to the geophysical exploration of layered strata, *Studia Mathematica*, **77** (1983), 163–181.

[186] A.H. Zemanian and P. Subramaniam, A solution for an infinite electrical network arising from various physical phenomena, *Internat. J. Circuit Theory and Applications*, **11** (1983), 265–278.

[187] A.H. Zemanian, R.P. Tewarson, C.P. Ju, and J.F. Jen, Three dimensional capacitance computations for VLSI/ULSI interconnections, *IEEE Trans. Computer-Aided Design*, **8** (1989), 1319–1326.

[188] A.H. Zemanian and T.S. Zemanian, Domain contractions around three-dimensional anomalies in spherical finite-difference computations of Poisson's equation, *SIAM J. Appl. Math.*, **46** (1986), 1126–1149.

Index of Symbols

This is a list of some of our symbols – along with the pages where they are defined. A few symbols have more than one meaning.

A, 4, 9
$b_{k,n}$, 195
$\mathcal{B}$, 5, 7
$c_{\mathbf{M}}$, 125
$c_{\mathbf{M}}^*$, 126
$\mathbf{c}$, 67
C^n, 2
C^∞, 47
E, 88
ess inf, 4
ess sup, 4
$f_{k,n}$, 195
f^ω, 203
f^{ω^p}, 206
F, 4, 11
$\mathcal{F}$, 6, 41, 135
$\mathcal{G}^0$, 8
$\mathcal{G}^1$, 70
$\mathcal{G}^p$, 141
$\mathcal{G}^\omega$, 148
$\mathcal{G}^{\bar{\omega}}$, 147
H, 4, 11
$\mathbf{H}$, 88
I_c, 172
$\mathcal{I}$, 63, 74, 154
$\mathcal{K}$, 64, 77, 154
$\mathcal{K}^0$, 64, 77, 154
l_c, 116
$l_{\mathbf{M}}$, 121
$l_{\mathbf{M}^*}$, 126
$l_{\mathbf{M}}^\#$, 122
l_2, 3
l_{2r}, 3
l_2^n, 217
l_{2r}^n, 220
$l_2^{n,2}$, 246
$l_{2r}^{n,2}$, 246
lim inf, 4
lim sup, 4
$\mathrm{Lip}_\alpha[\mathcal{H}, \mathcal{H}]$, 113
L_2^n, 218
$L_2^{n,2}$, 246
$\mathcal{L}$, 133, 157
$\mathcal{L}^0$, 133, 157
$\mathcal{N}$, 5

$\mathcal{N}^0$, 8, 70
$\mathcal{N}^1$, 70
$\mathcal{N}^q$, 142
$\mathcal{N}^\omega$, 148
$\mathcal{N}_a$, 111
N, 44
$\mathbf{N}_c$, 93
$\mathbf{N}_c^{(\nu)}$, 96
o, 4
O, 4
Q_j, 168
$\mathbf{R}$, 88
R^n, 2
$\mathrm{sgn}(\cdot)$, 3
T^0, 68
T^{p-1}, 141
$T^{\bar{\omega}}$, 148
V, 4, 9
V_c, 172
$\mathcal{V}$, 84, 155
W_1, 179
$W_{1,k}$, 195
W^f, 185, 191
W^b, 185, 191
W_k^f, 195
W_k^b, 195
Z^n, 3, 217
Δ_2, 127
Θ^n, 218
ω, 147
Ω, 4, 9
$\mho$, 4, 9
$\emptyset$, 2
$\aleph_0$, 68
$\cdot^o$, 2
$\bar{\cdot}$, 2
$\|\cdot\|_1$, 2
$\|\!\|\cdot\|\!\|$, 122
$\cdot \backslash \cdot$, 2
$\cdot \rightsquigarrow \cdot$, 3
$\cdot \mapsto \cdot$, 3
$\cdot \sim \cdot$, 4
$\cdot \cong \cdot$, 123
$(\cdot, \cdot)$, 3
$[\cdot ; \cdot]$, 4, 217

Index of Symbols

Index

algebraic sum, 12
apparent resistivity, 269, 271
  discrete, 272
asymptotic behavior, 4

basic current, 75, 133, 154
bijection, 4
binary tree, 67, 132
borehole resistivity measurements, 271
boundary conditions, 248
  at infinity, 248
  beyond infinity, 260, 264
branch, 5, 7
  adjacent, 6
  boxed, 230
  end, 7
  floating, 220
  grounding, 220
  incident, 5
  parallel, 5
  parallel to a coordinate, 220
  resistive, 9
  unboxed, 230
Brouwer's theorem, 58

capacitance, 11
  computation of, 275
Cartesian product, 2
cascade, 158
  lattice, 159, 186
  swiveled, 192
  uniform, 159
  ω-cascade, 203
  $\vec{\omega}$-cascade, 203
  ω^p-cascade, 206
chainlike graph, see graph, chainlike
characteristic immittance, 179
  backward, 185, 191
  forward, 185, 191
characteristic resistance, 164
chord, 47
chord dominance, 51
component, 6
concavity of driving-point
  immittance, 101
conductance
  branch, 9

shunt, 168
  voltage-controlled, 53
connected graph, 6
  finitely, 71
  0-connected, 8, 71
  1-connected, 73
  q-connected, 146, 150
  ω-connected, 149
connected set in R^2, 172
continued fraction, 235
contraction mapping principle, 56, 161
correspondence, 3
coupling between branches, 11
current-transfer ratio, 163

diameter (q-diameter), 154
direct problem, 270
disjoint, totally, see totally disjoint
domain, 3
domain contraction, 274
driving-point immittance, 195
dual analysis, 83, 155

embrace, 8, 69, 72, 141, 143
Euclidean norm, 2
Euclidean space, 2
exceptional element, 69, 141, 148
exterior problem, 20, 268

family, 5
forest, 6
finite-difference approximation, 21
Foster's averaging formula, 228
Fourier expansion, 116
function, 3
  complementary, 120
  decreasing, 167
  increasing, 167
  inverse, 3
  multivalued, 3
  restriction of, 3
  single-valued, 3
  strictly convex, 120
functional, 3
  strictly convex, 134

Gateaux differential, 134

graph
  chainlike, 33
  chainlike, divergently, 34
  chainlike, finitely, 34
  connected, *see* connected graph
  countable, 5
  finite, 5
  locally finite, 5
  of a function, 3
  reduced, 8, 71, 143, 148
  transfinite, 20
  0-graph, 8
  1-graph, 70
  k-graph, 149
  p-graph, 142
  ω-graph, 148
  $\vec{\omega}$-graph, 147
grid
  cubic, 221
  cylindrical resistive, 24
  el-section, 260
  grounded, 220, 223, 230
  infinity imperceptible in, 233
  one-dimensionally nonuniform, 230
  partially grounded, 230
  semi-infinite, 230, 245
  semi-infinite, transfinite, 260
  square, 221
  ungrounded, 221, 225, 230
  ungrounded, driving-point
    resistance in, 227, 245
  uniform, 221, 223, 225
  k-grid, 151
  n-dimensional rectangular, 220
ground node, 86, 220

Halin's theorem, 37
Hilbert's coordinate space, 3
homeomorphism, 168

immittance, 167
  characteristic, 179, 185, 191
  driving-point, 195
impedance, 44, 274
imperceptible infinity, 193, 233
incidence matrix, 110
incident, terminally, 72
induced current, 46
inductance, 11
injection, 3
integer, n-dimensional, 3, 217
isomorphism, topological linear, 218

joint, 46
Jordan arc, 171
Jordan curve, 171
Jordan's theorem, 172

Kirchhoff's current law, 12, 80, 155
Kirchhoff's voltage law, 12, 82, 155
König's lemma, 36

ladder, 159
  operator, 232
  transfinite, 19
Laplace's equation, 21
lattice cascade, 159, 186
lattice network, 187
lattice point, 3, 217
least power principle, 99
leftmost subpath, 145
limb, 41
limb analysis, 24, 49
  elementary, 239
Lipschitz condition, 57, 113
loop, 6
  chord-tree, 47
  transfinite, 152
  transfinite, perceptible, 154
  transfinite, proper, 154
  1-loop, 73
  1-loop, perceptible, 75, 82
  1-loop, proper, 74
  p-loop, 144
  ω-loop, 149
loop current, 64, 74, 154

mapping, 3
  backward, 167
  forward, 167
  one-to-one, 3
  onto, 3
matrix
  invertible in blocks, 48
  Laurent, 218
  row-finite, 48
Maxwell's equations, 26
Menger's theorem, 28
modular sequence space, 121
monotonicity law, 103
MOSFET threshold voltage, 278

network
  converted, 93
  electrical, 9

lattice, 187
  regular, 111
  three-terminal, 158
  transfinite, 20
  0-network, 74
  1-network, 74
  k-network, 153
node, 5, 8, 65
  adjacent, 6
  degree of, 5
  end, 6, 7
  finite, 5
  floating, 220
  $\mathcal{G}$-finite, 32
  $\mathcal{G}$-infinite, 32
  incident, 5
  infinite, 5
  isolated, 5
  ordinary, 69, 81, 154
  restraining, 81
  0-node, 8
  1-node, 68
  p-node, 141
  ω-node, 148
node voltage, 104, 222
normal log, 24
Norton's form, 9
Norton's load, 209
numerical range, 217

Ohm's law, 9
open circuit, 11
operator, 3
  adjoint, 117
  admittance, 111
  conductance, 84
  Laurent, 218
  Laurent, principal array in, 219
  monotone, 114
  positive, 217
  resistance, 64, 110, 131, 154
  resistance, of an ∞-port, 231
  surface, 270, 274, 277
  voltage-transfer, 236
operator network, 232, 291
orb, 43
orientation
  of a branch, 9
  of a loop, 49, 64
  of an orb, 45
orthonormal basis, 116

Parseval's equation, 116
path, 6
  endless, 6, 72, 144
  finite, 6, 72, 144
  nontrivial, 6, 144
  one-ended, 6, 72, 144
  perceptible, 92
  transfinite, 152
  0-path, 8, 71
  1-path, 72
  p-path, 144
  ω-path, 148
  $\vec{\omega}$-path, 147
  $\vec{\omega}$-path, endless, 148
  (k,q)-path, 151
perceptible infinity, 199, 233, 245
Poisson's equation, 21
port, 161
  ∞-port, 230
port condition, 231, 290
power, total isolated, 74, 83, 93, 153

quadratically summable, 3

random walk, 285
  recurrent, 286
  transfinite, 287
  transient, 286
range, 3
rank, 140
Rayleigh's monotonicity law, 103
reactive element, 11
reciprocity principle, 80
regime, 11
relation, 3
resistance
  branch, 9
  characteristic, 164
  current-controlled, 53
  driving-point, 227, 245
  input, 17
  series, 168
resistance parameters, 162
Riesz's representation theorem, 64
rightmost subpath, 145

section
  0-section, 8
  1-section, 73
  q-section, 146, 150
  ω-section, 149
self impedance, 44

self-loop, 5
semicontinuous, 134
separating node set, 36
separatrix, 182
series resistance, 168
set
  compact, 2
  compact, weakly sequentially, 134
  countable, 1
  denumerable, 1
  partition of, 2
  summable, 208, 262
short circuit, 9
shorted at infinity, 70
shorted subnetwork, 98
shunt conductance, 168
singleton, 2
source
  branch current, 9
  branch voltage, 9
  dependent, 11
  mixed, 247
  nodal current, 221
  pure current, 10
  pure voltage, 9
sonde, 24
space
  Banach, dual of, 126
  Hilbert's coordinate, 3
  isomorphic, 123
  modular sequence, 121
  reflexive, 130, 134
span, 4
spine, 34, 39
stage of a cascade, 159
star network, 88
subgraph, 5
  branch-induced, 5
  intersection of, 5
  node-induced, 5
  partition of, 6
  spanning, 6
  union of, 5
subnetwork, 12
  expanding sequence of, 65
  opened, 65
  shorted, 98
summable set in R^2, 208
summable, quadratically, 3
support, 2, 63
surface electromagnetic
  measurements, 272

surface resistivity measurements, 269
surjection, 3

Tellegen's equation, 65, 79, 84
terminals of a cascade, 161
Thevenin's form, 9
Thevenin's load, 209
Thomson's least power principle, 99
tie, 43
tip
  effectively shorted, 105
  elementary, 7
  0-tip, 67
  0-tip, disconnectable, 105
  0-tip, nondisconnectable, 104
  $(p-1)$-tip, 140
  $\vec{\omega}$-tip, 148
tip voltage, 104
totally disjoint, 6, 72, 73, 141, 143
  except terminally, 36, 144
trajectory, 177
  origin-evading, 193
  transfinite, 205
transference of a current source, 91
  proper, 93
transference of a voltage source, 89
  proper, 91
transformation, 3
  finite Fourier, 218
  Fourier-series, 218
transient behavior, 27
transimpedance, 44
tree, 6
  binary, 67, 132
Turam configuration, 272, 280
two-port, 158

uniform Δ_2 condition, 127

voltage-current regime, 11
voltage-transfer function, 236
voltage-transfer ratio, 163

weak convergence, 134
weakly sequentially compact set, 134
Wenner configuration, 269

Young's inequality, 121
Yukawa's potential equation, 20, 279